Vol. 10

Frontiers in
Nanobiomedical
Research

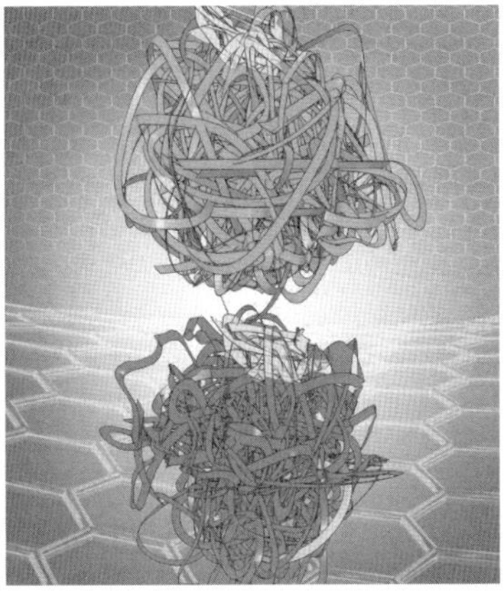

TISSUE ENGINEERING AND NANOTHERANOSTICS

Frontiers in Nanobiomedical Research

ISSN: 2251-3965

This new book series "Frontiers in Nanobiomedical Research" (FNBMR) effectively facilitates interactions and collaborations between the nano and medical research communities, who are focused on solving key problems based on emerging nanoscience and technologies, by publishing scientific monographs, graduate level texts, and reference materials, including handbooks and major reference works. These books will address the applications of a broad spectrum of materials, chemical, biological and medical sciences, and engineering. In addition, the book series publications summarize the wealth of experimental results in both nanomaterials and biomedicine and introduce new aspects of nanoscience relevant to biological and medical applications. The series also demonstrates new trends in the field and cultivates a new research community in both frontier research and higher education.

Published

Vol. 11 *Cancer Therapeutics and Imaging: Molecular and Cellular Engineering and Nanobiomedicine*
edited by Kaushal Rege and Sheba Goklany (Arizona State University, USA)

Vol. 10 *Tissue Engineering and Nano Theranostics*
edited by Donglu Shi (University of Cincinnati, USA) and Qing Liu (Tongji University, China)

Vol. 9 *The World Scientific Encyclopedia of Nanomedicine and Bioengineering II: Bioimplants, Regenerative Medicine, and Nano-Cancer Diagnosis and Phototherapy (A 3-Volume Set)*
edited by Donglu Shi (University of Cincinnati, USA), Maoquan Chu (Tongji University, China) and Jiang Chang (Chinese Academy of Sciences, China)

Vol. 8 *Bioengineering in Wound Healing: A Systems Approach*
edited by Martin L. Yarmush (Rutgers University, USA & Harvard Medical School, USA) and Alexander Goldberg (Tel Aviv University, Israel)

Vol. 7 *Multiscale Technologies for Cryomedicine: Implementation from Nano to Macroscale*
edited by Xiaoming He (The Ohio State University, USA) and John C. Bischof (University of Minnesota, USA)

For further details, please visit: http://www.worldscientific.com/series/fnbmr

(Continued at the end of the book)

Vol. 10
Frontiers in Nanobiomedical Research

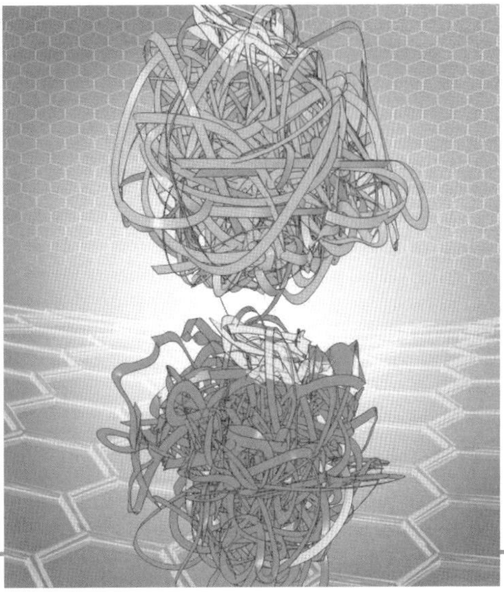

TISSUE ENGINEERING AND NANOTHERANOSTICS

editors

Donglu Shi
University of Cincinnati, USA

Qing Liu
Tongji University, China

World Scientific

NEW JERSEY • LONDON • SINGAPORE • BEIJING • SHANGHAI • HONG KONG • TAIPEI • CHENNAI • TOKYO

Published by

World Scientific Publishing Co. Pte. Ltd.
5 Toh Tuck Link, Singapore 596224
USA office: 27 Warren Street, Suite 401-402, Hackensack, NJ 07601
UK office: 57 Shelton Street, Covent Garden, London WC2H 9HE

Library of Congress Cataloging-in-Publication Data
Names: Shi, Donglu, editor. | Liu, Qing, 1962 June 22– editor.
Title: Tissue engineering and nanotheranostics / edited by Donglu Shi & Qing Liu.
Other titles: Frontiers in nanobiomedical research ; v. 10. 2251-3965
Description: New Jersey : World Scientific, 2017. | Series: Frontiers in nanobiomedical research ; vol. 10 | Includes bibliographical references and index.
Identifiers: LCCN 2016056675 | ISBN 9789813149182 (hardcover : alk. paper)
Subjects: | MESH: Tissue Engineering | Theranostic Nanomedicine
Classification: LCC R857.T55 | NLM QS 525 | DDC 612/.028--dc23
LC record available at https://lccn.loc.gov/2016056675

British Library Cataloguing-in-Publication Data
A catalogue record for this book is available from the British Library.

Desk Editors: Dr. Sree Meenakshi Sajani/Joy Quek

Typeset by Stallion Press
Email: enquiries@stallionpress.com

Printed in Singapore

Preface

This book is about underlying concepts and recent developments in nanotechnology, tissue engineering, and nanomedicine. In short, the title terms "Nano Theranostics" refer to diagnosis and therapy using nanotechnology.

The authors of the chapters have attempted to address the subject matter and the corresponding issues in a tutorial manner so that they are readily accessible to technical nonspecialists and university students. Many new concepts are introduced and explained in terms of established theories and experimental data.

The book is divided into two parts: (1) Tissue engineering and (2) Nanomedicine. Tissue engineering refers to biomaterials development for skeletal regeneration and stem cell cased translational medicine. Nanomedicine refers to drug delivery imaging, biosensing, and localized treatment of cancer via various chemical and physical means.

Interestingly tissue engineering and nanomedicine are both "biomaterials", but they are better characterized by their biological outcomes. While the objective of tissue engineering is cell wellbeing and soft tissue regeneration, the objective of nanomedicine is the destruction of cells (cancer cells). That is tissue engineering promotes "good cells" while nanomedicine suppresses "bad cells".

Tissue engineering has advanced into clinical settings such as stem cell neuron repair, bio-resorbable stent, and tissue-engineered cartilage. Current tissue engineering research uses computer simulation and 3D printing technologies advancing the state-of-the-art in an unprecedented fashion.

The first chapter of the book discusses the status of obtaining skeletal muscle that may be lost due to diseases and injuries. Some of the research highlights are described in Chapter 2 on biomaterial-based patches as fetal surgery implants. The authors give detailed experimental results on implant development for open fetal surgery and fetoscopic applications. They also discuss the internal environment surrounding the fetus in terms of pH changes, mechanical impact, and degradation profiles. Chapter 3 updates the status in 3D printing for cartilage regeneration. Articular cartilage is introduced with various unique biological features in comparison to most common tissues. Chapter 4 focuses on a fascinating topic regarding differentiation of human pluripotent stem cells. In particular, this chapter deals with the aspects of human pluripotent stem cells for therapeutic applications. Critical issues in this area are discussed including the difficulties involved in differentiating human pluripotent stem cells from diseases-related cell lineages, including neurons, cardiac muscle cells and pancreatic beta cells.

The remainder of the book considers issues in nanomedicine with an emphasis on bio-imaging, bio-sensing, nanodelivery, and multifunctional nanocarriers for cancer thereanostics. Some effective means of nano therapy involving the photothermal effect is also introduced.

Chapter 5 describes recent progress on the development of magnetic nanohybrids for MRI and phototherapy. The authors have carried out extensive experiments on developing novel nanomaterials are particularly effective for early detection of premature tumor and proper treatment. Chapter 6 summarizes some of the key developments in biosensing and imaging. Several conducting nanoparticles have been observed to exhibit the so-called localized surface plasmonic resonance (LSPR), responsible for the pronounced photothermal effects. LSPR of nanoparticles is also tunable via the particle size, shape, and microenvironment. The LSPR nanaoparticles can therefore be utilized for design of biosensors that respond to cell protein expression and DNA hybridization.

Chapter 7 deals with nanoparticle delivery to cancer cells. In nano therapeutics, the key is the nanoparticle delivery to cancer cells in a targeted and controlled fashion. This chapter introduces the methodology that is tailored to a specific characteristic of cancer cells: an excess of centrioles. Centrioles are structurally composed of small cylindrical organelles lying adjacent to the nucleus, creating weak electromagnetic field in a cell. The electromagnetic field is proportional to the excess of centrioles providing a biomarker for magnetically charged toxic nanoparticles. Chapter 8 gives an overview on the development of multifunctional nanoparticles for cancer theranostics. In dealing with complicated issues of cancer therapy, nano carriers have to be designed according to the clinical requirements and oncological characteristics. For a particle to function both as an imaging contrast agent and lethal anticancer entity, various functional moieties have to be installed in a single carrier at nanoscale. These include anticancer drugs, fluorescent quantum dots, tumor specific ligands, and physical means for cancer cell killing such as photothermal nanoparticles (fold and Fe3O4). As such, the design and development of a multifunctional carrier can be complicated by various conflicting effects. These critical issues are addressed with currently developed new approaches.

This book is intended to serve as an informative tutorial for non-technical specialists from fields of medicine, chemistry, materials science, biomedical engineering, and biology. It will shed new light on the various strategies adopted for both tissue engineering and nanomedicine. Both undergraduate and graduate students will find the book a valuable reference not only on biomedicine, but also on materials synthesis. We hope these chapters will provide timely and useful information for the progress of tissue engineering and nanomedicine. We are grateful to all invited authors for their excellent contributions to this book.

Dr. Donglu Shi
Adjunct Professor
The Institute for Translational Nanomedicine
Shanghai East Hospital
The Institute for Biomedical Engineering & Nano Science
Tongji University School of Medicine, Shanghai 200092, China

Professor and Chair
The Materials Science and Engineering Program
Department of Mechanical and Materials Engineering
College of Engineering and Applied Science
University of Cincinnati
Cincinnati, OH 45221, USA

Dr. Qing Liu
Adjunct Professor
The Institute for Translational Nanomedicine
Shanghai East Hospital
The Institute for Biomedical Engineering & Nano Science
Tongji University School of Medicine, Shanghai 200092, China

CEO
Jiangyin Recongene Biomedical Technologies, Co. Ltd.
Jiangyin, Jiangsu 214437, China

Contents

Chapter 1

Engineering Approaches for Creating Skeletal Muscle

Caleb Vogt, Mitchell Tahtinen and Feng Zhao

Department of Biomedical Engineering
Michigan Technological University
Houghton, MI 49931, USA

Abstract: Engineered skeletal muscle grafts have made great progress during the past decades, benefiting from a growing understanding of mechanobiology and stem cell differentiation. Current techniques are widely varied, ranging from *in vitro* methods following the classical tissue engineering paradigm to *in situ* approaches such as host cell recruitment. In different ways, all of these try to supply mechanical toughness while providing the necessary signals for differentiation and maturation of the engineered skeletal muscle.

Keywords: Skeletal muscle, volumetric muscle loss, myogenesis, muscle regeneration, stem cell engineering.

1. Clinical Need for Skeletal Muscle

The biomedical research community has long held an interest in engineering replacements for skeletal muscles. Diseases and injuries that result in muscle loss can lead to significant weakness and deformity, and it is hoped that regenerating or replacing the muscle tissue will mitigate these negative outcomes. Of the many opportunities for intervention, two classes are particularly prominent in the clinic. The first of these is wasting disorders, which result in muscle loss on a large scale throughout the body. The second is volumetric muscle loss (VML), which can be the result of injury or surgery.

1.1. Wasting Diseases and Disorders

There are several diseases and disorders that can cause muscular wasting, including a class known as muscular dystrophy (MD). Duchenne's muscular dystrophy (DMD), for example, is caused by a defective DMD gene, which encodes the protein dystrophin. Dystrophin is an integral member of the dystrophin-associated protein complex, which connects a myocyte's actin cytoskeleton to the extracellular matrix (ECM).[1] It is thought that this connection shields the sarcolemma from excessive stress, especially during eccentric contraction.[2,3] In DMD, mutations result in a non-functional dystrophin protein, which is unable to fulfill its mechanical role. This, in turn, is thought to lead to higher sarcolemmal stresses, eventually resulting in elevated intracellular Ca^{2+}.[4] Although work is still underway to understand all of the underlying processes, the end result on the organ level is myocyte death, fat deposition, and fibrosis. Due to these changes to the muscle tissue, patients with DMD experience pronounced muscle weakness. The diagnosis is generally made at a young age as the children are unable to keep up with their peers, and most of them are confined to wheelchairs by the age of 10.[5,6] Although treatment for DMD patients is improving, many succumb to respiratory or cardiac complications by the age of 20. DMD is recessive, and because the DMD gene is located on the X chromosome, it affects men more often than women. Female carriers can experience similar symptoms to males in some cases.[7] In 2015, it was determined that the combined prevalence of DMD and Becker's muscular dystrophy (BMD) (another prevalent dystrophy) among boys aged 5–9 was about 2 in 10,000.[8] Although DMD is the most common, there are many other dystrophies. BMD is very similar to DMD, and results in partially functional dystrophin. Therefore, the symptoms of BMD are generally not as serious as those of DMD. Other notable dystrophies include myotonic, congenital, Emery–Dreifuss, facioscapulohumeral, limb-girdle, distal, and oculopharyngeal. Other conditions of clinical concern are cachexia and sarcopenia. Cachexia is a loss of body weight due to chronic disease, such as cancer. Much of this lost weight is the result of muscle wasting.[9] Sarcopenia is age-related muscle loss and weakness.[10] Each of these diseases and disorders are systemic, affecting the entire body through faulty genetics or disease (although the primary symptoms may be more specific). Because of this, treatment strategies will differ from muscle loss caused in only one location.

1.2. Volumetric Muscle Loss

Significant amounts of muscle tissue can be lost through either traumatic injury or surgery. Often classified as VML, such a wound can be very difficult for the body to heal.[11] Severe injuries such as those sustained in vehicular accidents or combat that cause substantial damage to the soft tissues are a major source of this problem. Bullet wounds can also cause VML injury, both in combat and in civilian conflict. Because of the inherent nature and unpredictability of such events, patients sustaining such traumatic wounds are generally handled on a case-by-case basis.[12] Various surgical procedures can also result in muscle loss, such as tumor ablation and limb salvage. These procedures are intended to remove soft tissue sarcomas in muscle, and necessitate injury to the affected organ.[11,13] VML has a large potential for negative impact on a patient's quality of life. Even if the wound is closed quickly and without infection, permanent deformity and disability may result. Ideally, wound healing in muscle proceeds in a similar manner to other connective tissues. The mechanical forces tear apart the tissue, rupturing the myoblasts and capillaries in the wound area.[14,15] A hematoma forms as the body attempts to contain the injury. Inflammation follows over the course of several days, and immune cells are recruited to remove debris and prevent spread of any infection.[15] To prevent further mechanical trauma and restore integrity to the tissue, fibroblasts begin depositing dense ECM within the wound.[15] The immune cells recruit resident satellite cells (SCs) to begin forming new myoblasts. The ideal healing resolution of remodeling the scar tissue, fusion of new myocytes, and reinnervation are prevented in VML injuries due to the large amount of scar tissue.[15] As a result, the scar remains intact and any tissue distal of the wound is denervated.[15] Primary concerns in the clinical treatment of VML injuries are wound closure, pain reduction, and promotion of healing. A common surgical method to achieve these goals utilizes muscle flaps.[11] If the injury is small enough, a muscle flap can be cut in the nearby muscle and repositioned in numerous ways to cover the wound.[16] Larger defects may require harvesting thin muscle grafts from a different donor site.[16] This treatment provides the wound site with a covering to help protect the injury, preformed vasculature to ensure adequate blood supply, and an additional source of cells for regenerating the lost muscle tissue.[16] Although this method is well established and often effective for closing wounds, lack of suitable donor sites and the necessity of causing another injury make it less than ideal. Skin grafting procedures may also be necessary once the underlying muscle damage has been addressed. When the wound is closed and has developed mechanical integrity, bracing and physical therapy can be used to mitigate any resulting disability.[17] Cellular engineering strategies could be applied to many of these problems, but the strategies used will be dependent on the type of disease or injury. For example, VML may eventually be treated by off-the-shelf muscle grafts engineered *in vitro*. However, this technique will require surgery, and will likely not be effective in systemic disorders such as DMD. Instead, dystrophies may be addressed through a cell injection approach, conditioning and expanding cells *in vitro* and injecting them into circulation to home in on their target sites throughout the body. Current tissue engineering approaches from 3D constructs to cell injection are discussed later in this chapter.

2. New Muscle: Myogenesis and Muscle Regeneration

The body has a remarkable ability to not only structure its connective tissues during embryogenesis, but also maintain and repair them as an adult. These processes can be broadly divided into two categories: myogenesis and regeneration. Myogenesis encompasses the development of muscle tissue in the embryo, and continues in some capacities until adulthood. Its role is generating new muscle, adding to the size and strength of an organism. On the other hand, regeneration is responsible for repairing muscle tissue from many types of injuries. Both of these modes of generating muscle tissue are worth considering when deciding on viable tissue engineering approaches. For example, the myogenic processes may be best suited for generating muscle from early stem cells. Regeneration will likely make a better model for muscle generated from adult stem cells such as the monopotent SCs.

2.1. Myogenesis

Myogenesis is a process that begins in the embryonic somite and is generally thought to occur in four distinct stages: embryonic, fetal, neonatal, and adult (Fig. 1). The first two stages, embryonic and fetal, occur prenatal. The third stage, neonatal, displays extensive proliferation and maturation of muscles. Finally, the fourth stage, adult, primarily exhibits quiescent stem cells termed SCs.[18,19] Each of these stages can be identified by the type of cells present, expression of specific surface proteins, active signaling pathways, and the cells proliferative capacity. However, there is some overlap seen between the stages and there have not been clearly established guidelines as to when each stage begins or ends.

The first stage, embryonic myogenesis, initiates in the somite. At this stage, myogenic progenitors express Pax3 and Pax7 transcription factors, which are induced through the Wnt, Shh, and BMP signaling pathways.[20] Pax3 is expressed first during the somite formation, but is progressively restricted as myogenic progenitors proliferate.[21] As Pax3 expression reduces, myogenic progenitors begin to express more Pax7 and start differentiation into myoblasts; initiating the second stage, fetal myogenesis. Upon differentiation, Pax7 expression is greatly reduced or absent, and expression of myogenic regulatory factors (MRFs) begin. The earliest MRFs to be expressed are Myf5 and

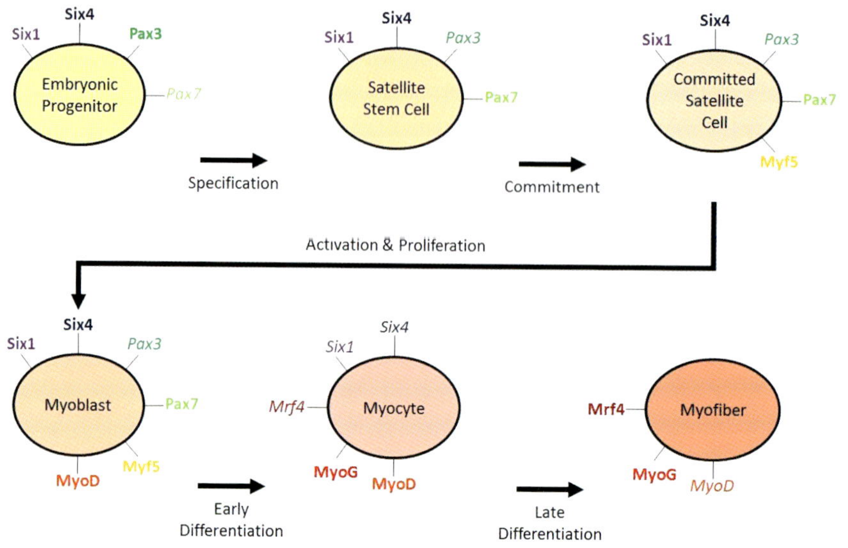

Fig. 1. Proteins expressed during the time course of myogenesis beginning with embryonic progenitors. Proteins that are strongly expressed are bold, while proteins that are weakly expressed are italicized.

MyoD, which are expressed in myoblasts during fetal development.[18] Proliferation of the myoblasts continues during this stage as well as differentiation into mononuclear myocytes. During this stage the myoblasts MRFs expressed are Myf5, MyoD, and Mrf4.[18] However, the mononuclear myocytes will express MyoD, Mrf4, and MyoG MRFs.[18] The differentiation of myoblasts is the first step to muscle fibril formation. During both the fetal and neonatal stage, mononuclear myocytes will group together to form multinuclear myofibrils and myotubes, which group together to form muscles. The final stage, adult myogenesis, involves adult progenitor cells that lie near the basement membrane of adult myofibers. These progenitor cells are quiescent stem cells that are termed SCs. These cells most commonly express the Pax7 marker along with Myf5, CD34, and M-Cadherin.[18,22] Recently, other receptors and transcription factors have been identified to play a critical role in myogenesis; such as, regulation from the G protein-coupled receptor kinase 2[23] or ephrine receptors and zinc finger protein transcription factors.[24]

2.2. Muscle Regeneration

Adult progenitor stem cells, or SCs, are the primary source of muscle regeneration. These cells reside in the basement membrane of adult myofibers and remain mitotically quiescent until needed. During the quiescent period the cell-intrinsic factors expressed by SCs have been identified to be Pax7, Foxo3, Notch, TTP, miR-31, and $p27^{KIP1}$.[25] Following stress to muscle tissue, which can be induced by weight bearing or trauma from an injury, SCs will become activated and begin differentiation into myoblasts.[26] These myoblasts will undergo multiple rounds of division prior to terminal differentiation into mononuclear myocytes that fuse together to form myofibrils. During activation many cell-intrinsic factors are expressed including Myf5.[25] After injury a large majority of SCs differentiate into myoblasts; therefore, the remaining cells must undergo self-renewal to replenish the progenitor cell population. The cells in this stage display few cell-intrinsic factors (Pax7, Spry1, and Foxo3).[25]

3. Stem Cell Candidates for Muscle Regeneration

Stem cells are defined as being undifferentiated cells that have the potential to differentiate into functional tissues of the body. Under the right conditions they are known to be highly proliferative, although some often remain dormant in the body's tissues until a need arises for generation of new tissue. Many types of stem cells have been identified, each with varying abilities in terms of the mature cell types they can give rise to, as well as their potential to proliferate. Embryonic stem cells (ESCs), induced pluripotent stem cells (iPSCs), and mesenchymal stem cells (MSCs) have all been shown to differentiate into myoblast-like cells and may be able to release beneficial cytokines to enhance wound healing.[27–35] These cell types are currently at the forefront of muscle tissue engineering.

3.1. Embryonic Stem Cells

The embryonic body is the home for many stem cells that have been shown to be pluripotent (capable of differentiation into almost any tissue type found in the body). Their role in the development of new organisms requires that they have a theoretically unlimited potential for self-replication and proliferation.[36] Both of these traits make ESCs an attractive source for tissue engineering purposes. Theoretically, implantation of ESCs into a wound site would provide everything necessary for complete replacement of all damaged tissues. However, a number of significant disadvantages are associated with this cell type. First are the ethical concerns with the harvesting of such cells, which necessarily deprives the embryonic body of the ability to develop into an adult.[27] Furthermore, due to the cells' proliferative and pluripotent nature, the risk of teratoma formation is high.[36] Finally, use of ESCs to form tissue engineered constructs for allografts can result in a detrimental immune response.[28]

3.2. Induced Pluripotent Stem Cells

Recently, it was discovered that it is possible to genetically reprogram somatic cells to behave as pluripotent stem cells.[29] This is conducted with genes that have been identified as essential in the maintenance of the embryonic environment, and has been best demonstrated on fibroblasts.[29] The benefit of this cell type, besides the pluripotency, is the ability to use cells harvested from the patient. The cells could be obtained from a small skin biopsy, and would theoretically be completely compatible with the patient's immune system if used as an autograft. iPSCs are not without their drawbacks, however, which include the risk of teratoma formation and low yield of the induction process.[30,31] Furthermore, the full ramifications of genetic manipulation are not yet known.[31] Myogenic differentiation has been accomplished through overexpression of critical genes including Pax7 or MyoD.[37,38]

3.3. Mesenchymal Stem Cells

The MSCs are resident stem cells in many tissues of the body, and are best known for their ability to differentiate into osteogenic, adipogenic, and chondrogenic lineages.[32] Several criteria for the identification of MSCs have been identified by the International Society for Cellular Therapy.[33] These criteria include adherence to plastic during culture and the ability to differentiate into osteoblasts, adipocytes, and chondrocytes.[33] In addition, the cells must be CD105, CD73 and CD90 positive but CD45, CD34, CD14 or CD11b, CD79alpha, CD19 and HLA-DR negative.[33] More recently, MSCs have been shown to support muscle cell growth, and can differentiate and fuse to form multinucleated myofibers.[34] There is increasing evidence that MSCs are involved in the upkeep of various tissues, especially through the release of growth factors and regulation of the behavior of monopotent tissue stem cells.[35] MSCs are also known to play a role in modulating the immune response through various mechanisms including contact inhibition of T-cells and B-cells.[39] During angiogenesis, MSCs act as pericytes, playing a significant role in stabilizing the stalks of new vessels.[40] They can be easily isolated from bone and adipose tissue, and expand easily *in vitro*. All of these aspects make these cells attractive for tissue engineering of skeletal muscle, especially since immunogenicity and lack of vascularization are issues that continue to plague the field. Recent attempts to utilize MSCs for restoring muscle tissue have shown great promise, although the descriptions for the mechanism of healing varies between some tests that find MSCs directly contributing to the myofiber population, and other tests that show MSCs only seeming to support growth through secretion of growth factors.[41] Although MSCs can differentiate into myoblast lineages *in vitro*, doubt has been casted on the ability of the *in vivo* environment to also induce such a change.[42] *In vitro* differentiation of MSCs can be accomplished through growth in media designed for skeletal muscle or culture media conditioned by skeletal muscle cells.[43] While these techniques have been shown to cause differentiation, more work is needed to fully replicate the adult myocyte phenotype in a culture.[43] MSCs are used in a wide variety of treatment models, including both injectable and volumetric tissue engineered deliveries.[44,45]

3.4. Satellite Cells

SCs have been identified as monopotent stem cells that reside alongside the myofibers in skeletal muscle.[46] These cells can be activated when there is need for new muscle fiber generation.[47] Following stimulation, they differentiate into myoblast cells, then fuse into multinucleated myofibers.[46] Some research has shown that the SCs will remain in their undifferentiated, inactive state for long periods until a signal disrupting the stem cell environment such as tissue damage is introduced.[48] Unfortunately, SCs do not maintain their potential to proliferate or differentiate well *in vitro*, although with careful culturing they may prove useful in tissue engineering of volumetric constructs.[42,49]

4. Engineering Approaches

There is much debate on which cells and approaches should be considered ideal. One thing that holds true for all treatment options is that they must all account for the development of myofibers and motor neurons with adequate blood supply from capillary structures to be effective replacements for damaged tissue. Tissue engineers have found that there are many parameters of cell culture that can alter the behavior of cells. Ideal substrate stiffness and topography, mechanical stimulation, and electrical stimulation can all be culture inputs that encourage differentiation, proliferation, or maturation of many stem cells and somatic cell tissues. Generally, it is accepted that bioreactors mimicking *in vivo* conditions can lead cells toward forming the desired tissues. Because there are many important cell types in skeletal muscle, it is important to design bioreactors to encourage development of nerves and vasculature, as well as muscle fibers. The tissue engineering strategies for attaining muscle growth can be divided into two main groups, those that develop muscle tissue *in vitro* and those that implant or inject the necessary materials to develop the new muscle *in vivo*.

4.1. *In Vivo* Approaches

Many traditional tissue engineering techniques involve developing a tissue (such as a cell sheet) *in vitro*, then implanting it to a target site. However, rather than making cellularized constructs, some researchers are attempting to use the wound environment itself as a driver for regeneration (Fig. 2). Possibly the simplest of these approaches involves implanting of acellular matrices in an attempt to recruit native cells into the wound area.[50] For example, it was found that implanting a commercially available gelatin scaffold containing bovine serum albumin was sufficient to recruit SCs from the surrounding tissues in a tibialis anterior wound model in Sprague-Dawley rats.[50] Other papers argue that scaffolding materials are not enough to substantially improve muscle function.[51] This may be due to size discrepancies between wound models, and it is possible that, although small wounds can be repaired by recruitment alone, large wounds will not have enough available stem cells to sufficiently populate the scaffold. The next step in this kind of treatment could be subsequent injection of stem cells following the implantation of an acellular scaffold.[45] A decellularized ECM scaffold from the muscle of Lewis rats was implanted and sutured into another rat of same species in a gastrocnemius VML wound model.[45] These researchers also found that treatment with ECM alone was not enough to cause recovery, even after 42 days.[45] However, injection of MSCs derived from Lewis rat bone marrow 1 week after ECM implantation did result in recovery of functionality, reaching statistical significance above the ECM-only treatment after 6 weeks.[45] Development of vascularization into the ECM was also found to be significantly enhanced by the MSC injection.[45] Rather than implantation of rigid scaffolding materials, some researchers are investigating the use of injectable, cell-loaded materials such as biomaterial glues and hydrogels.[44] In one such study, fibrin glue, gelatin-based FloSeal® hemostat gel, and hyaluronate-alginate hydrogels were all tested both with and without loading of MSCs.[44]

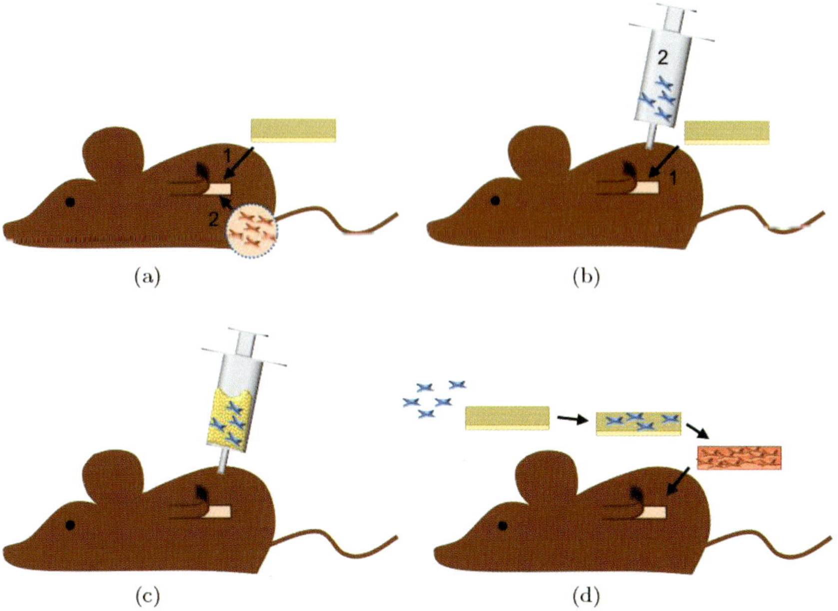

Fig. 2. Depiction of current approaches to engineering skeletal muscle. (a) Implant scaffold for host cell recruitment.[50] (b) Implant scaffold then inject stem cells.[45] (c) Cell-loaded injectable scaffold.[44] (d) Cells added to scaffold and differentiated *in vitro*.[49,52]

It was found that all of the treatments were able to restore functionality to the muscle wound model, but it should be noted that the excised volume was very small compared to other similar studies.[44,45,50] Interestingly, it was found that the injectables that were loaded with MSC-conditioned media alone resulted in less irritation than those that contained MSCs.[44] It should be noted that this treatment model involved xenografting from human stem cells into a mouse model, which may be a cause for this finding.

4.2. *In Vitro* Approaches

Muscle tissue formed more traditionally *in vitro* has had success in mouse models, as demonstrated in 2014.[52] Mouse ecto-mesenchymal stem cells (EMSCs) were induced along the myogenic lineage with dimethyl sulfoxide (DMSO)-containing culture media.[52] A thin structure composed of collagen was used as a scaffold, and differentiated EMSCs were allowed to grow to form a cell sheet, eventually degrading the original collagen scaffold.[52] Following 14 days of culture, the cell sheet morphology was generally aligned, and composed of elongated cells.[52] The cell sheets were rolled to form a 3D structure, then implanted into an excised volume muscle defect model in the quadriceps of BALB/C mice.[52] The allograft was allowed to grow for 14 days, then removed for analysis.[52] It was found that MyoD, an important myogenic marker, was detected in the excised construct.[52] Furthermore, it was found the EMSCs fused and formed multinucleated cells.[52] Another study done on SCs *in vitro* to form 3D constructs through spontaneous detachment and 3D organization found that it was possible to make cylindrical muscle constructs that were contractile under electrical stimulation.[49] Most recently, the emerging technique of 3D bioprinting was used to print mouse myoblasts into tube structures.[53] Many of these approaches may prove to be useful in the future of tissue engineered muscle.

4.2.1. Vascularization

Transport of gasses, nutrients, and waste are critical to any tissue. Consequences due to the lack of functional vasculature become much more apparent when tissue engineered constructs have a thickness greater than about 200 μm.[54] The diffusion limit of oxygen is one of the primary constraints when engineering volumetric muscle tissue, and must be addressed if the tissue is to avoid necrosis.[54] In skeletal muscle engineering, some promising approaches take a multicellular approach to the problem of forming capillary beds *ex vivo*. In one approach, wild-type mouse muscle was broken up and the cells were obtained by dissociation from the matrix.[55] The resulting cells were cultured in a dish to form a cell sheet, which was then immobilized at two opposing points to allow the cell sheet to spontaneously roll into a 3D spindle.[55] It was found that not only did the cells form myofibers, but also the endothelial cells in the culture organized into vascular-like structures.[55] When implanted into an injury model in the tibialis anterior of mice, the engineered muscle could integrate with the native tissue and it was shown to have a significant effect on the recovery of grip strength.[55] A similar study using a variety of cells, including fibroblasts, endothelial cells, and myoblasts, also showed similar results in terms of vascularization.[56] It is possible that the use of multiple cell types, or at least prior induction of stem cells along different lineages, may be necessary to obtain the complexity needed for formation of both vasculature and myofibers.

All of these approaches, both *in vitro* and *in vivo*, show some promise. However, to precisely control the outcomes of these types of experiments, many engineering factors must be considered. Cells are responsive to both chemical and physical stimuli, opening many possible routes to modifying their behavior. These engineering factors are summarized in Table 1 and also described below.

4.2.2. Scaffolding materials

Fully formed myofibers are very specifically arranged with a high degree of alignment within skeletal muscle tissue, so it comes as no surprise that beneficial effects can be generated by providing the cells with a patterned substrate.[57] It has been discovered that aligned

Table 1. Overview of engineering factors for tissue engineering of skeletal muscle. Various factors leading to improved differentiation or function are highlighted.

	Myoblasts	Endothelial cells	Neurons
Substrate properties	12 kPa stiffness nanofibers + microchannels → alignment and differentiation[58,59]	200 Pa → vasculature 2500 Pa → proliferation[60]	7 kPa stiffness + microchannels → proliferation[63]
Mechanical stimulation	Cyclic stretching or static strain → alignment and differentiation[64,66]	Cyclic stretching → perpendicular alignment and angiogenesis[67]	Equibiaxial stretching → neurite outgrowth[70]
Electrical stimulation	0.2 V/m 1 Hz 4 ms → increasing contractility[71]	100–300 mV/mm DC field → perpendicular alignment and angiogenesis[72]	NGF necessary + 100–200 mV/mm at 100 Hz (sinusoid) → neurite elongation[73]

nanoscale topography such as electrospun fibers can stimulate the parallel alignment of C2C12 murine myoblasts, a common model cell for myoblast culture.[58] Coincident with this alignment was an increase in proliferation and upregulation of myosin heavy chain (MYH) and myogenin (MyoG) expression, signaling the beginning of a progression toward myofiber fusion.[58] When the nanofiber topography was coupled with microgrooves, it was found that the grooves restricted myofiber movement and promoted further maturation.[58] Investigations into ideal substrate stiffness have been performed with murine SCs. The cells were cultured on polyethylene glycol (PEG) hydrogels of varying stiffness by controlling the percent weight of the polymer solution.[59] It was found that gels that mimicked the natural stiffness of skeletal muscle, roughly 12 kPa in this study, had the best effect on the SCs in terms of their final engraftment and spread in a murine wound model.[59] The soft hydrogels were also able to significantly reduce cell death commonly experienced by this cell type on rigid polystyrene culture plastic.[59] Keeping these results in mind, it appears that an ideal substrate for tissue engineering of muscle must provide stiffness close to that of the native tissue and ideally have a combination of topographies that lead to improved alignment, proliferation, and differentiation.

Vascular endothelial cells are the primary components of capillary beds in the muscle, and their unique mechanobiology must be addressed in conjunction with that of the myoblasts for formation of functional muscle. Much like with muscle, it has been determined that the substrates that most closely mimic the natural environment can lead to better outcomes in terms of vascularization. It has been shown that endothelial cell morphology changes drastically when cultured on substrates of increasing stiffness. Above 3 kPa, endothelial cells lose their normal shape and begin to elongate drastically.[60] Furthermore, the presence of thick bundles of stress fibers has been observed.[60] These stress fibers, and the other drastic morphological changes between the substrate stiffness disappear when the cells reach confluency.[60] Some authors have noticed that bovine aortic endothelial cells (BAECs) have stronger attachment and increased spreading on stiff (1.1 MPa) vs. soft (0.3 MPa) PDMS substrates.[61] Some authors believe that the effect of a soft substrate on endothelial cells comes from more emphasis on the cell–cell junctions rather than substrate adhesions. This seems to lead to the beginning of more stable vascular networks.

Neurons can rely on a number of factors to guide their direction of regrowth. It has been shown many times that nanotopographies or nanofibers have the potential to direct and enhance neuronal axon growth.[62,63] In a paper demonstrating the effect of substrate mechanics on cell behavior, researchers showed that substrate stiffness can have a large effect on both neurons and glial cells.[63] The substrates were made out of a tunable poly(ε-caprolactone) triacrylate (PCLTA) polymer and patterned with various dimensions of microscale grooves.[63] Rat PC12 pheochromocytoma cells, chosen to represent neurons, were cultured on these substrates for up to 7 days.[63] It was found that neuron proliferation was enhanced on substrates with a lower stiffness of 7 kPa when compared to the substrates at 10 kPa.[63] Interestingly, rat Schwann cell precursor line cells (SpL201) preferred the higher stiffness.[63] It was proposed that, in the body, Schwann cells may have a mechanical cushioning effect on the neurons.[63] This leads to the possibility that the best approach for engineering skeletal muscle involves choosing a particular order of introducing cells to the culture to best utilize their stiffness and other properties.

4.2.3. Mechanical stimulation

Numerous studies have been done on the effects of mechanical forces on myoblasts and their precursors. The mechanical load is often applied to a deformable substrate on which the cells are grown. Two primary modes of uniaxial stretching have been investigated, cyclic and static. It was discovered that a 10% periodic cyclic strain could enhance the differentiation of human skeletal myoblasts, as demonstrated by vast increases in miR-1, miR-133a, miR-206, and myogenic transcription factors Mef-2C.[64] The strain waveform of 10% strain at 0.5 Hz was applied for 1 h at a time with 5 h in between each session for 14 days.[64] Interestingly, another study has shown that myogenic differentiation of bone-derived human MSCs was enhanced by periodic strain, although the differentiation was accompanied by a decrease in proliferation.[65] Static strain has also been shown to have beneficial effects on myoblast cultures, leading to enhanced myofiber fusion.[66] Strain was applied to a 3D fibrin gel culture of C2C12 myoblasts for 6 days. Each day consisted of 6 h at 10% strain followed by 18 h at 3% strain. The stimulation also increased expression of crucial markers like MyoG and desmin, and increased cell alignment parallel to the strain direction.[66] In summary, both cyclic and static strains have the potential to increase differentiation and fusion of myoblasts into myofibers, although cyclic stretching may better mimic the *in vivo* conditions.

Although endothelial cells are not primarily load-bearing cells in the skeletal muscle, various researchers have found that they respond to mechanical stretching. When BAECs were cyclically and uniaxially stretched they aligned perpendicularly to the direction of stretch.[67] Some researchers suggest that this perpendicular orientation reduces the strain on the cells.[68] Stretching the collagen gel on which the BAECs were cultured was enough to increase the number of vascular sprouts into the gel as much as vascular endothelial growth factor (VEGF) treatment.[67] This response was determined to occur through Rho-associated protein kinase (ROCK)-dependent pathways.[67] Interestingly, VEGF and cyclic strain had a negative interaction with each other in the 3D environment, resulting in less sprouting than when either was done alone.[67] Conversely, in a 2D monolayer environment, VEGF plus the stretching had an additive effect.[67] This points to the necessity of testing both types of environments, as the 2D culture may not necessarily behave the same way as a 3D system. Overall, mechanical stimulation was able to improve vascular sprouting.

Neurons are not typically thought of as being highly involved with mechanical forces, however, they have the ability to be very responsive to applied strain. It has been demonstrated that when neurons are cultured in such a way that the nerve body and the axon are attached to different, sliding surfaces, the axons can be lengthened by slowly moving the surfaces apart.[69] This phenomenon is known as axon stretch growth. The neurons could be stretched at 42 μm per hour.[69] Therefore, neurons are definitely capable of compensating for applied strains through axonal elongation. Other researchers have found that neurite outgrowth is increased by cyclic equibiaxial stretching.[70]

Human neuroblastoma (SH-SY5Y) cells were cultured on BioFlex culture plates and mechanically stimulated to 10% at 0.25 Hz.[70] This culture was performed either with or without the addition of retinoic acid (RA), a well-known mediator of neurite outgrowth.[70] It was found that by simply applying this biaxial strain, neurite length and numbers per cell was significantly increased.[70] Stretch also tended to increase the expression of neurofilament light chain (NFL), which is an important part of the neural cytoskeleton, but this change was not statistically significant.[70] This cyclic stretching is more relevant to skeletal muscle tissue than the axonal stretch growth, however, both studies demonstrate that carefully applied forces are able to improve outcomes for neuron cultures.

4.2.4. Electrical stimulation

It is well established that muscles that are not used will atrophy, and muscles that becomes denervated for any number of reasons will often undergo the same. Building on this knowledge, tissue engineers have attempted to use electrical stimulation as a biomimicking signal to enhance the efficacy of muscle constructs. A study done with C2C12 cells found that application of a sine-wave stimulus led to more pronounced striation, larger myofibers, and more MYH expression.[57] The electrical current was oscillating at 1 Hz and had a current of 22 mA. Another study found that periodic stimulation by application of a square voltage waveform has the potential to increase the total protein content and force generated by a construct.[71] It also found that there is a potential for overstimulation in terms of field strength.[71] Best results were obtained when cells were cultured at a field strength of 0.3 V/mm.[71] This work also found a good response at the 1 Hz frequency, in agreement with the first paper. Electrical stimulation, therefore, should have a role in promoting better muscle tissue development. However, it must be done in a manner that prevents overstimulation.

Much like mechanical strain, electrical fields are able to cause changes in endothelial cell behavior and preferential alignment. When human umbilical vein endothelial cells (HUVECs) were exposed to 100–300 mV/mm DC fields, they responded with elongation and aligned along a vector perpendicular to the applied field.[72] This realignment of the cells was found to be VEGFR dependent.[72] Along with the morphological changes, the electrical stimulation caused an increased release of VEGF from the HUVECs.[72] Fields with strengths less than 100 mV/mm did not have as profound of an effect on cells, but the effect was still statistically significant.[72] In general, electrical stimulation led to increased endothelial cell orientation and increased VEGF production, both of which are beneficial for angiogenesis.

As their primary role in the body is electrical signal transduction, it would make sense that electrical stimulation would have an effect on neurons. However, electrical stimulation alone has not been enough to cause changes in neurite outgrowth.[73] PC12 cells grown on collagen-coated culture plastic was stimulated with indium tin oxide electrodes with a 100 Hz alternating signal generating 50, 100 or 200 mV/mm oscillating fields.[73] While simple electrical stimulation was not able to cause any significant effect in cell behavior, it had a strong effect on the cells when the culture was supplemented with nerve growth factor (NGF).[73] It was determined that the electrical stimulation with NGF could increase average neurite length in cells.[73] Optimum effect was reached with field strength of 100 mV/mm and 100 ng/mL NGF supplementation.[73] In summary, electrical stimulation can be beneficial to neurite development of neurons. However, addition of other factors like NGF are necessary to see the full effect.

4.2.5. Oxygen tension

In terms of stem cells used in tissue engineering, oxygen tension also may play a role in development. It has been previously shown that low oxygen content can preserve MSCs in an undifferentiated state, and a more recent study shows that hypoxia has a similar effect on muscle SCs. Decreased oxygen led to faster proliferation, and still allowed SCs to be differentiated in induction media.[74]

5. Challenges and Outlook

Although exciting progress is being made in this field, there are still a number of challenges that will need to be addressed. These include the complexity of the muscle's natural structure, which contains many cell types and is organized into a very specific structure.[75] Although it has been shown that use of multiple cell types can be beneficial in forming vascularized tissue, very little work has been done on innervating the tissue *in vitro*.[75] Another problem, which is more specific to VML, is the use of fresh wound models.[76] These wounds do not really reflect what happens in clinical practice, where it is more likely that the body will have already gone through some of the healing and remodeling process before a tissue engineered intervention is introduced.[11,77] To move forward, it will be necessary to begin integrating knowledge about the primary cell types in skeletal muscle. Although the three cell types discussed here have very different roles and structures, it is clear that they all have the potential to respond to the same mechanobiological stimuli. It is hoped that this work will serve as a tool for future tissue engineering endeavors as more biologically relevant muscle constructs continue to be developed. Questions yet to be answered include, details such as whether cyclic or static strain are able to provide better results for myoblast proliferation or differentiation. Also, what is the best way to reconcile the differences in the electrical stimulation methods used by researchers? All of these challenges will need to be addressed for the field to move forward to a point where skeletal muscle tissue engineering for VML is clinically useful.

References

1. Ehmsen J, Poon E, Davies K. The dystrophin-associated protein complex. *J. Cell Sci.* **115**(14), 2801–2803 (2002).
2. Godfrey C, Muses S, McClorey G, Wells KE, Coursindel T, Terry RL, Betts C, Hammond S, O'Donovan L, Hildyard J, El Andaloussi S, Gait MJ, Wood MJ, Wells DJ. How much dystrophin is enough: the physiological consequences of different levels of dystrophin in the mdx mouse. *Hum. Mol. Ge.* **24**(15), 4225–4237 (2015).

3. Heydemann A, McNally EM. Consequences of disrupting the dystrophinsarcoglycan complex in cardiac and skeletal myopathy. *Trends Cardiovasc. Med.* **17**(2), 55–59 (2006).
4. Deconinck N, Dan B. Pathophysiology of duchenne muscular dystrophy: Current hypotheses. *Pediatr. Neurol.* **36**(1), 1–7 (2007).
5. Bushby K, Finkel R, Birnkrant DJ, Case LE, Clemens PR, Cripe L, Kaul A, Kinnett K, McDonald C, Pandya S, Poysky J, Shapiro F, Tomezsko J, Constantin C. Diagnosis and management of duchenne muscular dystrophy, part 1: Diagnosis, and pharmacological and psychosocial management. *The Lancet Neurology* **9**(1), 77–93 (2010).
6. Kohler M, Clarenbach CF, Bahler C, Brack T, Russi EW, Bloch KE. Disability and survival in duchenne muscular dystrophy. *J. Neurol. Neurosurg. Psychiatry.* **80**(3), 320–325 (2009).
7. Florian A, Rösch S, Bietenbeck M, Engelen M, Stypmann J, Waltenberger J, Sechtem U, Yilmaz A. Cardiac involvement in female duchenne and becker muscular dystrophy carriers in comparison to their first-degree male relatives: a comparative cardiovascular magnetic resonance study. *Eur. Heart J. Cardiovasc. Imaging* **17**(3), 326–333 (2016).
8. Romitti PA, Zhu Y, Puzhankara S, James KA, Nabukera SK, Zamba Gideon KD, Ciafaloni E, Cunniff C, Druschel CM, Katherine D, Dennis MJ, Matthews F, Meaney J, Andrews JG, Caspers Conway KM, Fox DJ, Street N, Adams MM, Bolen J. Prevalence of duchenne and becker muscular dystrophies in the United States. *Pediatr.* **135**(3), 513–521 (2015).
9. Morley JE, Thomas DR, Wilson M-MG. Cachexia: Pathophysiology and clinical relevance. *Am. J. Clin. Nutr.* **83**(4), 735–743 (2006).
10. Malafarina V, riz Otano F, Iniesta R, Gil-Guerrero L. Sarcopenia in the elderly: Diagnosis, physiopathology and treatment. *Maturitas.* **71**(2), 109–114 (2012).
11. Grogan BF, Hsu JR. Volumetric muscle loss. *J. Am. Acad. Orthop. Surg.* **19**, s35+ (2011).
12. Aurora A, Garg K, Corona BT, Walters TJ. Physical rehabilitation improves muscle function following volumetric muscle loss injury. *BMC Sports Sci. Med. Rehabili.* **6**, 41 (2014).
13. Kraybill WG, Pollock RE, Chao AH, Mayerson JL, Chandawarkar R, Scharschmidt TJ. Surgical management of soft tissue sarcomas: Extremity sarcomas. *J. Surg. Oncol.* **111**, 540+ (2015).
14. Baoge L, Van Den Steen E, Rimbaut S, Philips N, Witvrouw E, Almqvist KF, Vanderstraeten G, Vanden Bossche LC. Treatment of skeletal muscle injury: A review. *ISRN Orthop.* **2012**: 7 (2012).

15. Turner NJ, Badylak SF. Regeneration of skeletal muscle. *Cell Tissue Res.* **347**, 759+ (2012).
16. Mathes SJ, Levine J. *Muscle Flaps and Their Blood Supply*, 6th Edition. Lippincott Williams and Wilkins, Philadelphta, PA, 2007, Book Section 5, pp. 42–51.
17. Garg K, Ward CL, Hurtgen BJ, Wilken JM, Stinner DJ, Wenke JC, Owens JG, Corona BT. Volumetric muscle loss: Persistent functional deficits beyond frank loss of tissue. *J. Orthop. Res.* **33**(1), 40–46 (2015).
18. Murphy M, Kardon G. Origin of vertebrate limb muscle: The role of progenitor and myoblast populations. In Grace K. Pavlath's, *Myogenesis*, volume 96 of *Current Topics in Developmental Biology*, Academic Press, Chapter 1, pp. 1–32 (2011).
19. Tajbakhsh S. Skeletal muscle stem cells in developmental versus regenerative myogenesis. *J. Intern. Med.* **266**(4), 372–389 (2009).
20. Zhao P, Hoffman EP. Embryonic myogenesis pathways in muscle regeneration. *Dev. Dynam.* **229**(2), 380–392 (2004).
21. Eng D, Ma H-Y, Gross MK, Kioussi C. Gene networks during skeletal myogenesis. *ISRN Dev. Biol.* **2013**: 8 (2013).
22. Tran T, Andersen R, Sherman SP, Pyle AD. Insights into skeletal muscle development and applications in regenerative medicine. In Kwang W. Jeon's, *International Review of Cell and Molecular Biology*, volume 300 of *International Review of Cell and Molecular Biology*. Academic Press, Cambridge, MA, pp. 51–83 (2013).
23. Garcia-Guerra L, Vila-Bedmar R, Carrasco-Rando M, Cruces-Sande M, Martn M, Ruiz-Gmez A, Ruiz-Gmez M, Lorenzo M, Fernndez-Veledo S, Mayor F, Murga C, Nieto-Vzquez I. Skeletal muscle myogenesis is regulated by g protein-coupled receptor kinase 2. *J. Mol. Cell Biol.* **6**(4), 299–311 (2014).
24. Alonso-Martin S, Rochat A, Mademtzoglou D, Morais J, de Reynis A, Aurad F, Chang TH-T, Zammit PS, Relaix F. Gene expression profiling of muscle stem cells identifies novel regulators of postnatal myogenesis. *Front. Cell Dev. Biol.* **4**(6), 58 (2016).
25. Almada AE, Wagers AJ. Molecular circuitry of stem cell fate in skeletal muscle regeneration, ageing and disease. *Nat. Rev. Mol. Cell. Biol.* **17**(5), 267–279 (2016).
26. Peault B, Rudnicki M, Torrente Y, Cossu G, Tremblay JP, Partridge T, Gussoni E, Kunkel LM, Huard J. Stem and progenitor cells in

skeletal muscle development, maintenance, and therapy. *Mol. Ther.* **15**(5), 867–877 (2007).
27. Doerflinger RM. Old and new ethics in the stem cell debate. *J. Law Med. Ethics.* **38**(2), 212–+ (2010).
28. Calderon D, Planat-Benard V, Bellamy V, Vanneaux V, Kuhn C, Peyrard S, Larghero J, Desnos M, Casteilla L, Pucat M, Menasch P, Chatenoud L. Immune response to human embryonic stem cell-derived cardiac progenitors and adipose-derived stromal cells. *J. Cell. Mol. Med.* **16**(7), 1544–1552 (2012).
29. Takahashi K, Yamanaka S. Induction of pluripotent stem cells from mouse embryonic and adult fibroblast cultures by defined factors. *Cell* **126**(4), 663–676 (2006).
30. Heslop JA, Hammond TG, Santeramo I, Tort Piella A, Hopp I, Zhou J, Baty R, Graziano EI, Proto Marco B, Caron A, Skld P, Andrews PW, Baxter MA, Hay DC, Hamdam J, Sharpe ME, Patel S, Jones DR, Reinhardt J, Danen EHJ, Ben-David U, Stacey G, Bjrquist P, Piner J, Mills J, Rowe C, Pellegrini G, Sethu S, Antoine DJ, Cross MJ, Murray P, Williams DP, Kitteringham NR, Goldring CEP, Park BK. Concise review: Workshop review: Understanding and assessing the risks of stem cell-based therapies. *Stem Cells Transl. Med.* **4**(4), 389–400 (2015).
31. Medvedev SP, Shevchenko AI, Zakian SM. Induced pluripotent stem cells: Problems and advantages when applying them in regenerative medicine. *Acta Naturae* **2**(2), 18–28 (2010).
32. Chamberlain G, Fox J, Ashton B, Middleton J. Concise review: Mesenchymal stem cells: Their phenotype, differentiation capacity, immunological features, and potential for homing. *Stem Cells* **25**(11), 2739–2749 (2007).
33. Dominici M, Le Blanc K, Mueller I, Slaper-Cortenbach I, Marini FC, Krause DS, Deans RJ, Keating A, Prockop DJ, Horwitz EM. Minimal criteria for defining multipotent mesenchymal stromal cells. the international society for cellular therapy position statement. *Cytotherapy* **8**(4), 315–317 (2006).
34. Stern-Straeter J, Bonaterra GA, Juritz S, Birk R, UR. Goessler, Bieback K, Bugert P, Schultz J, Hormann K, Kinscherf R, Faber A. Evaluation of the effects of different culture media on the myogenic differentiation potential of adipose tissue- or bone marrow-derived human mesenchymal stem cells. *Int. J. Mol. Med.* **33**(1), 160–170 (2014).

35. Kalinina NI, Sysoeva VYu, Rubina KA, Parfenova YeV, Tkachuk VA. Mesenchymal stem cells in tissue growth and repair. *Acta Naturae* **3**(4), 30–37 (2011).
36. Martello G, Smith A. The nature of embryonic stem cells. *Annu. Rev. Cell Dev. Biol.* **30**, 647–675 (2014).
37. Salani S, Donadoni C, Rizzo F, Bresolin N, Comi GP, Corti D. Generation of skeletal muscle cells from embryonic and induced pluripotent stem cells as an *in vitro* model and for therapy of muscular dystrophies. *J. Cell. Mol. Medi.* **16**(7), 1353–1364 (2012).
38. Abujarour R, Bennett M, Valamehr B, Tong Lee T, Robinson M, Robbins D, Le T, Lai K, Flynn P. Myogenic differentiation of muscular dystrophyspecific induced pluripotent stem cells for use in drug discovery. *Stem Cells Transl. Medi.* **3**(2), 149–160 (2014).
39. Ma S, Xie N, Li W, Yuan B, Shi Y, Wang Y. Immunobiology of mesenchymal stem cells. *Cell Death Differ.* **21**(2), 216–225 (2014).
40. Duffy RM, Feinberg AW. Engineered skeletal muscle tissue for soft robotics: Fabrication strategies, current applications, and future challenges. *Wiley Interdiscipl. Revi. Nanomed. Nanobiotechnol.* **6**(2), 178–195 (2014).
41. De Bari C, Dell'Accio F, Vandenabeele F, Vermeesch JR, J-Raymackers M, Luyten FP. Skeletal muscle repair by adult human mesenchymal stem cells from synovial membrane. *J. Cell Biol.* **160**(6), 909–918 (2003).
42. Fishman JM, Tyraskis A, Maghsoudlou P, Urbani L, Totonelli G, Birchall MA, De Coppi P. Skeletal muscle tissue engineering: Which cell to use? *Tissue Engineering Part B-Reviews* **19**(6), 503–515 (2013).
43. Merrison AF. Human bone marrow-derived stem cells: Myogenic potential and the factors influencing skeletal muscle differentiation. *Journal of Neurology Neurosurgery and Psychiatry* **78**(2), 206–206 (2007).
44. Pereira T, Armada-da Silva PAS, Amorim I, Rema A, Caseiro AR, Gartner A, Rodrigues M, Lopes MA, Bartolo PJ, Santos JD, Luis AL, Mauricio AC. Effects of human mesenchymal stem cells isolated from wharton's jelly of the umbilical cord and conditioned media on skeletal muscle regeneration using a myectomy model. *Stem Cells International* (2014).
45. Merritt EK, Cannon MV, DW. Hammers, Le LN, Gokhale R, Sarathy A, Song TJ, Tierney MT, Suggs LJ, Walters TJ, Farrar RP. Repair of traumatic skeletal muscle injury with bone-marrow-derived mesenchymal stem cells seeded on extracellular matrix. *Tissue Engi. Part A.* **16**(9), 2871–2881 (2010).

46. Seale P, Rudnicki MA. A new look at the origin, function, and stem-cell status of muscle satellite cells. *Dev. Biol.* **218**(2), 115–124 (2000).
47. Brack AS, Rando TA. Tissue-specific stem cells: Lessons from the skeletal muscle satellite cell. *Cell Stem Cell* **10**(5), 504–514 (2012).
48. Pallafacchina G, Franois S, Regnault B, Czarny B, Dive V, Cumano A, Montarras D, Buckingham M. An adult tissue-specific stem cell in its niche: A gene profiling analysis of *in vivo* quiescent and activated muscle satellite cells. *Stem Cell Res.* **4**(2), 77–91 (2010).
49. Lam MT, Y-Huang C, Birla RK, Takayama S. Microfeature guided skeletal muscle tissue engineering for highly organized 3-dimensional freestanding constructs. *Biomaterials* **30**(6), 1150–1155 (2009).
50. Ju YM, Atala A, Yoo JJ, Lee SJ. In situ regeneration of skeletal muscle tissue through host cell recruitment. *Acta Biomater.* **10**(10), 4332–4339 (2014).
51. Garg K, CL. Ward, Rathbone CR, Corona BT. Transplantation of devitalized muscle scaffolds is insufficient for appreciable *de novo* muscle fiber regeneration after volumetric muscle loss injury. *Cell and Tissue Res.* **358**(3), 857–873 (2014).
52. Nie X, Xing YJ, Deng MJ, Gang L, Liu R, Zhang YJ, Wen XJ. Ecto-mesenchymal stem cells from facial process: Potential for muscle regeneration. *Cell Biochem. Biophys.* **70**(1), 615–622 (2014).
53. Kang H-W, Jin Lee S, In Ko K, Kengla C, Yoo JJ, Atala A. A 3d bioprinting system to produce human-scale tissue constructs with structural integrity. *Nat. Biotech.* **34**(3), 312–319 (2016).
54. Kaully T, Kaufman-Francis K, Lesman A, Levenberg S. Vascularizationthe conduit to viable engineered tissues. *Tissue Eng. Part B Rev.* **15**(2), 159–169 (2009).
55. Carosio S, Barberi L, Rizzuto E, Nicoletti C, Del Prete Z, Musar A. Generation of ex vivo-vascularized muscle engineered tissue (x-met). *Sci. Rep.* **3**, 1420 (2013).
56. Dana E, Pavlov Daria A, Kaufman-Francis K, Koffler J, Landesberg A, Levenberg S, Yulia S. Improved vascular organization enhances functional integration of engineered skeletal muscle grafts. *Proceedings of the National Academy of Sciences of the United States*, **108**, 14789+ (2011).
57. Hosseini V, Ahadian S, Ostrovidov S, Camci-Unal G, Chen S, Kaji H, Ramalingam M, Khademhosseini A. Engineered contractile skeletal muscle tissue on a microgrooved methacrylated gelatin substrate. *Tissue Eng. Part A.* **18**(2324), 2453–2465 (2012).

58. Jana S, Leung M, Chang JL, Zhang MQ. Effect of nano- and micro-scale topological features on alignment of muscle cells and commitment of myogenic differentiation. *Biofabrication* **6**(3) (2014).
59. Gilbert PM, Havenstrite KL, Magnusson KE, Sacco A, Leonardi NA, Kraft P. Substrate elasticity regulates skeletal muscle stem cell self-renewal in culture. *Science* **329** (2010).
60. Yeung T, PC. Georges, Flanagan LA, Marg B, Ortiz M, Funaki M, N Zahir, Ming W, Weaver V, Janmey PA. Effects of substrate stiffness on cell morphology, cytoskeletal structure, and adhesion. *Cytoskeleton.* **60**(1), 24–34 (2005).
61. Ataollahi F, Pramanik S, Moradi A, Dalilottojari A, Pingguan-Murphy B, Abas WAW, Abu Osman NA. Endothelial cell responses in terms of adhesion, proliferation, and morphology to stiffness of polydimethylsiloxane elastomer substrates. *J. Biomed. Mater. Res. Part A.* **103**(7), 2203–2213 (2015).
62. Discher DE, Janmey P, Y-l Wang. Tissue cells feel and respond to the stiffness of their substrate. *Science* **310**(5751), 1139–1143 (2005).
63. Cai L, Zhang L, Dong J, Wang S. Photocured biodegradable polymer substrates of varying stiffness and microgroove dimensions for promoting nerve cell guidance and differentiation. *Langmuir* **28**(34), 12557–12568 (2012).
64. Cheng CS, El-Abd Y, Bui K, Hyun YE, Hughes RH, Kraus WE, Truskey GA. Conditions that promote primary human skeletal myoblast culture and muscle differentiation *in vitro. Am. J. Physiol.-Cell Physiol.* **306**(4), C385–C395 (2014).
65. Egusa H, Kobayashi M, Matsumoto T, Sasaki JI, Uraguchi S, Yatani H. Application of cyclic strain for accelerated skeletal myogenic differentiation of mouse bone marrow-derived mesenchymal stromal cells with cell alignment. *Tissue Eng. Part A.* **19**(5–6), 770–782 (2013).
66. Heher P, Fuchs C, Pruller J, Maleiner B, Kollmitzer J, Runzler D, Teuschl A, Wolbank S, Redl H. Static mechanical strain improves myogenic differentiation in a bioreactor-based 3d skeletal muscle engineering approach. *Wound Repair Regen.* **22**(5), A83–A83 (2014).
67. Wilkins JR, Pike DB, Gibson CC, Li L, Shiu YT. The interplay of cyclic stretch and vascular endothelial growth factor in regulating the initial steps for angiogenesis. *Biotechnol. Progr.* **31**(1), 248–257 (2015).
68. Jufri NF, Mohamedali A, Avolio A, Baker MS. Mechanical stretch: physiological and pathological implications for human vascular endothelial cells. *Vascular Cell* 7, 12 (2015).

69. Loverde JR, Tolentino RE, Pfister BJ. Axon stretch growth: The mechanotransduction of neuronal growth. *J. Vis. Exp: (JoVE)*. **54**, 2753 (2011).
70. Higgins S, Soon Lee J, Ha L, Yul Lim J. Inducing neurite outgrowth by mechanical cell stretch. *BioResearch Open Access*. **2**(3), 212–216 (2013).
71. Ito A, Yamamoto Y, Sato M, Ikeda K, Yamamoto M, Fujita H, Nagamori E, Kawabe Y, Kamihira M. Induction of functional tissue-engineered skeletal muscle constructs by defined electrical stimulation. *Sci. Rep*. **4** (2014).
72. Zhao M, Bai H, Wang E, Forrester JV, McCaig CD. Electrical stimulation directly induces pre-angiogenic responses in vascular endothelial cells by signaling through vegf receptors. *J. Cell Science*. **117**(Pt 3), 397–405 (2004).
73. Chang Y-J, Hsu C-M, Lin C-H, Shiang-Cheng Lu M, Chen L. Electrical stimulation promotes nerve growth factor-induced neurite outgrowth and signaling. *Biochim. Biophys. Acta*. **1830**(8), 4130–4136 (2013).
74. Koning M, Werker PMN, van Luyn MJA, Harmsen MC. Hypoxia promotes proliferation of human myogenic satellite cells: A potential benefactor in tissue engineering of skeletal muscle. *Tissue Eng. Part A*. **17**(13–14), 1747–1758 (2011).
75. Perniconi B, Coletti D. Skeletal muscle tissue engineering: Best bet or black beast? *Frontiers in Physiology* **5**, 255 (2014).
76. Cittadella Vigodarzere G, and Mantero S. Skeletal muscle tissue engineering: Strategies for volumetric constructs. *Front. Phys*. **5** (2014).
77. Li MTA, Willett NJ, Uhrig BA, Guldberg RE, Warren GL. Functional analysis of limb recovery following autograft treatment of volumetric muscle loss in the quadriceps femoris. *J. Biomech*. **47**(9), 2013–2021 (2014).

Chapter 2

Characterization of Biomaterial Patches as Fetal Surgery Implants

Rigwed Tatu* and Chia-Ying Lin*,†,‡

**Department of Biomedical Engineering*
†Department of Orthopedic Surgery
‡Department of Neurosurgery
University of Cincinnati
Cincinnati, OH 45221, USA

Abstract: Fetal research has a considerable impact on the life of a child and his family, making it extremely critical for the scientific community. Biomaterial implants used in hernia repair and aortic arch reconstruction are also employed for open fetal surgery and fetoscopic applications. The internal environment around the fetus is drastically different compared to an abdominal or cardiac setting, and needs to be studied further with regards to the pH changes, mechanical impact during fetal growth, and degradation profiles. These studies will in-turn affect the desired properties of polymeric implants used in fetal surgery, and also the characterizations to be performed. What we learn from them will not only help us understand the property requirements of the implants, but also aid in realizing the relationship between the fetal tissues and the implants. This chapter reviews the range of characterizations carried out on polymeric implants used in abdominal and cardiac applications, which need to be applied for fetal implants for a more confident approach towards fetal surgery.

Keywords: Fetal surgery, fetoscopy, tissue engineering, polymeric implants, degradation, mechanical properties, inflammatory response.

1. Introduction

1.1. Fetal Surgery

Being an extremely sensitive field of surgery, fetal research has garnered immense interest from the scientific community. Over the years, there has been a revolution of research efforts to surgically treat birth defects without posing any risk to the unborn baby and the mother. These efforts range from open fetal surgery to minimally invasive interventions to stem cell transplantation, but some consequences such as preterm labor has not yet been solved.[1]

Neural tube defects are one of most common birth defects, and involve incomplete development of the neural elements. Subsequently, improper development of adjacent bone and mesenchymal structures are also observed. Fetal surgeries are carried out to rectify neural tube defects, which are detected by ultrasound techniques in the first few weeks of pregnancy. As these defects are detected, they need to be repaired at the earliest to eliminate the risk of post-birth problems for the child. The two most commonly occurring neural tube defects are spina bifida and anencephaly.[2]

1.2. Spina Bifida

Spina Bifida is the most frequently occurring neural tube defect in the United States, affecting 0.05% of babies born annually (Fig. 1). Precisely known as "split spine", this defect leads to deficiencies in the development of the brain, spinal cord and meninges (the protective covering around the brain and spinal cord). Of the various spina bifida forms myelomeningocele (MMC) is the most severe and emerges when neural elements are unsealed due to opening in the spine, inducing partial or complete paralysis of the body parts beneath the spinal aperture.[3,4]

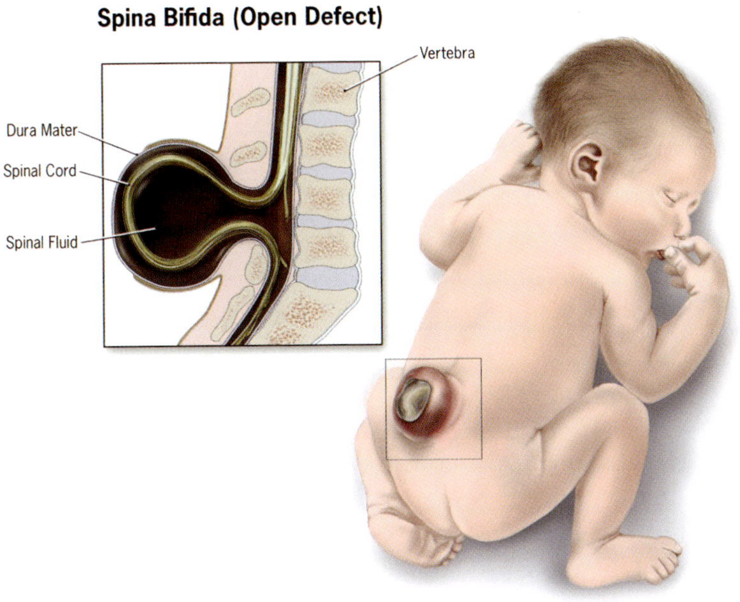

Fig. 1. Spina Bifida defect.[5]

1.3. Myelomeningocele Repair

Prenatal or postnatal surgery is the only choice for alleviating MMC.[6] It is impossible to restore activity in the damaged nerve tissues. Specific organizations in the United States possess the competency to execute fetal surgery for treatment of MMC through National Institute of Health (NIH) experimented protocol.[7] Open surgery involves dural closure followed by suturing of spinal defect, but it poses a potential risk for both mother and fetus.[8] Recent studies have indicated benefits of a fetoscopic approach, which consists of suturing the skin over a polymeric patch.[8] Recently, researchers have augmented a coverage method comprising patch placement over the defect and adhesion using a sealant, which displayed encouraging outcomes in fetal sheep and two human procedures.[8] The properties of patches used are of utmost importance, since they would govern the performance after implantation and in turn affect the development of neural elements and healing of surgical lesion. Some surgeons prefer the use of patches made from extracellular matrix and pericardium to facilitate tissue ingrowth and acceptance by the body, but these have certain manufacturing restrictions.

1.4. Polymeric Patches in Fetal Surgery

Polymeric patches made of polypropylene (PP), polyethylene terephthalate and polytetrafluoroethylene (PTFE) are commonly used in surgeries such as aortic arch reconstruction and hernia repair.[9] Patches employed in these surgeries are also used in fetal surgery, but there are numerous issues associated with this technique. There is negligible prior art available on characterization of patches used for fetal surgery, and this subject needs to be delved into deeper and explored further. Fetal surgery refers to a complex domain, since the after-effects of surgery are directly linked to the life of a child.

Surgical patches have been categorized on the basis of long-term physical behavior, weight, size and elasticity.[9] We will discuss the range of characterizations carried out for patches used for hernia repair, aortic arch reconstruction, etc. which can also be applied to the field of fetal research.

2. Literature Review

2.1. Inflammatory Response

Inflammation can be described as local tissue reaction to an injury or implantation. An inflammatory response is initiated with clot formation and contraction. Any tissue-material interaction is accompanied by protein adsorption to the biomaterial, but small extent of research has been done on this phenomenon.[10] Scar formation is associated with an inflow of defense cells between mesh pores and creation of connective tissue with deposition of collagen.[11] Fibrotic scar shrinkage causes mesh retraction, which can be avoided by possessing optimum mechanical properties.[11] The shape and structure of the surgical implant govern the biocompatibility of the implant, which depends on the shrinkage and protein adsorption characteristics.[10]

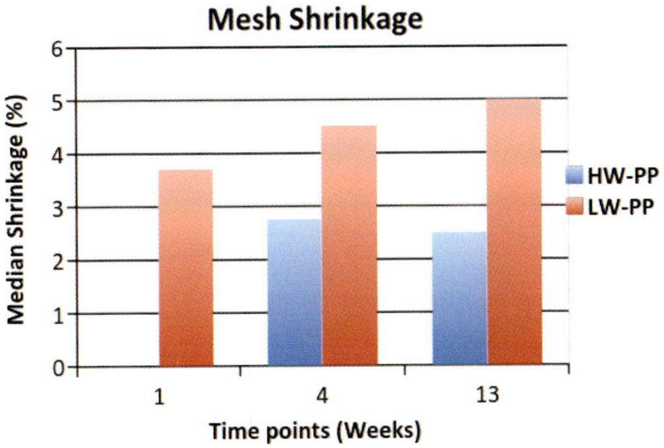

Fig. 2. Median mesh shrinkage vs. time points for LW–PP and HW–PP meshes.[11]

Zogbi and coworkers compared the shrinkage of heavyweight PP (HW–PP) and lightweight PP (LW–PP) meshes after 7, 28 and 90 days of fixation on abdominal surfaces of 25 Wistar rats (Fig. 2). A HW–PP mesh (*Marlex*®) measuring 95 g/m^2 was implanted on the right side of each rat and a LW–PP mesh (*Parietene*®) on the left side.[11] The statistical analysis was carried out using the Statistical Package for Social Sciences (SPSS) 18.0 and shrinkage was computed using the Kruskal–Wallis test, with a 95% level of significance ($P \leq 0.05$). The significance for each mesh shrinkage at 7, 28 and 90 days was 0.647, 0.020 and 0.013, respectively for HW–PP mesh, and 0.040, 0.014 and 0.001, respectively for LW–PP mesh.[11]

This report by Zogbi and coworkers hypothesize that less HW–PP mesh shrinkage at 90 days (2.5%) than at 28 days (2.7%) is due to the progressive weight gain of the rats, and links the mesh weight with the shrinkage and thus the inflammatory response.[11]

An investigation by Jacob and coworkers distinguished the shrinkage of a lightweight large pore PP mesh (*Optilene*® *LP*) and mediumweight large pore knitted PTFE mesh (*GORE*® *INFINIT*® *mesh*), after porcine implantation (Fig. 3). An analysis of variance was performed with a statistical significance in *p*-values below 0.05, using a SPSS 14.0 statistics program. Measurements after 3 months of

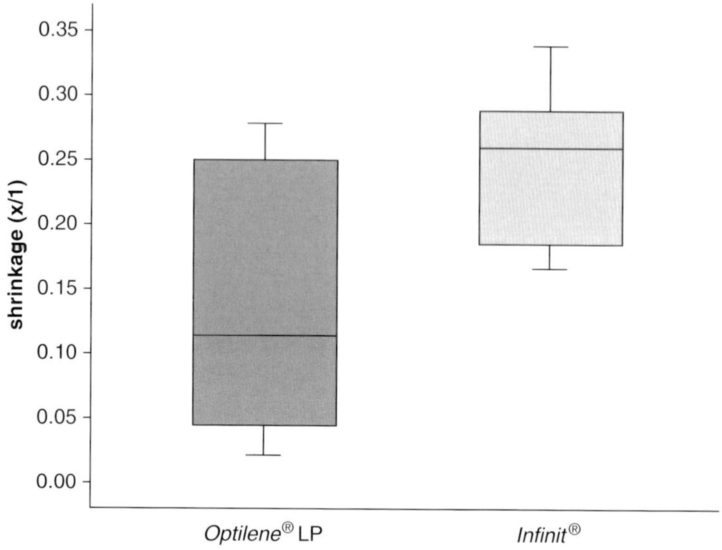

Fig. 3. Boxplots showing shrinkage rate for *Optilene® LP* and *INFINIT®* meshes.[12]

implantation showed a reduction in total surface area with an average shrinkage of 14 ± 10% for *Optilene® LP* against 25 ± 6% for *INFINIT®*, which was significant (p = 0.017).[12] The maximum reduction in length and width was 1.2 cm and 1.0 cm, respectively for *Optilene® LP*, and 2.0 cm and 1.0 cm, respectively for *INFINIT®*.[12]

This study also establishes the relationship between mesh weight and inflammatory response. In case of patches used for fetal surgeries such as MMC repair, the patch weight would be of great importance not only due to inflammatory response but also due to the possibility of a negative effect on the growth of fetus. Hence, the optimal weight of a surgical patch needs to be studied further for fetal research.

2.2. Mechanical Properties

Mechanical properties of surgical patches are of primary importance, since they impart requisite virtues such as elasticity, stiffness, long-term durability, strength and modulus. Fatigue failure or deformation of an implanted mesh can lead to numerous complications and eventual revision surgery.

Li and coworkers studied the biomechanical behavior of *Prolene*® PP mesh, which is commonly used for surgical repair procedures. Dogbone-shaped samples of *Prolene*® mesh were cut coursewise (0°), walewise (90°), and in the diagonal (45°) direction (Fig. 4) using a laser cutter. Instron™ was used to carry out uniaxial tensile testing.[13]

To simulate large pressure changes during coughing and walking upstairs, cyclic tests were conducted to observe the mechanical response of the mesh under large amplitude responsive loading.[13] The strength necessary for fascial closure of the abdominal wall is dependent on the intraabdominal pressure, varying from 0.2 kPa to 20 kPa in healthy humans. Considering the abdominal wall as a thin-walled cylinder, a tensile strength of 1.6 kN/m is needed to bear the maximum intraabdominal pressure. Using this value and the mesh sample width (40 mm), the maximum force encountered by the mesh was estimated.[13] The force to bear resting pressure was gauged in a similar manner, and used to decide the loading pattern for cyclic tests.[13]

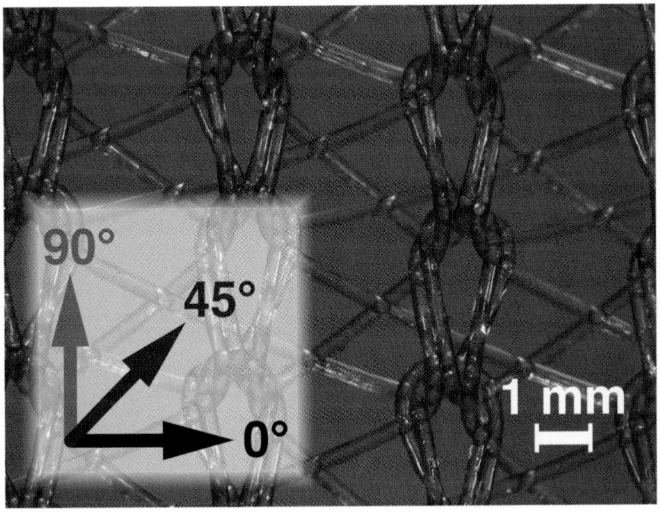

Fig. 4. *Prolene*® mesh directions.[13]

Table 1. Stiffness values for cyclic testing in 0°, 45° and 90° directions.

Direction	Stiffness C1 (MPa)	Stiffness C2 (MPa)	Stiffness C3 (MPa)
0°	14.17	21.72	20.05
45°	17.47	26.62	25.75
90°	21.23	26.19	26.43

The samples were initially loaded to 5 N prior to the cyclic tests, and 5–30 N (Cycle 1) was considered as the normal physiological loading range, whereas 5–60 N (Cycle 2) was taken as the physiological loading limit. Cycle 3 was also carried out between 5 N and 30 N. The stiffness was calculated as the slope of loading curve of the 10th cycle.[13] Results from cyclic testing show that the stiffness increased with successive loading cycles, which suggests strain hardening in the material (Table 1). A considerable increase in C1 and C2 values is visible for all three directions, hinting at the possibility of strain hardening at the physiological loading limit, experienced during extreme pressure variation.[13] Studying the effects of cyclic loading after exposure to phosphate buffered saline (at 37°C) for time of intended implantation would provide a more accurate understanding of the stiffness response and strain hardening of the polymer.

Failure tests were conducted on three samples from each of the three directions, where each sample was elongated to failure at 1 mm/s using a 1 kN load cell (Fig. 5).[13] The stiffest response was unveiled by samples loaded in 90° direction, whereas the 0° direction proved to be the most compliant. Results of failure tests clearly highlight the anisotropic variation of the meshes, which needs to be examined in greater detail by biaxial tests.[13]

In totality, Li and coworkers did an excellent job of failure analysis and cyclic testing for patches, to mimic real-time conditions. As the abdominal pressures carry weightage while designing a patch for hernia repair, similarly the patches used for fetal surgeries need to be designed taking into account the growth of the fetus. The patch should have optimal elastic properties to accommodate radial stretching due to fetal growth over a period of few months.

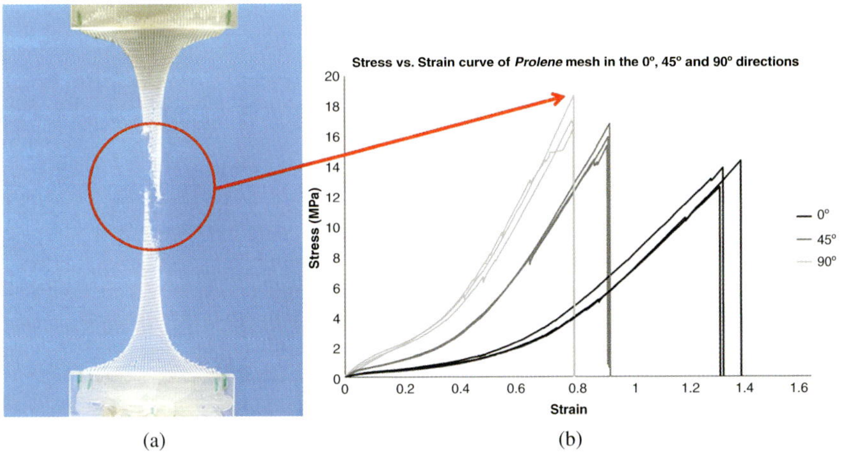

Fig. 5. (a) Failure of *Prolene*® samples cut in coursewise (90°) direction; (b) stress vs. strain curves for failure tests on *Prolene*® mesh.[13]

Another technique to interpret mechanical properties and anisotropy was reported by Pott and Schwarz.[14] Six different meshes were assessed in view of longitudinal and transverse directions through uniaxial tensile tests carried out on a Zwick™ 020 universal testing machine, to compute maximum loads using force-displacement data.[14] The meshes studied were *Dynamesh-Ipom*® (PVDF), *Parietene*® (PP), *Prolene*® (PP), *Surgipro Pro*® (PP), *Ultrapro Mesh*® (absorbable polyglecaprone-25 and non-absorbable PP filaments), and *Vicryl*® (resorbable polyglactin filaments). A student's t-test with a confidence level of 95% was used for intramaterial comparisons (longitudinal vs. transverse), whereas ANOVA variance analysis with significant differences for $p < 0.001$ was employed for intermaterial comparison of different mesh types (SPSS 18.0).[14]

For tensile testing, a dogbone-shaped die (ISO 527-1) was used to cut samples in longitudinal (warp) and orthogonal (weft) directions. Prior to testing, the specimens were immersed in isotonic saline for 30 min.[14] Test conditions included a strain rate of 50 mm/min and were ended when recorded load fell below 90% of the maximum load. The test results of maximum load were compared to human data pertaining to maximum forces in the abdominal wall (Fig. 6).[14–17]

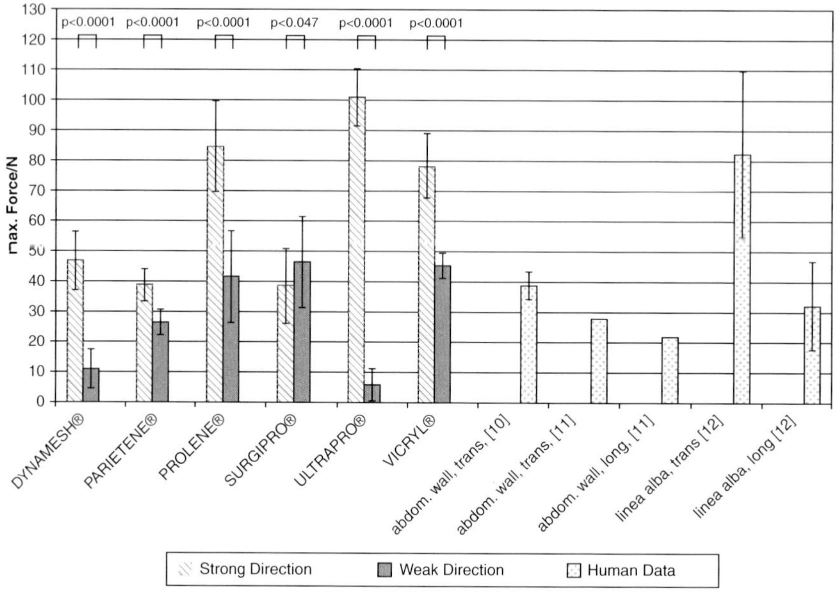

Fig. 6. Bar graphs depicting the maximum load of the mesh materials.[14]

Uniaxial testing would be of equal importance in the case of fetal patches, since they would give an idea about the patch response and integrity during fetal growth. The placement of the patch at the right position would also carry value if the patch demonstrates anisotropy.

An experimental analysis carried out by Velayudhan and Martin[18] evaluate the dynamic creep properties of four commonly used hernia meshes (*Prolene®*, *Ultrapro®*, *Vypro®* and *Vypro® II*), primarily consisting of non-absorbable PP.[18] Dynamic creep tests were executed on an EnduraTEC BioDMA, using a 250-N load cell.[18] Mesh specimens were cut along machine (warp) and perpendicular (weft) directions; and subjected to force controlled sinusoidal oscillation for 100,000 cycles at a frequency of 1 Hz.[18] The chosen mean load for each experiment was 16 N/cm, which was based on physiological load values and force amplitude of 4 N was retained.[19] Secant modulus was determined from the slope of the hysteresis loop. The dynamic creep behavior was probed in phosphate buffer saline (PBS, pH = 7.3) at 37°C, which effectively simulated physiological environments. The secant moduli increased gradually for all

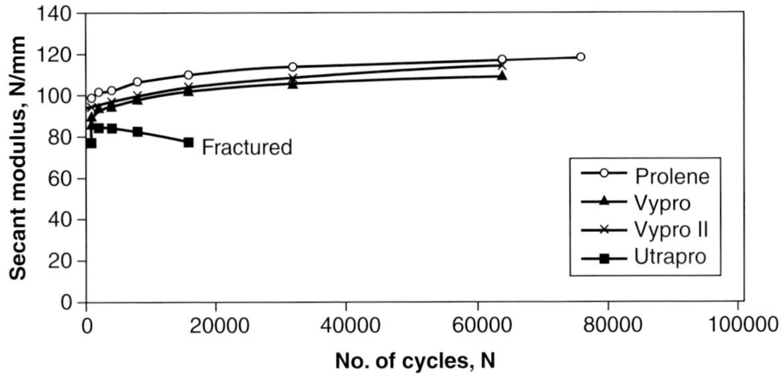

(a)

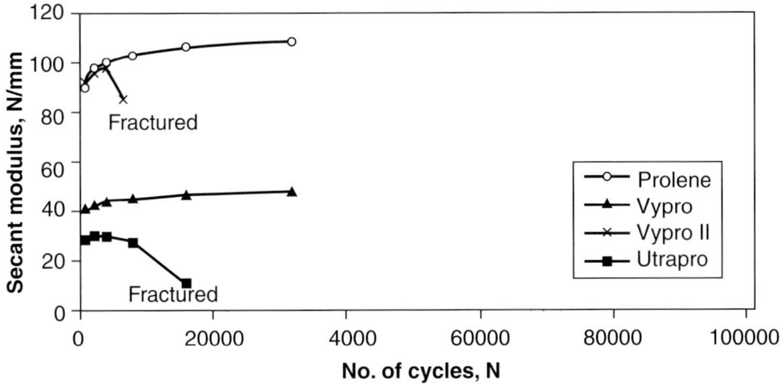

(b)

Fig. 7. Variation of secant modulus of various meshes in (a) warp and (b) weft directions with the number of cycles tested in PBS at 37°C.[18]

meshes in both directions, except for *Ultrapro*® where it declined after an initial rise (Fig. 7).[18]

From the results, Velayudhan and Martin concluded that all meshes strain hardened and underwent permanent plastic deformation, which can lead to complications for long-term implantations.[18] The authors pointed out that lower secant moduli of meshes at 37°C in PBS than that at 23°C in air was due to higher polymer chain mobility at elevated temperatures.[18] This claim can be further consolidated by providing glass transition temperature values of the polymers used, and debating whether a temperature of 37°C would affect the chain mobility. Though the test conditions in this article better mimic the physiological conditions encountered by the mesh, biaxial testing and *in vivo* models would provide superior insights into mesh performance. It would be interesting to study the effect on fetal patches during the breathing cycle of the mother. If the breathing cycle of the mother exerts any kind of pressure before the incision in the placenta has completely healed, dynamic tests might be of importance.

Deeken and Abdo assessed the physico-mechanical behavior of surgical meshes used in hernia repair.[20] Nine FDA-approved meshes were assessed, including four uncoated, one PTFE, one polyester and three with partially absorbable filaments. Each specimen was made using a 2.5 cm × 5.1 cm template, thickness was measured using laser micrometry and density was calculated by dividing specimen mass by specimen area. Tear resistance testing was established in compliance with ASTM specification #D2261-07a. A 2.5 cm slit was cut from the edge to the center of the mesh to create 2 tabs and each tab was clamped in the upper and lower grips of an Instron machine (Fig. 8(a)). Out of the 12 mesh specimens considered, six were oriented perpendicular and other six were parallel to longest dimension of the mesh interstices. Tear strength was computed as the maximum load sustained by the mesh, when the specimen was stretched in tension at 300 mm/min.[20] For ball burst testing, five specimens measuring 9 cm × 9 cm were prepared and mesh orientation was not taken into account owing to the biaxial nature of the test.[20] A test fixture was constructed

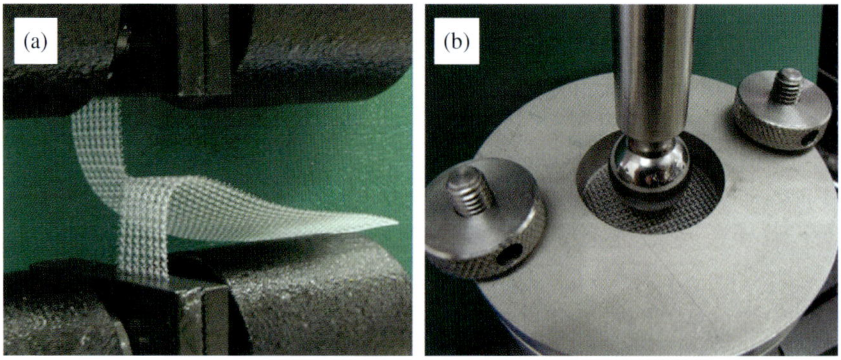

Fig. 8. Biomechanical test fixtures for (a) tear resistance and (b) burst testing.[20]

based on ASTM specification #D3787-07. After clamping the mesh specimen between two circular rings, a 2.5 cm diameter ball was compressed on the specimen at a speed of 300 mm/min until failure (Fig. 8(b)).[20]

Except for the *ProLite® Ultra* (uncoated PP mesh), *INFINIT® Mesh* (uncoated PTFE mesh), and *C-QUR® Lite Small and ULTRAPRO®* (meshes with partially absorbable coatings), all others resisted tearing and displayed tear strengths greater than 20 N. All meshes had tensile strengths greater than 50 N/cm during ball burst testing, except the *INFINIT® Mesh* (uncoated PTFE mesh) and *ULTRAPRO®* (PP mesh with absorbable coating). The results indicated that certain meshes must not be used in conditions that require tensile strength above 50 N/cm, such as in an extremely obese patient. Also, certain meshes demonstrated strains below the physiological range of approximately 10–30% for the human anterior abdominal wall,[19,20] which can lead to the mesh stretching lesser than the abdominal wall. This study has the advantage of carrying out ball burst strength testing, which encompasses the biaxial forces experienced by the mesh *in vitro*.[20]

In the case of fetal implants, ball burst strength tests would perfectly simulate the forces encountered by the patch during the growth process of the fetus. Based on the burst strength test data, the viscoelastic properties of fetal implants can be programmed. This would ensure that the growth of the fetus is not hampered by the patch stiffness, and the mechanical integrity of the implant is sustained over the complete gestation period.

2.3. Degradation Studies

Material degradation needs to be investigated in detail for surgical patches, since it can have a deep impact on the biocompatibility and tissue response. Byproducts released during degradation of polymers also need to be analyzed, to confirm whether a particular mesh is ideal for clinical use.[10] Sometimes, mesh fixation leads to damage to adjacent tissues, triggering phagocytes and macrophages that degrade foreign materials before wound healing begins. In the midst of this process, oxygen is metabolized, and lysosomes containing oxidants like hydrogen peroxide are released. Since a mesh can never be fully degraded and dispatched from the body, constant activation of phagocytic cells and production of oxidants can occur, stimulated by an inflammatory response.[21]

Characterization of explanted PP hernia meshes was carried out by Costello and Bachman, using scanning electron microscopy (SEM), differential scanning calorimetry (DSC) and thermogravimetric analysis (TGA).[21] The sample population ($n = 14$) was composed of PP components from PP/expanded PTFE composite hernia meshes such as *Composix*® *E/X* or *Kugel*® *Composix*. PP meshes were extracted from patients in need of revision surgery due to chronic pain, recurrence and other complications. Adherent tissue was separated from the meshes before testing, by immersion in a sodium hypochlorite solution for 2 h at 37°C, followed by repeated rinse cycles in distilled water and overnight drying.[21]

SEM was performed on sputter-coated pristine and explanted mesh samples using a Hitachi S-4700 scanning electron microscope, by a secondary electron detector at an accelerating voltage of 5 keV.[21] Using a PyrisTM1 DSC was conducted on mesh samples weighing 1–3 mg with an empty aluminum pan acting as the reference. Samples were heated from 50°C to 400°C at 10°C/min under nitrogen flow. For thermogravimetric analysis, a TGAQ50 was programmed to heat the sample from room temperature to 600°C at 10°C/min under nitrogen flow, where samples were cut to approximately 5 mm × 5 mm.[21] Statistical analyses were performed using GraphPad Prism version 4.0 where significant differences in means of pristine versus explanted specimens were determined by a one-way analysis of variance with a 95% confidence interval and Dunnett's post-test ($n = 3$, $p < 0.05$).

SEM micrographs of explanted PP specimens divulged cracks, fissures and blisters were compared to micrographs of pristine specimens (Fig. 9), and such surface abnormalities were observed for 85% of the explanted specimens. The DSC results are shown in Fig. 9, where subject #2 and subject #9 represent the results for all 14 explants. A lower melting temperature (T_m) of approximately 163°C was

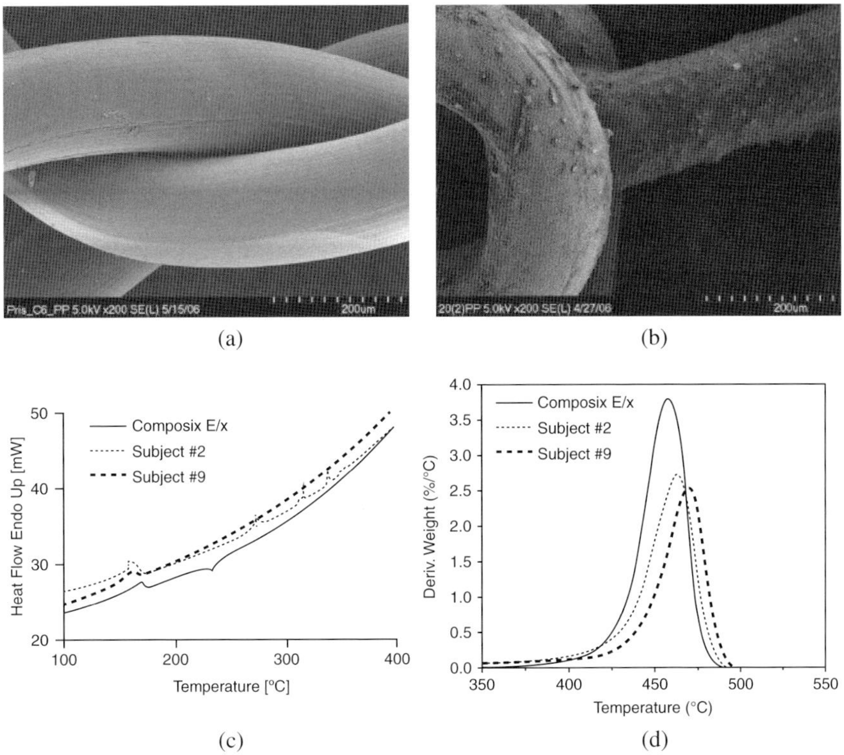

Fig. 9. (a) SEM of the PP component of a pristine Composix E/X mesh; (b) SEM of an explanted PP mesh with blisters on fibers; (c) DSC; (d) TGA thermograms of explanted and pristine meshes.[21]

displayed by explanted specimens, in comparison with 166°C for pristine specimens, and five of the 14 specimens had statistically significant T_m values (Fig. 9). A shift is observed in the peak temperature, from 458°C for pristine specimens to approximately 465°C for explanted meshes. The author hints at the possibility of *in vivo* oxidation of PP, based on SEM, DSC and TGA results. The broader melting peak in DSC was attributed to increased polydispersity of the material, but no gel permeation chromatography analysis was performed to study the change in molecular weight distribution. The author states that oxidized materials usually undergo some weight loss, which could have been supported by gravimetric analysis.

An established degradation profile is of prime importance for fetal patches as well, for the ones made from biodegradable materials. Even in case of inert implants, studying the *in vitro* degradation is extremely critical, owing to the effect of monomer leaching and surface erosion on biocompatibility and toxicity.

3. Conclusions and Future Work

From the literature review section, it is evident that there has been considerable research done on characterization of surgical patches and meshes used in abdominal and cardiac domains. Many patch materials have hydrophobic surfaces, which might lead to inflammatory reactions in the body, and this hurdle needs to be overcome by using hydrophilic coatings.[10,22] There was no literature on surface energy of patches, measured by contact angle tests, which could determine the nature and chemistry of the surface. The mechanical characterization also has numerous limitations since no research article encompasses all the properties that need to be considered for selection of a mesh, namely the anisotropy, elongation, strain, stiffness, dynamic creep, and modulus. Evidence of *in vivo* oxidation is supported by morphology and thermal studies, but no literature presents Fourier transform infrared spectroscopy data about the chemical species on the surface.[23] Degradation studies are limited to few weeks,[21,24] and do not simulate actual times of implantation and only focus on molecular weight degradation of the polymer over time, but ignores the degradation of mechanical and thermal properties. These additional tests need to be carried out for surgical implants, and also need to be carefully modified and implemented for fetal surgery implants.

There exists a grave necessity to carry out relevant tests to characterize fetal surgery implants, to encompass the specific requirements for the physiological environment and simulate the conditions the implant would be subjected *in vivo*. For example, we need to carry out studies on monomer leaching over successive time points to assess if the implant chemically reacts with physiological components to release toxic substances. These studies can be conducted by liquid chromatography-mass spectroscopy tests, or *in vitro* transwell degradation tests at different time points of exposure. Another major aspect to be considered is the volumetric expansion of the fetus over the weeks of gestation till birth. The visco-elastic properties of the implants need to coincide with this volumetric expansion, without exerting too much force to hinder fetal growth. A thorough analysis of cell viability and proliferation on the implants needs to be performed for the time period of implantation. This would govern the performance of the implant, and dictate the degradation profile in case of non-inert implants. Degradation tests over a range of times can define the trigger points at which the mechanical integrity of the implant starts to deteriorate. These experiments can determine the life of the implant *in vivo*, until when the mechanical behavior is consistent.

Based on the above discussion, the future work would involve relevant characterizations such as Fourier transform infrared spectroscopy for surface functional groups, and contact angle studies for the knowledge of hydrophobicity. Permeability studies on the implant would also be necessary if the application cannot permit fluid transfer. Gel permeation chromatography tests after implantation would be helpful to gauge changes in molecular weight distribution due to hydrolytic or enzymatic degradation.

References

1. Bruner JP, Tulipan, NE. Endoscopic coverage of fetal open myelomeningocele in utero. *Am. J. Obstetr. Gynecol.* 153–158 (1997).
2. Kaufman BA. *Neural Tube Defects. National Center for Biotechnology Information.* Elsevier Saunders (2004).
3. Mitchell L, Adzick N. Spina bifida. *Lancet* **364**(9448), 1885–1895 (2004).
4. Northrup H, Volcik K. Spina bifida and other neural tube defects. *Curr. Probl. Pediatr.* Elsevier, 317–332 (2000).
5. McCormack J. Facts for spina bifida life expectancy and unexpected death. www. Medichere.com. 18 Mar. (2016).
6. Scully S, Mallon M. Fetal myelomeningocele repair: A new standard of care. *AORN J.* **96**(2), 175–195 (2012).
7. Burrows P, Johnson M. A randomized trial of prenatal versus postnatal repair of myelomeningocele. *N. Engl. J. Med.* 993–1004 (2011).

8. Fontecha C, Peiro J. Fetoscopic coverage of experimental myelomeningocele in sheep using a patch with surgical sealant. *Eur. J. Obstetr. Gynecol. Reprod. Biol.* **156**(2), 171–176 (2011).
9. Poussier M, Denève E. A review of available prosthetic material for abdominal wall repair. *J. Visc. Surg.* **150**(1), 52–59 (2013).
10. Anderson J. Inflammatory response to implants. *ASAIO Journal* 101–107 (1988).
11. Zogbi L, Trindade E. Comparative study of shrinkage, inflammatory response and fibroplasia in heavyweight and lightweight meshes. *Hernia.* **17**(6), 765–772 (2013).
12. Jacob D, and Schug-Pass C. Comparison of a lightweight polypropylene mesh (Optilene® LP) and a large-pore knitted PTFE mesh (GORE® INFINIT® mesh) — Biocompatibility in a standardized endoscopic extraperitoneal hernia model. *Langenbecks Arch. Surg. Deutsche Gesellschaft Für Chirurgie* **397**(2) (2012).
13. Li X, Kruger J. Characterizing the *ex vivo* mechanical properties of synthetic polypropylene surgical mesh. *J. Mech. Behav. Biomed. Mater.* **37**, 48–55 (2014).
14. Pott P, Schwarz M. Mechanical properties of mesh materials used for hernia repair and soft tissue augmentation. *PloS One* 7(10), e46978, (2012).
15. Klinge U, Klosterhalfen B. Modified mesh for hernia repair that is adapted to the physiology of the abdominal wall. *Eur. J. Surg.* **164**, 951–960 (1998).
16. Cobb W, Burns J. Normal intraabdominal pressure in healthy adults. *J. Surg. Res.* **129**, 231–235 (2005).
17. Hollinsky C, Sandberg S. Measurement of the tensile strength of the ventral abdominal wall in comparison with scar tissue. *Clin. Biomech.* **22**, 88–92 (2007).
18. Velayudhan S, Martin D. Evaluation of dynamic creep properties of surgical mesh prostheses — uniaxial fatigue. *J. Biomed. Mater. Res. B.* **91**(1), 287–296 (2009).
19. Junge K, Klinge U. Elasticity of the anterior abdominal wall and impact for reparation of incisional hernias using mesh implants. *Hernia* **5**, 113–118 (2001).
20. Deeken C, Abdo M. Physicomechanical evaluation of polypropylene, polyester, and polytetrafluoroethylene meshes for inguinal hernia repair. *J. Am. Coll. Surg.* **212**(1), 68–79 (2011).

21. Costello C, Bachman S. Materials characterization of explanted polypropylene hernia meshes. *J. Biomed. Mater. Res. B.* **83B**, 44–49 (2007).
22. Bridges AW, Garcia, AJ. Anti-inflammatory polymeric coatings for implantable biomaterials and devices. *J. Diabetes. Sci. Technol.* (2008), 1 November 2008 (2015).
23. Slee Joshua B, Abigail J. Christian. Addressing the inflammatory response to clinically relevant polymers by manipulating the host response using ITIM domain-containing receptors. *Polymers*, MDPI, 29 Sept. 2014. 26 Feb. 2015.
24. Akihito M, Nishibe T. Patches for carotid artery endarterectomy: Current materials and prospects. *J. Vasc. Res.* **206**(13). NIH Public Access, 1 July 2010. (2015).

Chapter 3

Three-dimensional Bioprinting for Cartilage Regeneration

Hailin Zhu*,†, Hui Gong*, Qing Liu‡
and Harry Huimin Chen*,†,§

**StemEasy Biotech, Ltd. BridgeBio Park*
Jiangyin, Jiangsu 214434, P. R. China
†*State Key Laboratory of Biotherapy, Sichuan University*
Chengdu, Sichuan 610041, P. R. China
‡*Jiangyin Recongene Biomedical Technologies, Co. Ltd.*
Jiangyin, Jiangsu 214437, China
§*huimin.chen@stemeasy.com*

Abstract: Articular cartilage is hyaline cartilage which has very limited self-repairingcapacity after its degeneration or injury. Recently, three-dimensional (3D) bioprinting provides a promising method for repair and regeneration of articular cartilage. Significant progress has been made in 3D bioprinting for cartilage regeneration, particularly in printing hydrogels in combination of cells and growth factors. In this chapter, we reviewed recent progress in cartilage 3D bioprinting, including the use of various cell sources and growth factors for cartilage formation. We also discussed the challenges and the future research directions of cartilage regeneration.

Keywords: Bioprinting, bioink, cartilage, tissue engineering, regenerative medicine.

1. Introduction

Articular cartilage is hyaline cartilage, which is rich in collagen type II and proteoglycan, and plays an important role in joint activities through bearing the mechanical load or lubricating joints. Unlike most tissues, articular cartilage does not have blood vessels, nerves, or immune response, and shows limited capacity for self-repair after degeneration or injury.[1] Currently, there are four main surgical approaches to treat articular cartilage lesions: microfracture, autologous chondrocyte implantation (ACI), mosaicplasty and osteochondral allograft.[2,3] These treatments often result in fibrous repair tissue that is rich in collagen type I. As the hyaline cartilage lacks mechanical properties, their fibrous repair may lead to degenerative changes and arthritis.[4,5]

A conventional scaffold-based tissue engineering method is based on random cell-seeding and growth factor administration.[6] In this approach, cells can only attach on the surface of the scaffold; their distribution inside and the inner composition of the product cannot be controlled.[7] Conventional scaffolds have obvious disadvantages that affect their clinical applicability, e.g. limited cell-seeding efficiency and control over spatial distribution and localization.[6,7]

Recently, cartilage tissue engineering (CTE) provides a promising method for cartilage repair and regeneration. Many reports have demonstrated the success of these methods in growing chondrocytes or undifferentiated cells alone or in combination with various types of three-dimensional (3D) scaffolds and hydrogels fabricated from natural or synthetic materials.[8–10] However, it is still difficult to obtain the long-term outcome of cartilage repair. With the boom of 3D bioprinting and new engineering technologies to create scaffolds of different materials and shape, there has been a wide development of printers and machines. Several "additive manufacturing" technologies that allow the fabrication of customized parts and devices with geometrically complex structures have been applied in the field of biofabrication.[11] These include fused deposition modeling (FDM),[12,13] pneumatic extrusion printing, stereolithography,[14–16] extrusion printing gels,[17] inkjet printing,[18–21] and selective laser sintering (SLS).[22,23] With regards to cartilage regeneration, hydrogel-based scaffolds are the main materials used, given their inherent compatibility with chondral tissue. And inkjet and pneumatic extrusion printers are the most commonly used machines in this field of tissue engineering.

3D bioprinting is a promising biofabrication method for cartilage regeneration. This emerging technology has overcome many limitations of current CTE method. This process combines cells and biomaterials in an ordered and predetermined way. Because the cells are immerging into the printing composition, it allows for the accurate positioning of cells and fabricating the construct in a layer-by-layer deposition process. In this chapter, we review the recent advances in cartilage bioprinting, classify the cell sources for cartilage formation, identify the current challenges and discuss the directions for future developments in cartilage regeneration.

2. Advances in Cartilage Bioprinting

2.1. Thermal Inkjet-Based Bioprinting

Inkjet printing is a non-contact printing technique that reproduces digital pattern information onto a substrate with tiny ink drops. Air bubbles generated by heating in the printhead collapse to provide pressure pulses to eject ink drops with various volumes from 10 pL to 150 pL. Bioink prepared for thermal inkjet printing is usually water based in order to minimize the clogging of the printhead. In 2012, Cui *et al.* developed an inkjet-based bioprinting system with simultaneous photopolymerization capable for 3D CTE (Fig. 1).[24]

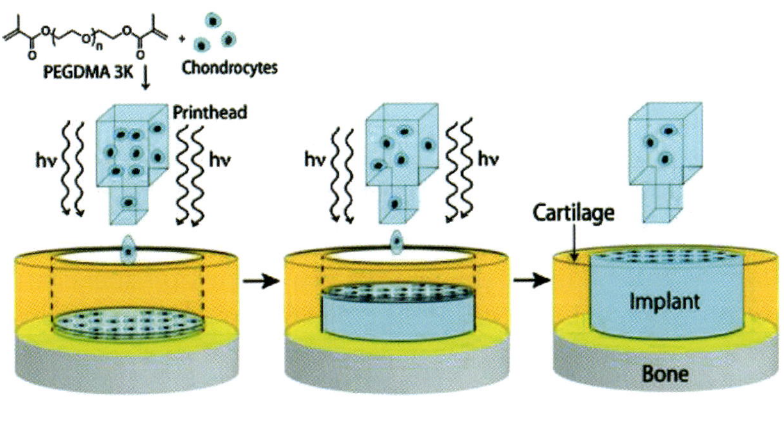

Fig. 1. Schematic of bioprinting cartilage with simultaneous photopolymerization process. PEGDMA, poly (ethylene glycol) dimethacrylate; hv, UV light energy.[24]

A polymerizable bioink was prepared by combing a polyethylene glycol dimethacrylate (PEGDMA) with a photoinitiator and a suspension of human chondrocytes. Printed cell-laden hydrogel firmly integrated with native tissue in 3D biopaper maintaining cell phenotype with consistent gene expression analysis and biochemical data.[24]

2.2. Extrusion-Based Bioprinting

Extrusion-based printing allows the deposition of cell-laden filaments and is regarded as the most suitable technique for the 3D bioprinting of viable constructs of several centimeters in size and with high cell densities. Consequently, for the printing of cartilage constructs, extrusion-based printing techniques are most often considered. Hydrogel prepared for extrusion printing must be viscous enough to keep its shape during printing and must have cross-linking abilities allowing for it to retain the 3D structure after printing. Cross-linking can be induced chemically (e.g. calcium ion to cross-link alginate), thermally, or using UV or visible light with the addition of appropriate initiators. For bioprinting, these cross-linking methods can be used separately or combined with each other.

Cartilage contains predominantly collagen, proteoglycans, water, and low numbers of chondrocytes. To mimic the environment for chondrocytes growth, a lot of studies are focused on natural polymers, such as collagen (Col), gelatin (Gel), hyaluronic acid (HA), chondroitin sulfate (CS) and alginate (AL). Gelatin is a water-soluble protein obtained by the denaturation of collagen. Functionalization of gelatin with unsaturated methacrylamide groups results in gelatin–methacrylamide (GelMA), which can form covalently cross-linked hydrogels in the presence of a photoinitiator and light. However, GelMA solutions have a low viscosity at 37°C which is incompatible with extrusion-based bioprinting. In 2013, Schuurman *et al.*[25] improved the printability of GelMA by adding HA. HA and CS are two of the most abundant glycosaminoglycans (GAGs) in cartilage Schuurman *et al.* They can be functionalized with methacrylic anhydride to become photocrosslinkable.[25] In 2014, Levett *et al.* developed a biomimetic extracellular matrix (ECM) for CTE centered on photocurable gelatin, HA and CS. They encapsulated human chondrocytes in GelMA-based hydrogels, and showed that with the incorporation of small quantities of photocrosslinkable hyaluronic acid methacrylate (HAMA), and to a lesser extent chondroitin sulfate methacrylate (CSMA), chondrogenesis and mechanical properties can be enhanced.[26] In the study by Levett PA *et al.* they investigated in detail the role of HAMA in the developed mechanical properties of engineered cartilage constructs. Their result showed that combinations of GelMA and HAMA are promising candidates for CTE. Encapsulated chondrocytes display a predominantly rounded morphology, and secreted ECM that increases the compressive modulus by up to three-fold over eight weeks culture.[27] In 2016, Costantini *et al.* presented an innovative method based on a coaxial-needles extruder for 3D printing and bioprinting alginate and ECM analogues-based bioinks.[28] They showed that by blending alginate with photocurable polymers such as GelMA, CSMA and HAMA, it was possible to formulate ECM biomimetic inks that can be used for CTE. All the employed hydrogels exhibited an enhanced chondrogenic differentiation of bone marrow-mesenchymal stem cells (BM-MSCs) after 3 weeks of culture in chondrogenic medium. Among the formulated bioinks, the one composed of alginate, GelMA and CSMA turned out to be the best candidate in neocartilage formation with the highest collagen type II/collagen type I and collagen type II/collagen type X ratios.[28]

2.3. Hybrid Bioprinting

Despite the ability to mimic the native properties of tissues, printed 3D constructs that are composed of naturally-derived biomaterials still lack structural integrity and adequate mechanical properties for use *in vivo*, thus limiting their development for use in load-bearing tissue engineering applications, such as cartilage.

The use of synthetic polymers such as poly (e-caprolactone) (PCL) and poly (D,L-lactic-co-glycolic acid) (PLGA) for scaffolding has yielded higher mechanical strengths, higher process ability, and controllable degradation rates. These synthetic polymer scaffolds can provide a biologically favorable, highly hydrated 3D structure similar to natural cartilage matrix. Therefore, combining both hydrogel and polymeric components into a hybrid construct can mimic the biologically and structurally supportive properties of cartilage, offering promise for optimizing CTE strategies.

In 2015, Kundu *et al.* used a multihead deposition system (MHDS) to fabricate 3D cell-printed scaffolds through layer-by-layer deposition of PCL and chondrocyte cell-encapsulated alginate hydrogel.[29] The 3D cell-printed scaffolds of PCL–alginate gel were implanted in the dorsal subcutaneous spaces of female nude mice. Histochemical (Alcian blue and haematoxylin and eosin (H&E) staining) and immunohistochemical (collagen type II) analyses of the retrieved implants after 4 weeks revealed enhanced cartilage tissue and type II collagen fibril formation in the PCL–alginate gel (+TGFβ) hybrid scaffold.[29]

In 2016, Izadifar *et al.* demonstrated that bioprinting 3D hybrid constructs of PCL and cell-impregnated alginate hydrogel is a promising approach for CTE (Fig. 2).[30] They evaluated the heat distribution of printed PCL strands and the rheological property and structural stability of alginate hydrogels at various temperatures and

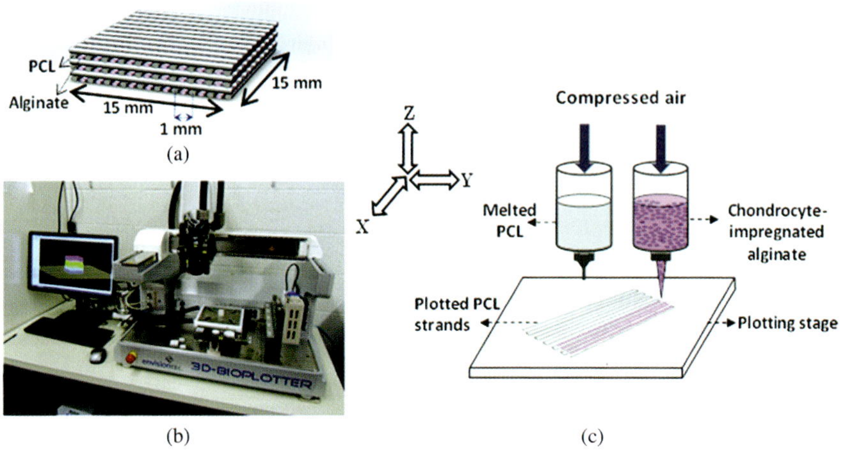

Fig. 2. Design and 3D bioprinting of hybrid constructs with structural and biological features. (a) Schematic of designed 3D hybrid construct with alternating strands of polycaprolactone (PCL) and chondrocyte-impregnated alginate in each layer, (b) 3D bioplotter system employed for biofabrication of designed hybrid constructs, and (c) hybrid biofabrication using pneumatic dispenser heads.[30]

concentrations. The cell viability, proliferation, and cartilage differentiation were observed at high levels in hybrid constructs.[30] Their data suggest that this hybrid fabrication method may improve engineering of cartilage and other tissues. Complex structural and biological properties could be designed into constructs that mimic the zonal characteristics of articular cartilage. Such biomimetic tissue constructs may promote more natural ECM formation.[30]

2.4. Optimization of Cartilage Bioprinting, from Photoinitiator to Formulation of Photocurable Bioinks

The most popular system for cartilage 3D printing is the combination of GelMA with other materials (HA, HAMA, Alginate...) in the presence of a photoinitiator to cross-link under UV light (320–365 nm). 2-hydroxy-1-[4-(2-hydroxyethoxy) phenyl]-2-methyl-1-propanone (I2959) is the most commonly used photoinitiator. However, the use of UV light has shown that it can influence chromosomal and genetic instability in cells.[31] Therefore, other photoinitiators that absorb in the visible light range may offer significant advantages for bioprinting. Fairbanks *et al.* have synthesized a initiator, lithium phenyl-2,4,6-trimethylbenzoylphosphinate (LAP), and explored its potential for application to photo-encapsulation of living cells.[32] The initiator demonstrated remarkable advantages over I2959, including greater water solubility, increased polymerization rates with 365 nm wavelength light, and absorbance above 400 nm that enables efficient visible light polymerization. Cell survival of fibroblasts encapsulated in LAP-initiated PEG diacrylate hydrogels was 95% or greater for every condition evaluated.[32] The survival rate of cells during the printing is affected by many factors. Billiet *et al.* examined the influence of needle type (conical vs. cylindrical), needle internal diameter, and dispensing pressure on the viability of cells.[33] They found that the highest cell viabilities, >97%, were observed at low dispensing pressures (1 bar) using a conical needle type ($\Phi = 200\ \mu m$). In addition, ALMA has maintained the characteristic of alginate in the formulation, which can be cross-linked by calcium ion.[33] In our own study, we have investigated different formulations of photocurable bioinks made of GelMA+HAMA, GelMA+ALMA and GelMA+HAMA+CSMA. Our primary data showed that viscosity of 5% GelMA+3% ALMA is suitable for our bioprinting system.

2.5. Bioprinting *In Situ*

A major advantage of the bioprinting approach is the ability to tailor implants to the anatomy of the defect and/or specific lesion by using medical imaging data to inform implant design. However, surgical approaches to chondral injury repair nominally require an initial debridement step to remove excess fibrous tissue around the defect. This means the size and shape of the final defect to be filled may not be accurately known prior to surgery and so prefabrication of the construct may not represent the best approach. Potentially, cartilage defects could be filled *in situ*, by printing the implant directly into the lesion.[34] In 2016, O'Connell *et al.* described the development of a handheld biofabrication tool, dubbed the "biopen", which enables the deposition of living cells and biomaterials in a manual, direct-write fashion.[35]

2.6. Bioprinting with Bioactive Material

Although 3D print scaffolds with cells offer much promise for articular cartilage repair, this kind of cell-based procedure is much more complex and must go through a very strict approval process from drug regulatory authority before applying in the clinic. Bioprinting with bioactive material, without live cells for cartilage regeneration, may simplify the process and get into clinical practice earlier.

Microfracture, the most widely used procedure for the repair of cartilage defects, so far, enhances migration of MSCs from bone marrow to the site of a cartilage defect, which provides a good cell source for the bioactive scaffold. However, microfracture often results in the formation of fibrocartilage that is biochemically and biomechanically inferior to hyaline articular cartilage. To induce MSCs to differentiate to hyaline cartilage, print scaffolds with sustained release of growth factors will provide favorable condition.[36] In 2010, Lee *et al.*

fabricated an anatomically correct bioscaffold using a composite of poly-ε-caprolactone and hydroxyapatite, infused with transforming growth factor $\beta 3$ (TGF$\beta 3$) to induce MSCs to chondrocytes. Their results suggest that the entire articular surface of the synovial joint can regenerate without cell transplantation.[37] In 2016, Shi *et al.* described a novel strategy by utilizing an ultraviolet (UV) light-reactive, rapidly cross-linkable matrix integrated with KGN-loaded nanoparticles to obtain the natural hyaline cartilage with a simple procedure. Their data shows that after a convenient one step procedure, this KGN-based release strategy could efficiently and persistently promote chondrogenesis.[38]

3. Cells for Cartilage Bioprinting

3.1. Cell Source

3.1.1. Chondrocyte

In 1994, Brittberg *et al.* introduced the autologous chondrocyte transplantation for the first time. They obtained healthy chondrocytes from the injured knee in the process of arthroscopy and cultured the chondrocytes for 14–21 days, and then injected chondrocytes to the defective area. This method reduced pain and swelling in all patients, and 2 years later, 14 of the 16 patients showed good to excellent results.[39] This result consists with Peterson *et al.* who evaluated 94 patients with 2–9 years follow-up after autologous chondrocyte transplantation. They found that 53 patients showed good repair tissue fill, and histologic analysis of 37 biopsies showed a correlation between hyaline like tissue and good to excellent clinical results.[40]

3.1.2. Mesenchymal stem cells

Human MSCs can be isolated relatively easily from a variety of adult mesenchymal tissues, have extensive proliferation potential and are easily expanded without loss of their multilineage differentiation potential within several passages. Therefore, MSCs are perceived as a

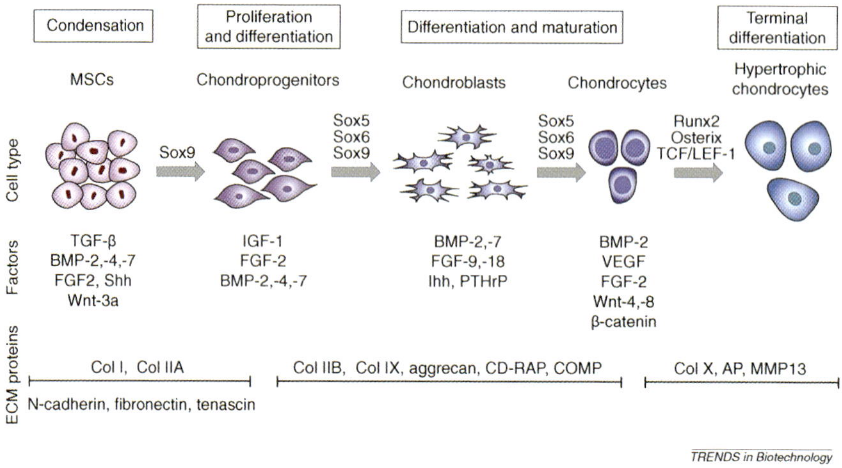

Fig. 3. Sequence of events leading to the differentiation of MSCs toward chondrocytes. alkaline phosphatase (AP), cartilage-derived retinoic acid-sensitive protein (CD-RAP), Col, collagen; cartilage oligomeric protein (COMP), matrix metalloprotease (MMP), vascular endothelial growth factor (VEGF).[47]

good cell source for regeneration of cartilage.[41] Recently, there are more and more studies focusing on the chondrogenic potential of MSCs from bone marrow,[42–45] adipose,[43–45] synovium,[44,45] periosteum,[44,47] umbilical cord[46] and muscle.[44,45] In 2005, one group compared the yield, expandability, differentiation potential, and epitope profile among MSCs from five different tissue sources (bone marrow, synovium, periosteum, muscle and adipose). Their results indicate that there are significant differences in MSC properties according to tissue source, further than donor and experimental variations. Synovium-derived cells particularly had the greatest ability for chondrogenesis.[44] Figure 3 illustrates the whole differentiation of MSCs towards chondrocytes.

3.1.3. Embrgonic stem cells/induced pluripotent stem cells

Currently, there are three methods for the differentiation of embryonic stem cells (ESCs) and induced pluripotent stem cells (iPSCs) toward chondrocytes, each method has advantages along with disadvantages.

Many research groups have published articles on the coculture of human embryonic stem cells (hESCs) and chondrocytes; they found that chondrocyte-secreted morphogenetic factors can promote the differentiation of hESCs. Coculture with primary chondrocytes can induce hESCs to differentiate toward the chondrocyte lineage. This coculture system formed colonies and secreted ECM containing GAG.[48] Furthermore, this result is confirmed by gene expression and immunostaining analysis. In the meantime, during monolayer expansion of the chondrogenically-committed cells, a dynamic expression profile of chondrocyte-specific genes was observed.[49] One obstacle of human pluripotent stem cells (hPSCs) clinical application is its tumorigenicity, but Karlsson. *et al.* take demonstrated that no teratoma formation was detected after transplantation of cocultured hESCs under the kidney capsule of SCID mice.[50]

The second chondrogenic differentiation method involves the formation of EBs from ESCs/iPSCs. For example, human iPS cells from fetal neural stem (FNS) cells can be successfully subjected to *in vitro* chondrogenic differentiation by EBs formation to form functional cartilaginous tissue.[51] Comparison shows that self-assembly of cells obtained by enzymatic dissociation of EBs is superior to self-assembly of EBs.[52] When chondro-induced human iPSCs (hiPSCs) were implanted in osteochondral defects created on the patellar groove of immunosuppressed rats and evaluated after 12 weeks, the defects showed a significantly better quality of cartilage repair than the no-treatment control, and the majority of cells in the regenerated cartilage consisted of implanted hiPSCs.[53]

Recently, a three-stage protocol has been developed for the differentiation of hESCs toward chondrocytes, driving the differentiation of hESCs through primitive streak–mesendoderm and mesoderm intermediates to a chondrocyte population. Gene expression analysis suggests that the hESCs progress through primitive streak or mesendoderm to mesoderm, before differentiating into a chondrocytic culture comprising cell aggregates which also express cell surface CD44 and aggrecan, and deposit a

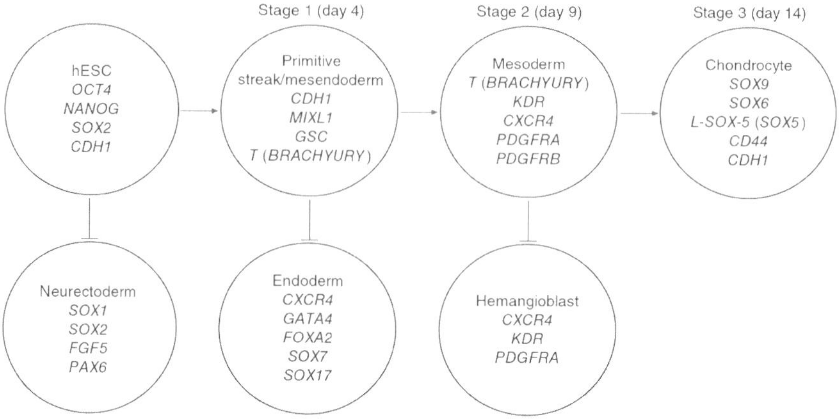

Fig. 4. Schematic of directed differentiation protocol in three stages. In stage 1, pluripotent hESCs are directed toward a primitive streak–mesendoderm population; in stage 2, differentiation proceeds to a mesoderm population; and in stage 3, toward chondrocytes. As some genes are expressed in different cell lineages and at different stages, the developmental status of each cell population was characterized by expression of panels of marker genes including SOX2, which is expressed by both pluripotent hESCs and cells derived from the neurectoderm germ layer, CDH1, expressed on pluripotent and mesendoderm cells and CXCR4, used to identify cell lineages from both the endoderm and mesodermal-derived hemangioblast.[54]

sulfated GAG and cartilage-specific collagen II matrix. At this final stage, 74% (HUES1 cells) and up to 95–97% (HUES7 and HUES8 cells) express the chondrogenic transcription factor SOX9.[54] Figure 4 schematically shows their differentiation protocol in three stages.

3.2. Growth Factors for Chondrogenic Differentiation

3.2.1. The transforming growth factor-β superfamily

The transforming growth factor-β (TGF-β) family comprises transforming growth factor-β, bone morphogenetic proteins (BMPs),[55] activins and inhibins.[56] TGF-β members direct the mesenchymal stem cell fate into the chondrogenic and osteogenic directions.[57]

3.2.1.1. *TGF-β*

The TGF-β family includes five members (TGF-β1–5) which are predominantly produced in bone and cartilage.

TGF-β1 at a concentration of 5 ng/mL has the most pronounced effect to stimulate bone marrow-derived MSCs for their chondrogenic differentiation in a dose-dependent relationship.[58] When transferred to the full thickness articular cartilage defects, it can be restored with hyaline cartilage from *in vivo* differentiated autologous MSCs, which was superior to implantation of *in vitro* differentiated autologous MSCs, as evidenced by a better surface zone repair and reconstitution. All results indicated TGF-1 was able to induce the MSCs into chondrocytes *in vivo* and prevent the deterioration of newly formed cartilage with time.[59]

hESCs also respond to TGF-β1, the hESC-derived cells exhibited growth factor–dependent matrix production in pellet culture but did not produce tissue characteristic of cartilage morphology. But when these cells were encapsulated in RGD modified poly(ethylene glycol) hydrogels, they formed neocartilage with basophilic ECM deposition within 3 weeks of culture, and produced cartilage-specific gene up-regulation and (ECM) production.[60]

3.2.1.2. *Bone morphogenetic proteins*

Bone morphogenetic proteins (BMPs) play a role in many stages of chondrogenic differentiation, initiating chondroprogenitor cell determination and differentiation of precursors into chondrocytes, and also at the stage of chondrocyte maturation and terminal differentiation.

Recombinant human BMP-2, -4 and -6 can enhance *in vitro* cartilage formation of MSCs from bone marrow. All the BMPs tested, increased chondrogenic differentiation as assayed by immunohistochemistry and by the size and weight of the cartilage synthesized.[61] Other group transfected BMP-4 into adipose-derived stem cells (ADSCs) by nanoparticles and evaluated the cartilage repair effect in a rabbit model. The results showed that the collagen type II protein and aggrecan expression was up-regulated *in vitro*.[62] The similar results were obtained when hESCs were cultured as an aggregate in a pellet culture system with BMP-7 treatment.[63]

3.2.2. Fibroblast growth factor family

The fibroblast growth factor (FGF) family comprises 22 structurally related proteins that bind one of four FGF receptors (FGFRs).

By forcing the expression of FGFR3in a pluripotent murine mesenchymal stem cell line (C3H10T1/2), Hoffmann *et al.* found that FGFR3 is adequate enough for chondrogenic differentiation, indicating an important role for FGF-signaling during the manifestation of the chondrogenic lineage in this cell line.[64] Meanwhile, another murine genetic-based model revealed that FGF18 signals through FGFR3 to promote cartilage construction.[65]

Allison *et al.* studied the effect of FGF-2 on bone marrow-derived MSCs and demonstrated that the use of 100 ng/mL of FGF-2 significantly increased MSCs pellet DNA and GAG content. Collagen type II content of the pellet was also increased by use of 10 ng/mL and 100 ng/mL of FGF-2. Collagen type II and aggrecan mRNA levels were increased by treatment with FGF-2 too.[66] FGF2 seems to have a positive effect on hESCs in the process embryoid bodies (EBs) formation and can induce greater numbers of osteogenic and chondrogenic lineage cells.[67]

3.2.3. Insulin-like growth factor family

Insulin-like growth factor-1 (IGF-1) significantly increased chondrogenesis in a dose-dependent manner when administered continuously throughout the culture period. *In situ* hybridization for type II collagen showed that continuous IGF-1 maintained type II collagen mRNA expression throughout the cambium layer from 2 to 6 weeks.[68] When entrapped in silk fibroin scaffolds, IGF-1 can stimulate chondrogenic differentiation of hMSC whereas no chondrogenic responses were observed on unloaded control scaffolds.[69]

3.2.4. Hedgehog family

The impact of Sonic hedgehog (Shh), a member of hedgehog family, on adult stem cells was tested on human bone marrow-derived MSCs. It showed expression of cartilage markers aggrecan, Sox9, CEP-68, and collagen types 2 and 10 within 3 weeks of Shh stimulation. Only r-Shh treated cells showed a very strong cell proliferation and much higher BrdU incorporation in cell assay systems.[70]

3.3. Biophysical Stimuli

3.3.1. Oxygen tension

Cartilages are mostly avascular tissues, with synovial fluid providing oxygen and nutrients that diffuse through the ECM. The lack of blood supply creates a hypoxic environment, with reports of oxygen levels that range from 1% to 8%, depending on the location of the tissue and depth inside.[71]

After exposure of mouse stromal ST2 stem cells to 1% oxygen, Robins *et al.* demonstrated that exposure to low oxygen levels induces genotypic and phenotypic changes consistent with differentiation along a chondrocyte pathway.[72] This conclusion agreed with Wang *et al.* who induced a two-fold increase in the rate of protein synthesis and a three-fold increase in total collagen synthesis in human adipose-derived adult stem cells.[73]

Hypoxic differentiation conditions enhanced the chondrogenic potential of hESCs. Koay *et al.* used the human ESC cell line H9, which showed significant increase in collagen II production by hypoxic conditions. Their result also indicates the possibility of generating a spectrum of different cartilages.[71]

3.3.2. Mechanical stimuli

Articular cartilage is primarily composed of type II collagen and proteoglycans,[74] the collagen type II provides stiffness to the tissue in tension, and the proteoglycans, which are negatively charged, attract water resulting in the provision of stiffness in compression.[75]

Dynamic compressive loading on chondrogenesis was performed on human MSCs. After 70 days of culture, dynamic compressive loading increased the mechanical properties of human MSCs encapsulated HA hydrogels, as well as the GAG and collagen contents.[76] The same result was obtained by Huang *et al.* who applied long-term dynamic compression to MSC-seeded constructs. Their results demonstrate that dynamic compressive loading initiated after a sufficient period of chondroinduction and with sustained TGF-β exposure enhances matrix distribution and the mechanical properties of MSC-seeded constructs.[77]

Shear force was another factor that had been widely investigated. The potential enhancing effect of surface shear on chondrogenic differentiation of hMSCs was studied by Schatti *et al.* They found that the application of shear superimposed upon dynamic compression led to significant increases in chondrogenic gene expression.[78]

4. Current Challenges and Future Directions

4.1. Choices of Cell Sources

In Sec. 2.1, multiple cell types have been discussed for their applications in bioactive cartilage implants. So far, in the field of cartilage bioprinting research, the focus has been predominantly on the use of chondrocytes. Nevertheless, when using autologous chondrocytes, obtaining sufficient cell numbers remains a challenge, since the differentiated chondrocytes have limited proliferative capacity. Allograft of cartilage has been proved to be safe due to cartilage immune privilege, and allogeneic juvenile chondrocytes was shown to produce more ECM than adult chondrocytes.[79] But it is hard to use juvenile chondrocytes as a major cell source for cartilage regeneration due to limitations of donors.

An alternative cell type for cartilage repair is MSC, which can be derived from multiple tissues, relatively easier to expand and can be differentiated into chondrocyte-like cells in the presence of specific growth factors (see details in Sec. 2.2). Scaffolds with growth factors are also being investigated in combination with the surgical method of microfracturing to coax the patient's own bone marrow MSCs to form articular cartilage. However, adequate cues to control MSC fate have to be provided, as these cells have the tendency to progress into hypertrophic chondrogenesis and to give rise to bone formation via the endochondral pathway once implanted *in vivo*.[80]

Furthermore, the induced pluripotent stem cells (iPSCs), which show unlimited self-renewal as ESCs and can be generated from numerous cell types (i.e. keratinocytes), remain an interesting cell source for cartilage regeneration. Although there are several publications which claimed iPSCs-derived chondrocytes' safety, more cases accretion and longer time observation are needed.

For clinical application of human iPSCs and ESCs, there is a requirement to minimize the risk of contamination with animal components. It requires that not only the culture media is chemically defined, but the culture plate coating material is xeno-free. In 2015, we have developed xeno-free coating substrate, which can be stored at 4°C and is ready to be used upon request, may serve as an easier way to amplify hESCs/iPSCs for clinical applications.[81]

4.2. Hybrid 3D Bioprinting

Cartilage 3D bioprinting can make the transition in the clinic from non-living personalized 3D printed implants toward biologically active living implants. Although chondrogenic cell-laden bioinks can be solidified in seconds, it may take months of transition time to function as articular cartilage. It is a challenge to remain the right shape, size and position for those soft cell-laden filaments during their transition. Absorbable and biocompatible polymers, e.g. PLA and PCL, can be the right choice for scaffold to provide a temporary support for chondrocyte-laden hydrogel. Kundu *et al.* have brought in an excellent example of the hybrid printing approach (see Sec. 1.3, Refs. 29,30). It is expected that this supporting scaffold will gradually retreat when printed cartilage grows and regains its function. However, to synchronize two speeds of scaffold degradation and cartilage growth, many factors need to be optimized on choices of cell type, cell density, and bioinks and its formulation.

4.3. Functional Repair of Cartilage Damage

Functional recovery of damaged cartilage is harder to achieve, but more beneficial for patients. To reach such a goal, unique cartilage architecture needs to be rebuilt.

It is believed that there is an appropriate zone-specific compositional and mechanical heterogeneity present in articular cartilage, and restoration of this zonal organization will improve integration and performance of the construct at the defect site.[82,83] 3D bioprinting may be a unique way to achieve this zonal construction, but it is a challenge to remain such organization after printing. We believe that mechanical loading regimes (i.e. periodical impact pressure) are important factors to rebuild the zonal organization *in vitro*, but to rebuild *in vivo*, a good rehabilitation strategy may play the trick after cartilage transplant surgery.

4.4. Cartilage 3D Bioprinting Through Bio-Pen

Many orthopedic surgeons may like the idea of using a hand-held 3D printing device (bio-pen) to cope with variable cartilage defects in operation rooms. Although printing directly into a defect is an exciting idea, it becomes a challenge to keep the 3D bioprinter in small size as a mobile device, and retain its complexity and accuracy as a standard 3D bioprinter. This approach has been exemplified by the direct *ex vivo* printing into osteochondral plugs or femurs.[34–36] We believe that for further development of biopen, the following factors should be taken in consideration: flexible single small tube combining bioink injection and light source, safety control for air pressure and light, and integration of defect scanning and bioprinting *in situ*.

4.5. Clinical Regulatory Standards for 3D Bioprinting Cartilage Implants

For clinical application, all materials for 3D bioprinting the cartilage implants, e.g. cells, growth factors and scaffolds must follow a set of regulations from drug administration authority. To guaranty the safety of bioprinted living cartilage implants, all the cells and bioinks have to meet their quality requirements, for example, sterility, endotoxin-free, safety and reproducibility, etc. For fabrication of bioinks, the whole process needs to be incorporated in a good manufacturing practice (GMP) facility and quality management system (ISO 9001:2000). For the final clinical application, the printer itself and all its components should be sterile and able to be handled in a sterile environment, such as in biosafety cabinets or operation rooms.

Although research on cartilage 3D bioprinting is growing rapidly, there are still lots of challenges. No approach to date has produced a regenerated hyaline cartilage with long-term stability and functional recovery. It is generally accepted that for stable long-term reconstruction and function repair, the therapy should not only address the cartilage but also focus on reconstructing the underlying bone and reestablishing joint homeostasis. In order to create constructs for successful cartilage regeneration, we need pay attentions not only on the mechanical strength, cell survival, and functionality, but also on deeper understanding of cartilage regeneration in general, including cell types, biological cues, and organization.

References

1. Fox AJS, Bedi A, Rodeo SA. The basic science of articular cartilage: structure, composition, and function. *Sports Health* **1**(6), 1461–1468 (2009).
2. Swieszkowski W. Repair and regeneration of osteochondral defects in the articular joints. *Biomol. Eng.* **24**(5), 489–495 (2007).
3. Redman SN, Oldfield SF, Archer CW. Current strategies for articular cartilage repair. *Eur. Cell. Mater.* **9**, 23–32 (2005).
4. Hayes DW, Brower RL, John KJ. Articular cartilage. Anatomy, injury, and repair. *Clin. Podiatr. Med. Surg.* **18**(1), 35–53 (2001).
5. Prakash D, Learmonth D. Natural progression of osteo-chondral defect in the femoral condyle. *Knee.* **9**(1), 7–10 (2002).
6. Langer R, Vacanti JP. Tissue engineering. *Science* **260**, 920–926 (1993).
7. Silva MM, Cyster LA, Barry JJ, Yang XB, Oreffo RO, Grant DM, Scotchford CA, Howdle SM, Shakesheff KM, Rose FR. The effect of anisotropic architecture on cell and tissue infiltration into tissue engineered scaffolds. *Biomaterials* **27**, 5909–5917(2006).
8. Caminal M, Peris D, Fonseca C, Barrachina J, Codina D, Rabanal RM, Moll X, Morist A, García F, Cairó JJ, Gòdia F, Pla A, Vives J. Cartilage resurfacing potential of PLGA scaffolds loaded with autologous cells from cartilage, fat and bone marrow in an ovine model of osteochondral focal defect. *Cytotechnology* **68**(4), 907–919 (2016).

9. Sharma B, Fermanian S, Gibson M, Unterman S, Herzka DA, Cascio B, Coburn J, Hui AY, Marcus N, Gold GE, Elisseeff JH. Human cartilage repair with a photoreactive adhesive-hydrogel composite. *Sci. Transl. Med.* **5**(167), 167ra6 (2013).
10. Wang DA, Varghese S, Sharma B, Strehin I, Fermanian S, Gorham J, Fairbrother DH, Cascio B, Elisseeff JH. Multifunctional chondroitin sulphate for cartilage tissue-biomaterial integration. *Nat. Mater.* **6**(5), 385–392 (2007).
11. Rosen W. Computer-aided design for additive manufacturing of cellular structures. *Comput. Aided. Des. Appl.* **4**, 585–594 (2007).
12. Hutmacher DW, Garcia AJ. Scaffold-based bone engineering by using genetically modified cells. *Gene.* **347**(1), 1–10 (2005).
13. Jung JW, Kang, HW, Kang, TY, *et al.* Projection image generation algorithm for fabrication of a complex structure using projection based microstereolithography. *Int. J. Precis. Eng. Manuf.* **13**, 445–449 (2012).
14. Lee JW, Ahn G, Kim JY, Cho DW. Evaluating cell proliferation based on internal pore size and 3D scaffold architecture fabricated using solid freeform fabrication technology. *J. Mater. Sci. Mater. Med.* **21**, 3195–3205 (2010).
15. Lee J-W, Kim J-Y, and Cho D-W. Solid free-form fabrication technology and its application to bone tissue engineering. *Int. J. Stem. Cells* **3**, 85–93 (2010).
16. Lee JW, Kang KS, Lee SH, Kim JY, Lee BK, Cho DW. Bone regeneration using a microstereolithography produced customized poly (propylene fumarate) diethyl fumarate photopolymer 3D scaffold incorporating BMP-2 loaded PLGA microspheres. *Biomaterials* **32**, 744–752 (2011).
17. Chung JH, Naficy S, Yue Z, Kapsa R, Quigley A, *et al.* Bioink properties and printability for extrusion printing living cells. *Biomater Sci* **1**, 763(2013).
18. Boland T, Xu T, Damon B, Cui X. Application of inkjet printing to tissue engineering. *Biotechnol. J.* **1**(9), 910–917 (2006).
19. Nakamura M, Kobayashi A, and Takagi F. Biocompatible inkjet printing technique for designed seeding of individual living cells. *Tissue Eng.* **11**(11–12), 1658–1666 (2005).
20. Ferris CJ, Gilmore KG, Wallace GG, Panhuis M. Biofabrication: an overview of the approaches used for printing of living cells. *Appl. Microbiol. Biotechnol.* **97**(10), 4243s–4258s (2013).

21. Ferris CJ Gilmore KG, Beirne S, McCallum D, Wallace GG, In het Panhuis M. Bioink for on-demand printing of living cells. *Biomater. Sci.* **1**, 224–230 (2013).
22. Wiria FE, Lim PN, Yeo JF, *et al.* Printing of titanium implant prototype. *Mater. Des.* **31**, S101–S105 (2010).
23. Simpson RL, Wiria FE, Amis AA, Chua CK, Leong KF, Hansen UN, Chandrasekaran M, Lee MW. Development of a 95/5 poly (L-lactide-co-glycolide)/hydroxylapatite and β-tricalcium phosphate scaffold as bone replacement material via selective laser sintering. *J. Biomed. Mater. Res. B.* **84**(1), 17–25 (2008)
24. Cui X, Breitenkamp K, Finn MG, Lotz M, D'Lima DD. Direct human cartilage repair using three-dimensional bioprinting technology. *Tissue. Eng. Part. A.* **18**(11–12), 1304–1312 (2012).
25. Schuurman W, Levett PA, Pot MW, van Weeren PR, Dhert WJ, Hutmacher DW, Melchels FP, Klein TJ, Malda J. Gelatin-methacrylamide hydrogels as potential biomaterials for fabrication of tissue-engineered cartilage constructs. *Macromol. Biosci.* **13**(5), 551–561 (2013).
26. Levett PA, Melchels FP, Schrobback K, Hutmacher DW, Malda J, Klein TJ. A biomimetic extracelluar matrix for cartilage tissue engineering centered on photocurable gelatin, hyaluronic acid and chondroitin sulfate. *Acta. Biomater* **10**(1), 214–223 (2014).
27. Levett PA, Hutmacher DW, Malda J, Klein TJ. Hyaluronic acid enhances the mechanical properties of tissue-engineered cartilage constructs. *PLoS One* **9**(12), e113216 (2014).
28. Costantini M, Idaszek J, Szöke K, Jaroszewicz J, Dentini M, Barbetta A, Brinchmann JE, Święszkowski W. 3D bioprinting of BM-MSCs-loaded ECM biomimetic hydrogels for *in vitro* neocartilage formation. *Biofabrication* **8**(3), 035002 (2016).
29. Kundu J, Shim JH, Jang J, Kim SW, Cho DW. An additive manufacturing-based PCL-alginate-chondrocyte bioprinted scaffold for cartilage tissue engineering. *J. Tissue. Eng. Regen. Med.* **9**(11), 1286–1297 (2015).
30. Izadifar Z, Chang T, Kulyk W, Chen X, Eames BF. Analyzing biological performance of 3D-printed, cell-impregnated hybrid constructs for cartilage tissue engineering. *Tissue. Eng. Part. C. Methods* **22**(3), 173–188 (2016).
31. Dahle J, Kvam E, Stokke T. Bystander effects in UV-induced genomic instability: antioxidants inhibit delayed mutagenesis induced by ultraviolet A and B radiation. *J. Carcinog.* **4**, 11(2005).

32. Fairbanks BD, Schwartz MP, Bowman CN, Anseth KS. Photoinitiated polymerization of PEG-diacrylate with lithium phenyl-2,4,6-trimethylbenzoylphosphinate: polymerization rate and cytocompatibility. *Biomaterials* **30**(35), 6702–6707 (2009).
33. Billiet T, Gevaert E, De Schryver T, Cornelissen M, Dubruel P. The 3D printing of gelatin methacrylamide cell-laden tissue-engineered constructs with high cell viability. *Biomaterials* **35**(1), 49–62 (2014).
34. Cohen DL, Lipton JI, Bonassar LJ, Lipson H. Additive manufacturing for in situ repair of osteochondral defects. *Biofabrication* **2**(3), 035004, (2010).
35. O'Connell CD, Di Bella C, Thompson F, Augustine C, Beirne S, Cornock R, Richards CJ, Chung J, Gambhir S, Yue Z, Bourke J, Zhang B, Taylor A, Quigley A, Kapsa R, Choong P, Wallace GG. Development of the Biopen: a handheld device for surgical printing of adipose stem cells at a chondral wound site. *Biofabrication* **8**(1), 015019 (2016).
36. Chen C, Bang S, Cho Y, Lee S, Lee I, Zhang S, Noh I. Research trends in biomimetic medical materials for tissue engineering: 3D bioprinting, surface modification, nano/micro-technology and clinical aspects in tissue engineering of cartilage and bone. *Biomater. Res.* **20**(10), (2016).
37. Lee CH, Cook JL, Mendelson A, Moioli EK, Yao H, Mao JJ. Regeneration of the articular surface of the rabbit synovial joint by cell homing: a proof of concept study. *Lancet* **376**(9739), 440–448 (2010).
38. Shi D, Xu X, Ye Y, Song K, Cheng Y, Di J, Hu Q, Li J, Ju H, Jiang Q, Gu Z. Photo-Cross-Linked scaffold with kartogenin-encapsulated nanoparticles for cartilage regeneration. *ACS. Nano.* **10**(1), 1292–1299 (2016).
39. Brittberg M, Lindahl A, Nilsson A, Ohlsson C, Isaksson O, Peterson L. Treatment of deep cartilage defects in the knee with autologous chondrocyte transplantation. *N. Engl. J. Med.* **331**(14), 889–895 (1994).
40. Peterson L, Minas T, Brittberg M, Nilsson A, Sjögren-Jansson E, Lindahl A. Two- to 9-year outcome after autologous chondrocyte transplantation of the knee. *Clin. Orthop. Relat. Res.* (374), 212–234 (2000).
41. Koga H, Engebretsen L, Brinchmann JE, Muneta T, Sekiya I. Mesenchymal stem cell-based therapy for cartilage repair: a review. *Knee Surg Sports Traumatol Arthrosc.* **17**(11), 1289–1297 (2009).
42. Colter DC, Sekiya I, Prockop DJ. Identification of a subpopulation of rapidly self-renewing and multipotential adult stem cells in colonies of

human marrow stromal cells. *Proc. Natl. Acad. Sci. USA.* **98**(14), 7841–7845 (2001).

43. Huang JI, Kazmi N, Durbhakula MM, Hering TM, Yoo JU, Johnstone B. Chondrogenic potential of progenitor cells derived from human bone marrow and adipose tissue: a patient matched comparison. *J. Orthop. Res.* **23**(6), 1383–1389 (2005).
44. Sakaguchi Y, Sekiya I, Yagishita K, Muneta T. Comparison of human stem cells derived from various mesenchymal tissues: Superiority of synovium as a cell source. *Arthritis. Rheum.* **52**(8), 2521–2529 (2005).
45. Yoshimura H, Muneta T, Nimura A, Yokoyama A, Koga H, Sekiya I. Comparison of rat mesenchymal stem cells derived from bone marrow, synovium, periosteum, adipose tissue, and muscle. *Cell. Tissue. Res.* **327**(3), 449–462 (2007).
46. Fong CY, Subramanian A, Gauthaman K, Venugopal J, Biswas A, Ramakrishna S, Bongso A. Human umbilical cord wharton's jelly stem cells undergo enhanced chondrogenic differentiation when grown on nano fibrous scaffolds and in a sequential two-stage culture medium environment. *Stem. Cell. Rev.* **8**(1), 195–209 (2012).
47. Vinatier C, Mrugala D, Jorgensen C, Guicheux J, Noël D. Cartilage engineering: a crucial combination of cells, biomaterials and biofactors. *Trends. Biotechnol.* **27**(5), 307–314 (2009).
48. Vats A, Bielby RC, Tolley N, Dickinson SC, Boccaccini AR, Hollander AP, Bishop AE, Polak JM. Chondrogenic differentiation of human embryonic stem cells: the effect of the micro-environment. *Tissue Eng.* **12**(6), 1687–1697 (2006).
49. Hwang NS, Varghese S, Elisseeff J. Derivation of chondrogenically-committed cellsfrom human embryonic cells for cartilage tissue regeneration. *PLoS One* **3**(6), e2498 (2008).
50. Bigdeli N, Karlsson C, Strehl R, Concaro S, Hyllner J, Lindahl A. Coculture of human embryonic stem cells and human articular chondrocytes results in significantly altered phenotype and improved chondrogenic differentiation. *Stem. Cells* **27**(8), 1812–1821 (2009).
51. Medvedev SP, Grigor'eva EV, Shevchenko AI, Malakhova AA, Dementyeva EV, Shilov AA, Pokushalov EA, Zaidman AM, Aleksandrova MA, Plotnikov EY, Sukhikh GT, Zakian SM. Human induced pluripotent stem cells derived from fetal neural stem cells successfully undergo directed differentiation into cartilage. *Stem Cells Dev.* **20**(6), 1099–1112 (2011).

52. Koay EJ, Hoben GM, Athanasiou KA. Tissue engineering with chondrogenically differentiated human embryonic stem cells. *Stem Cells* **25**(9), 2183–2190 (2007).
53. Ko JY, Kim KI, Park S, Im GI. *In vitro* chondrogenesis and *in vivo* repair of osteochondral defect with human induced pluripotent stem cells. *Biomaterials* **35**(11), 3571–3581 (2014).
54. Oldershaw RA, Baxter MA, Lowe ET, Bates N, Grady LM, Soncin F, Brison DR, Hardingham TE, Kimber SJ. Directed differentiation of human embryonic stem cells toward chondrocytes. *Nat. Biotechnol.* **28**(11), 1187–1194 (2010).
55. Wozney JM, Rosen V. Bone morphogenetic protein and bone morphogenetic protein gene family in bone formation and repair. *Clin. Orthop. Relat. Res.* (346), 26–37 (1998).
56. Centrella M, Horowitz MC, Wozney JM, McCarthy TL. Transforming growth factor-beta gene family members and bone. *Endocr. Rev.* **15**(1), 27–39 (1994).
57. Kingsley DM. The TGF-beta superfamily: new members, new receptors, and new genetic tests of function in different organisms. *Genes. Dev.* **8**(2), 133–146 (1994).
58. Worster AA, Nixon AJ, Brower-Toland BD, Williams J. Effect of transforming growth factor beta 1 on chondrogenic differentiation of cultured equine mesenchymal stem cells. *Am. J. Vet. Res.* **61**(9), 1003–1010 (2000).
59. Fan H, *et al.* Porous gelatin-chondroitin-hyaluronatetri-copolymer scaffold containing microspheres loaded with TGF-b1induces differentiation of mesenchymal stem cells in vivo for enhancing cartilage repair. *J. Biomed. Mater. Res. A.* 77, 785–794 (2006).
60. Hwang NS, Varghese S, Zhang Z, Elisseeff J. Chondrogenic differentiation of human embryonic stem cell–derived cells in arginine-glycine-aspartate modified hydrogels. *Tissue. Eng.* **12**(9), 2695–2706 (2006).
61. Sekiya I, Larson BL, Vuoristo JT, Reger RL, Prockop DJ. Comparison of effect of BMP-2, -4, and -6 on *in vitro* cartilage formation of human adult stem cells from bone marrow stroma. *Cell. Tissue. Res.* **320**(2), 269–276 (2005).
62. Shi J, Zhang X, Zhu J, Pi Y, Hu X, Zhou C, Ao Y. Nanoparticle delivery of the bone morphogenetic protein 4gene to adipose-derived stem cells promotes articular cartilage repair *in vitro* and *in vivo. Arthroscopy* **29**(12), 2001–2011.e2 (2013).

63. Nakagawa T, Lee SY, Reddi AH. Induction of chondrogenesis from human embryonic stem cells without embryoid body formation by bone morphogenetic protein 7 and transforming growth factor beta1. *Arthritis. Rheum.* **60**(12), 3686–3692 (2009).
64. Hoffmann A, Czichos S, Kaps C, Bächner D, Mayer H, Kurkalli BG, Zilberman Y, Turgeman G, Pelled G, Gross G, Gazit D. The T-box transcription factor Brachyury mediates cartilage development in mesenchymal stem cell line C3H10T1/2. *J. Cell. Sci.* **115**(Pt 4), 769–781 (2002).
65. Davidson D, Blanc A, Filion D, Wang H, Plut P, Pfeffer G, Buschmann MD, Henderson JE. Fibroblast growth factor (FGF) 18 signals through FGF receptor 3 to promote chondrogenesis. *J. Biol. Chem.* **280**(21), 20509–205015 (2005).
66. Stewart AA, Byron CR, Pondenis H, Stewart MC. Effect of fibroblast growth factor-2 on equine mesenchymal stem cell monolayer expansion and chondrogenesis. *Am. J. Vet. Res.* **68**(9), 941–945 (2007).
67. Lee TJ, Jang J, Kang S, Jin M, Shin H, Kim DW, Kim BS. Enhancement of osteogenic and chondrogenic differentiation of human embryonic stem cells by mesodermal lineage induction with BMP-4 and FGF2 treatment. *Biochem. Biophys. Res. Commun.* **430**(2), 793–797 (2013).
68. Fukumoto T, Sperling JW, Sanyal A, Fitzsimmons JS, Reinholz GG, Conover CA, O'Driscoll SW. Combined effects of insulin-like growth factor-1 and transforming growth factor-1 on periosteal mesenchymal cells during chondrogenesis *in vitro*. *Osteoarthritis. Cartilage* **11**(1), 55–64 (2003).
69. Uebersax L, Merkle HP, Meinel L. Insulin-like growth factor I releasing silk fibroin scaffolds induce chondrogenic differentiation of human mesenchymal stem cells. *J. Control. Release* **127**(1), 12–21 (2008).
70. Warzecha J, Göttig S, Brüning C, Lindhorst E, Arabmothlagh M, Kurth A. Sonic hedgehog protein promotes proliferation and chondrogenic differentiation of bone marrow derived mesenchymal stem cells *in vitro*. *J. Orthop. Sci.* **11**(5), 491–496 (2006).
71. Koay EJ, Athanasiou KA. Hypoxic chondrogenic differentiation of human embryonic stem cells enhances cartilage protein synthesis and biomechanical functionality. *Osteoarthritis Cartilage* **16**(12), 1450–1456 (2008).
72. Robins JC, Akeno N, Mukherjee A, Dalal RR, Aronow BJ, Koopman P, Clemens TL. Hypoxia induces chondrocyte-specific geneexpression in

mesenchymal cells in association with transcriptional activation of Sox9. *Bone.* **37**(3), 313–322 (2005).
73. Wang DW, Fermor B, Gimble JM, Awad HA, Guilak F. Influence of oxygen on the proliferation and metabolism of adipose derived adult stem cells. *J. Cell. Physiol.* **204**(1), 184–191 (2005).
74. Mow VC, Guo XE. Mechano-electrochemical properties of articular cartilage: their inhomogeneities and anisotropies. *Annu. Rev. Biomed. Eng.* **4**, 175–209 (2002). Epub 2002 Mar 22.
75. McMahon LA, O'Brien FJ, Prendergast PJ. Biomechanics and mechanobiology in osteochondral tissues. *Regen. Med.* **3**(5), 743–759 (2008).
76. Bian L, Zhai DY, Zhang EC, Mauck RL, Burdick JA. Dynamic compressive loading enhances cartilage matrix synthesis and distribution and suppresses hypertrophy in hMSC-Laden hyaluronic acid hydrogels. *Tissue. Eng. Part. A.* **18**(7–8), 715–724 (2012).
77. Huang AH, Farrell MJ, Kim M, Mauck RL. Long-term dynamic loading improves the mechanical properties of chondrogenic mesenchymal stem cell-laden hydrogel. *Eur. Cell. Mater.* **19**, 72–85 (2010).
78. Schätti O, Grad S, Goldhahn J, Salzmann G, Li Z, Alini M, Stoddart MJ. A combination of shear and dynamic compression leads to mechanically induced chondrogenesis of human mesenchymal stem cells. *Eur. Cell. Mater.* **22**, 214–225 (2011).
79. Adkisson HD, Martin JA, Amendola RL, Milliman C, Mauch KA, *et al.* The potential of human allogeneic juvenile chondrocytes for restoration of articular cartilage. *Am. J. Sports. Med.* **38**, 1324–1333 (2010).
80. Visser J, Gawlitta D, Benders KE, Toma SM, Pouran B, van Weeren PR, *et al.* Endochondral bone formation in gelatin methacrylamide hydrogel with embedded cartilage-derived matrix particles. *Biomaterials* **37**, 174–182 (2015).
81. Zhu Hailin, Yang Jinliang, Wei Yuquan, Chen H-M. Development of a xeno-free substrate for human embryonic stem cell growth. *Stem. Cell. International.* **2015**(1), 1–8 (2015).
82. Klein TJ, Malda J, Sah RL, Hutmacher DW. Tissue engineering of articular cartilage with biomimetic zones. *Tissue. Eng. Part. B. Rev.* **15**(2), 143–157 (2009).
83. Hollander AP, Dickinson SC, Kafienah W. Stem cells and cartilage development: Complexities of a simple tissue. *Stem. Cells* **28**(11), 1992–1996 (2010).

Chapter 4

Directed Differentiation of Human Pluripotent Stem Cells for Therapeutic Applications

Lauren N. Randolph*, Evan B. Witmer* and Xiaojun Lian*,†,‡

** Department of Biomedical Engineering*
The Huck Institutes of the Life Sciences
Pennsylvania State University
University Park, PA 16802, USA

† Department of Biology
The Huck Institutes of the Life Sciences
Pennsylvania State University
University Park, PA 16802, USA

‡ Lian@psu.edu

Abstract: Human pluripotent stem cells (hPSCs), including human embryonic stem cells and induced pluripotent stem cells, are a potentially inexhaustible supply of human cells because they can be propagated indefinitely while still retaining the capacity to differentiate into all somatic cell types. Hurdles facing utilization of hPSCs in regenerative medicine include lack of reliable and efficient methods to differentiate hPSCs into disease related cell lineages, including neurons, cardiac muscle cells and pancreatic beta cells.

During differentiation, hPSCs make a number of developmental decisions to generate first the three germ layers (ectoderm, mesoderm and endoderm), followed by subsequent decisions to generate all of the body's mature tissues. These differentiation processes often involve a limited set of developmental cell signaling pathways, which can be modulated by growth factors and small molecules. Therefore, common methods of differentiating hPSCs to therapeutic applicable cell lineages require application of specific combination of growth factors and/or small molecules at distinct differentiation stages. This chapter discusses recent progress in the production of therapeutically applicable cells via directed differentiation of hPSCs.

Keywords: Human pluripotent stem cells, induced pluripotent stem cells, stem cell culture, directed differentiation, ectoderm, beta cells, hepatocytes, lung progenitors, cardiomyocytes, hematopoietic cells, endothelial cells.

1. Stem Cell Totipotency and Pluripotency

Fertilization occurs when the sperm fertilizes the egg. The fertilized egg is imbued with the trait of totipotency. This totipotency allows the cells to proliferate and differentiate into all the cell types that make up the human body as well the extraembryonic cells that make up the placenta. However, this level of differential potential only lasts for the first few divisions of the zygote. Those cells, termed embryonic stem cells (ESCs), not associated with making the placental tissue, are said to be in a state of pluripotency and can be obtained from the inner cell mass during the blastocysts phase.

Pluripotent stem cells have the ability to become all cell types within the human body. Upon differentiation, a cell's fate is further limited to ectoderm, mesoderm, or endoderm tissue types. Before unraveling the production of each of these germ layers and their derivative cell types, it is important to understand the characteristics the characteristics of maintained pluripotency. Stem cell research depends on strategies to maintain the pluripotent state. Otherwise, these cells will lose pluripotency, due to spontaneous differentiation, before researchers can definitively test their potential applications.

1.1. Pluripotency Markers

In order to understand the state of pluripotency, researchers studied the genes that are upregulated in the undifferentiated ESCs. Oct4, also called Oct3, is a Pit-Oct-Unc (POU) transcription factor shown to be exclusively expressed in blastomeres, pluripotent stem cells, and the germ cell lineage. Embryos deprived of Oct3/4, lost the pluripotent quality of their inner-cell mass.[1] Further testing showed that repression of Oct3/4 in ESCs induced a loss of pluripotency and caused differentiation to trophectoderm, the tissue from the outer cell layer of the blastocyst.[2]

As with Oct3/4, Sox2 was also identified as a pluripotency marker gene. Within the early embryos, Oct3/4 expression has been shown to overlap with Sox2 in the pluripotent cells of the inner cell mass.[3] Many of the shared targets of Sox2 and Oct3/4 have therefore been established as markers for pluripotency, including Nanog and Zfp206.[4,5]

1.1.1. Tumorigenic pluripotency markers

Some genetic markers of stemness responsible for long-term pluripotency maintenance and rapid proliferation of ESCs have been identified as tumorigenic. Several of these genes have been identified and tested for their roles in ESC maintenance including STAT3, c-Myc, and KLF family.

The LIF/STAT3 pathway is necessary and sufficient in self-renewal of ESCs in mice, but not in humans.[6,7] Western blot analysis showed LIF activation of STAT3 is not as strong as in human ESCs as in murine ECSs.[8] In human ESCs, a constitutively active STAT3, STAT3C, was not capable of maintaining an undifferentiated state.[8]

Transcription factor c-Myc has been shown to be activated by STAT3 signaling in mice.[9] Its expression can independently adjudicate self-renewal and maintenance of pluripotency.[9] More recently, c-Myc is shown to play a role in generating induced pluripotent stem cells (iPSCs) from both mouse and human fibroblasts, which will be discussed later.

The KLF family has also been used in the process of creating iPSCs. KLF-4 overexpression was reported to reduce differentiation in cells.[10] Upon further examination of the KLF family, KLF-1, KLF-2, and KLF-5 have all been found to share KLF-4's ability to induce and maintain pluripotency.[11]

1.2. Serum Free Culture

Early culture conditions for human ESCs (hESCs) included the culture medium supplemented with fetal bovine serum. However the serum was a major source of undefined differentiation factors and therefore led to the promotion hESC differentiation. Subsequently, serum replacement was created by Price *et al.* in 1998, consisting of undisclosed albumins and albumin substitutes.[12] Two years later, Amit *et al.* reported a successful serum-free media with the addition of basic fibroblast growth factor (bFGF), also called FGF2. A nearly four-fold increase in cloning efficiency of hESCs was observed.[13]

FGF2's exact role in promoting hESC pluripotency was further investigated. A significant downregulation of Oct3/4 occurred upon the removal of FGF2's respective receptor from hESCs, suggesting a role for FGF2 signaling in regulating Oct3/4 expression.[14] Ding *et al.* showed that FGF2 starvation followed by subsequent reapplication led to rapid activation of PI3-K.[14] In the same study, PI3-K inhibition with LY294002, resulted in the downregulation of pluripotency markers Oct3/4, PODXL, and Tra-1-60, emphasizing the importance of FGF2 and downstream PI3-K signaling in maintaining pluripotency.[14]

1.3. Feeder Free Culture

Along with the culture media, the surface on which hESCs are grown is equally important for successful stem cell culture. Initial efforts to maintain hESCs *in vitro* required culture on a "feeder layer" made of mouse embryonic fibroblasts (MEFs).[15] In murine ESC culture, the addition of LIF to the media activated the LIF/STAT3 pathway, and, as discussed previously, is sufficient to maintain murine ESCs eliminating the need for the feeder layer.[16] However, for hESC, LIF/STAT3 is ineffective.

Xu *et al.* demonstrated a successful feeder-free system for hESC culture where cells were plated on Matrigel or laminin in MEF-conditioned medium (MEF-CM).[17] These feeder-free cells had a doubling time comparable to that of cells grown on a feeder layer.[17] While MEF-CM's combination with Matrigel and laminin spurred cell growth, using gelatin with the MEF-CM was incapable of supporting the feeder-free culture. This suggests that necessary signals for stem cell proliferation were most likely coming from the Matrigel/laminin layers in addition to the MEF-CM.[17]

Matrigel was closely examined for the factors it could be contributing to in the maintenance of pluripotency. Matrigel is a basement membrane made up of continuous layers of specialized extracellular matrix, forming an interface between cells and their adjacent connective tissue. Basement membranes also act as a storage depot for growth factors *in vivo*. Specifically, Matrigel contains laminin, collagen IV, and heparin sulfate proteoglycans as its main component.[18] Combinations of the MEF-CM with either laminin or whole Matrigel had much larger yields of undifferentiated cells than MEF-CM combined with any other component of Matrigel.[19] Xu *et al.* suggested laminin may be binding a receptor created from integrin $\alpha 6$ and $\beta 1$ in order to signal the pluripotency maintenance.[17]

1.4. Mouse Embryonic Fibroblast Conditioned Media

The contents of MEF-CM remain necessary for maintaining feeder-free, serum free hESC populations. The components of MEF-CM have been examined by numerous groups. An uncultured medium (UM) is often used in these analyses to determine if its application to hESCs, combined with the minimum essential additional factors, can substitute for MEF-CM.

The addition of bone morphogenetic protein 4 (BMP4) to mouse ESCs helped maintain stemness; this led to similar experiments with hESCs with contradictory results.[20] The BMP4 treatment did not maintain pluripotency but instead caused differentiation with enhanced expression of trophoblast markers.[21] Therefore, Noggin, a BMP inhibitor, was explored to counteract BMP4 induced differentiation.

In 2005, Wang *et al.* demonstrated that Noggin and FGF2 are the minimum essential factors required to sustain hESC pluripotency with UM and no feeder cell layer was used.[22,23] These results suggest FGF2 and Noggin are the two primary active factors in MEF-CM.

Activin A was identified as another active component in MEF-CM and a product of feeder cells. Beattie *et al.* reported that, in feeder-free, MEF-CM free conditions, Activin A application was able to maintain hESCs in the undifferentiated state for more than 20 passages.[24] A similar study combining FGF2 with Activin A in feeder-free, MEF-CM free conditions yielded equally positive results.[25] Similar results were acquired combining Nodal with FGF2, but other members of the TGFβ-superfamily did not show the same capacity for maintenance of pluripotency including transforming growth factor beta 1 (TGFβ1), Cripto, and BMP4.[25]

2. Induced Pluripotent Stem Cells

iPSCs are undifferentiated stem cells derived from adult fibroblasts through the sequential application of pluripotency inducing transcription factors. The first iPSCs were generated by Takahashi *et al.* from mouse embryonic and adult fibroblast cells in a study that used retrovirus-mediated introduction of four transcription factors: Oct3/4, Sox2, c-Myc, and KLF4.[26] These cells are indistinguishable from ESCs in morphology, proliferation, gene expression, and the ability to generate a teratoma consisting of all three germ layers. Using the same four factors, Takahashi *et al.* generated iPSCs from human fibroblasts the following year.[27] Thomson *et al.* contemporaneously demonstrated the generation of human iPSCs using Oct3/4, Sox2, NANOG, and LIN28.[28]

Further study by Nakagawa *et al.* in 2008, showed that retroviral induction of c-Myc was not necessary to form iPSCs from either mice or human fibroblasts.[11] Although retroviral induction of c-Myc may be unnecessary, endogenous c-Myc protein itself may still be required for iPSC generation. It has been suggested that the induction of Oct3/4, Sox2, and KLF4 may be enough to activate c-Myc transcription endogenously.

With or without c-Myc, after treatment with transcription factors, iPSCs were selected by Fbx15 expression. However, expression of another marker, Nanog, was later established as a better method for selection. iPSCs established through Fbx15 expression differ from authentic ESCs in three ways: gene expression, DNA methylation patterns, and failure to produce adult chimaeras. Nanog selected iPSCs demonstrated competent generation of adult germ line chimaeras.[29]

Yu *et al.* discovered another method to generate iPSCs using a different set of transcription factors: Oct3/4, Sox2, Nanog, and LIN28. These iPSCs could be derived from human somatic cells such as human foreskin and lung fibroblasts. The resulting cells met all of the same pluripotency qualifications as those iPSCs created by Takahashi *et al.*

Since the introduction of iPSC technology, whenever reporting new pluripotent maintenance or differentiation techniques for hESCs, the study is often repeated on iPSCs as well. If results are successful for hESC application, then similar results are expected with iPSCs.

3. Differentiation

Upon loss of signals promoting the undifferentiated state, a stem cell will make the transition from pluripotency to a more specialized cell type (neuron, beta cell, cardiomyocyte, etc.). This transition is called differentiation and results in the formation of tissue from one to the three germ layers: ectoderm, mesoderm, and endoderm. The activity of different cellular signaling mechanisms aids in the determination of which germ layer a cell will become. These signaling pathways regulate gene transcription to create a unique pattern of gene expression corresponding to a cell's fate. During differentiation, stem cells undergo gradual shifts in morphology and function passing through several intermediate stages before reaching a final specialized cell state.

In the embryo, differentiation begins in the blastula phase, where the embryo takes the shape of a hollow ball. The blastula begins a process called gastrulation, transitioning into three distinct germ layers. Gastrulation begins with an invagination of the hollow blastula leading to the designation of the two earliest germ layers, ectoderm and endoderm, followed by emergence of the mesoderm. Upon germ layer specification, cells differentiation into final somatic cell types.

In recent years, biomedical science and regenerative medicine coordinate synergistic efforts in technology development towards clinical

Table 1. Specialized cells derived from the ectoderm.

Subpopulation of ectoderm	Products
Surface ectoderm	Epidermis, hair, nails, lens of the eye, sebaceous glands, cornea, tooth enamel, the epithelium of the mouth and nose
Neural crest	Peripheral nervous system, adrenal medulla, melanocytes, facial cartilage, dentin of teeth
Neural tube	Brain, spinal cord, posterior pituitary, motor neurons, retina

implementation of stem cells or stem cell derived cell types. Successful implantation of these cells could help replace damaged or failing tissues. Other potential applications include treating diabetes, preventing heart disease, and creating antitumor drugs. The development of *in vitro* differentiation strategies that mimic *in vivo* events is crucial for the creation of stem cell derived cell therapies. The following sections provide a description of each germ layer and summarize directed differentiation methods for cell types derived from endoderm and mesoderm.

3.1. Ectoderm

The ectoderm is the outermost layer of cells developed during gastrulation. Specified ectodermal cells sit in the upper hemisphere, or animal pole, of the blastula in a region called the animal cap. As the embryo proceeds to develop, the ectoderm tissue will further specify into three more subsets: the external ectoderm, the neural crest, and neural tube. From these subsets, the ectoderm is integrated into a wide variety of organs and systems. Ectodermal products are listed in Table 1 with the subpopulation of ectodermal tissue from which they derive.

3.2. Endoderm

While the ectoderm forms from the outermost layer of the invaginating blastula, the tissue forming the innermost germ layer develops into the endoderm. Endoderm gives rise to cells making up the gastrointestinal and respiratory tracts, endocrine glands, and the urinary system. As opposed to ectoderm, endoderm induction does not rely as heavily on inhibiting the signals for other germ layers. Instead, it relies on a complex network of signaling unique to its own tissue. The directed differentiation techniques used to promote stem cell differentiation to beta cells, lung progenitors, and hepatocytes follow.

3.2.1. Beta cells

Beta cells, characterized by their specialized ability to secrete and store insulin for the regulation of blood glucose levels as well as other substances like C-peptide, compose the pancreatic islets of the pancreas. The loss of functional beta cells typifies type I diabetes. As a result, the potential for an inexhaustible source of functional beta cells for type 1 diabetes therapy is the primary reason researchers are interested in the directed differentiation of beta cells from hESCs. Several groups have developed and directed differentiation protocols to derive beta cells or functionally similar cells.

The first step in the differentiation from a hESC to a beta cell is dermal layer specification, in this case, endodermal specification. This requires cellular signaling to move first from pluripotency to mesendoderm and then to a definitive endodermal cell type characterized by high levels of Sox17 and FOXA2 expression.[30] In practice, this first transition is commonly accomplished by the addition of Activin A to a low serum content culture medium.[30] Activin A, a TGF-β related growth factor, mimics the function of the nodal ligand to trigger the same cell signaling cascade that would be endogenously activated.[31] From the definitive endoderm, the next step of differentiation requires pancreatic progenitor specification and further differentiation to a beta cell.

Several protocols have been published using a variety of different factors to induce pancreatic progenitor and later insulin producing cell specification. One of the common induction factors, retinoic acid, has been reported to be involved in endoderm differentiation and pancreatic bud formation.[32] In 2006, D'Amour *et al.* published a five-stage protocol for differentiating a 2D monolayer culture of pluripotent stem cells into endocrine hormone-expressing cells that secrete and store insulin.[33] However, the cells produced from this protocol lacked definitive beta cell function as they coexpressed other hormones such as glucagon. They also exhibited other functional differences from adult beta cells including lowered processing efficiency of proinsulin and minimal C-peptide secretion response to changes in glucose.[33] This method was later refined from a five-stage differentiation to a four-stage process resulting in pancreatic endoderm and endocrine precursors.[34]

In 2007, Shim *et al.* demonstrated a 3D embryoid body culture technique to derive pancreatic endodermal cells from human pluripotent stem cells (hPSCs) using the sequential additions of serum, Activin A, and retinoic acid to the culture.[35] They established that the cells resulting from grafts transplanted into mice produced insulin, proinsulin, C-peptide, and glucagon and rescued a hyperglycemic state suggesting the ability to further differentiate into beta-like cells *in vivo*.[35] However, the uncontrolled nature of the final differentiation stage made it difficult to develop a definitive understanding of the complete process directing beta cell differentiation. Jiang *et al.* utilized a three-stage protocol and suspension culture system to derive insulin-producing cells from hESCs.[36] Their methods employed Activin A to produce the definitive endoderm and retinoic acid to induce pancreatic commitment. Supplements added during the maturation stage of this differentiation included bFGF and nicotinamide.[36] The resulting insulin-producing cells exhibited expression of several beta cell markers, including Pdx1, insulin, C-peptide, glut2, glucagon, and amylase, and the cells secreted insulin in response to changes in glucose level.[36] The efficiency of this differentiation method was reported at 15%, leaving room for optimization.[36]

Two years later, Jiang *et al.* demonstrated that an improved version of their differentiation protocol could be used to derive insulin producing cells *in vitro* from both hESCs and iPSCs.[37] This protocol implemented a pancreatic progenitor expansion stage by adding epidermal growth factor (EGF) prior to the maturation stage, resulting in a 25% yield of insulin-producing cells.[37] The chemically defined protocol represented an advantage to this differentiation method; however, improvements in efficiency would be beneficial in conjunction with definitive functional characterization of the resulting cells to determine likeness for adult islet cells.

In 2012, Rezania *et al.* described a four-stage protocol to differentiate a highly enriched population of 98% PDX1 positive pancreatic progenitor cells.[38] They used Activin A to induce definitive endoderm formation and then sequentially treated the cells with additional factors including FGF7, Noggin, retinoic acid, SANT1, Alk5i, and TBP.[38] They showed that these cells could rescue hyperglycemic conditions when transplanted to a mouse with drug-induced diabetes and observed that the *in vivo* maturation of the transplanted pancreatic progenitors mimicked human fetal pancreatic development.[38] In 2014, Pagliuca *et al.* reported a three-dimensional scalable culture system for generating beta cells *in vitro* with 85% efficiency for deriving PDX1 positive cells (Fig. 1).[39] Using a growth factor mediated approach, they generated beta cells that not only responded to changes in glucose *in vivo*, but also *in vitro*.[39] They confirmed that these cells expressed mature beta cell markers, as opposed to fetal cell markers, and demonstrated subcellular structure similar to mature beta cells.[39] Further research is required to evaluate the similarity of

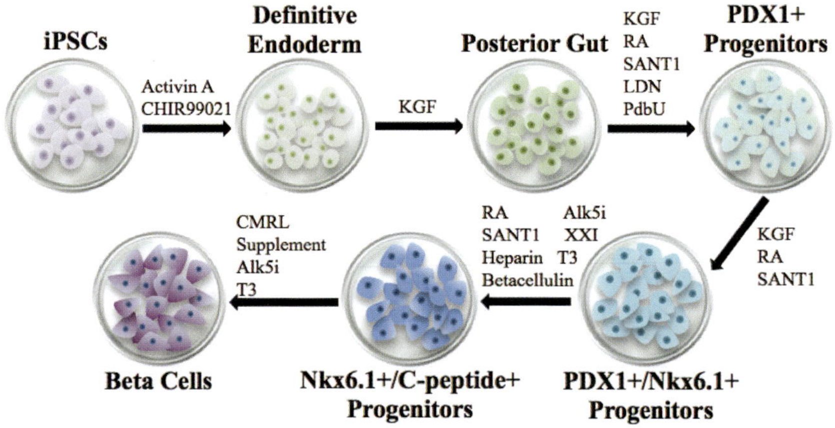

Fig. 1. Schematic illustrating the beta cell directed differentiation protocol developed by Pagliuca *et al.* using growth factor mediated transition from pluripotency through endodermal and progenitor states to functional beta cells.[39]

these stem cell derived beta cells to primary beta cells, but this scalable strategy is promising for clinical application.

3.2.2. Hepatocytes

Hepatocytes are the primary cells creating the liver. Hepatocyte functions include protein synthesis and storage, cholesterol, bile salt, and phospholipid synthesis, and detoxification and excretion of wastes or toxic materials. Exposure to toxic chemicals, alcohol abuse, or viral hepatitis can potentially cause irreversible damage to hepatocytes. This can lead to chronic liver failure, and liver transplantation represents the current clinical treatment to address functional tissue loss. As with any organ transplant, the demand exceeds the availability of viable organs. Challenges also arise with tissue compatibility and rejection. Stem cell based therapies for liver disease offer a potential alternative to restore functional hepatocytes. Additionally, stem cell derived hepatocytes could be utilized for drug toxicity testing.

Like beta cells, hepatic differentiation requires definitive endoderm specification followed by tissue, in this case hepatic, specification. In 2005, Schwartz *et al.* tested the ability of different cytokines to induce hepatocyte lineage specification in a serum free environment.[40] They found FGF4 and hepatocyte growth factor (HGF) significantly increase the expression of fetal hepatocyte markers, such as albumin, AFP, and CK18.[40] It has also been shown that the composition of extracellular matrix substratum used for differentiation culture may play a role in supporting hepatocyte specification. Culture on type 1 collagen during directed differentiation produced hepatocyte-like cells with albumin secretion closest to adult hepatocytes.[40,41]

In 2007, Cai *et al.* published a three-stage protocol for the differentiation of hESCs to hepatocyte or hepatocyte-like cells.[42] Their protocol used Activin A for definitive endoderm induction, recognized as particularly important in hepatic differentiation as extraembryonic endodermal cells express some of the same markers as hepatocytes.[42] FGF4 and BMP2 were utilized to induce hepatic progenitor specification with 70% efficiency for producing cells with fetal hepatocyte marker expression.[42] After maturation with HGF, oncostatin M, and Dex, the resulting cells expressed adult liver cell markers and displayed functional hepatocyte characteristics such as albumin secretion, LDL uptake, glycogen storage, and inducible cytochrome P450 activity.[42] In 2009, Cai *et al.* demonstrated that another similar protocol was effective in directing differentiation from iPSCs to hepatocyte-like cells.[43] While these cells exhibited many hepatocyte characteristics, the hESC and iPSC derived hepatocyte-like cells failed to produce the same functionality as adult hepatocytes. Albumin secretion, urea synthesis, and cytochrome P450 activities tested substantially lower than adult hepatocytes.[43] Similar results were obtained by Duan *et al.* using a similar differentiation strategy with the addition of BMP4 and DMSO to the culture.[44]

Basma *et al.* devised a three-stage differentiation using Activin A to induce the definitive endoderm, HGF and DMSO for hepatic specification, and Dex during the maturation stage.[45] To further enrich their hepatic-like cell population, they selected ASGPR positive cells by fluorescence activated cell sorting. The ASGPR positive cells displayed higher secretion of albumin, urea, and AAT, but the functionality still fell short of primary hepatocytes.[45] In 2013, Hannan *et al.* described a protocol that they used to generate hepatocyte-like cells from hESCs and iPSCs with 80% efficiency.[46] This strategy has five different stages and requires 25 days to generate the hepatocyte like cells.[46] These cells were observed to express rapidly decreasing levels of fetal hepatocyte markers as they matured; however about 10 days after the differentiation, the cells died off as functionality decreased.[46] As a result, hepatic differentiation methods need further refinement to improve the maturation stages in order to achieve improved mature hepatocyte functionality.

3.2.3. Lung progenitors

Lung epithelial cells line the respiratory tract. They participate in gas exchange, prevent infection, and remove foreign pathogens. These cells offer potential uses for disease modeling, drug testing, and clinical transplantation for conditions like cystic fibrosis (CF) where the lung endothelial cells lack normal function, producing a thicker mucous that impedes airway function. Unlike the pancreatic and hepatic lineages which derive from the posterior foregut or hindgut endodermal tissues, cells comprising the lungs originate from the anterior foregut endodermal tissue.[47] Lineages from the anterior foregut endoderm have proved more challenging to differentially recapitulate using directed *in vitro* culture techniques. Additionally, minimal knowledge exists about cell specific markers for lung epithelial cells or lung progenitor cells. One known progenitor marker is transcriptional regulator Nkx2-1; however, Nkx2-1 is also found in thyroid and fore brain tissue and is therefore not lung tissue specific.[47–49] This further complicates the isolation and characterization of lung progenitors in a differentiation culture.

Like pancreatic and hepatic differentiations, the first step requires definitive endodermal induction. This can be done with Activin A; however this approach has been shown to prefer posterior foregut and hindgut lineages with further differentiation.[48] In 2011, Green *et al.* determined that following Activin A treatment with inhibition of both BMP and TGF-β signaling, using Noggin and SB-431542, induced an enriched population of anterior foregut endodermal cells from the definitive endoderm population.[48] Using this strategy to derive anterior foregut endodermal cells from murine ESCs, Longmire *et al.* showed that subsequent stimulation of BMP and FGF signaling resulted in promotion of Nkx2-1 positive lung and thyroid progenitor cells with minimal neuro-ectoderm induction.[49] They activated BMP and FGF signaling by addition of FGF2, Wnt3a, FGF10, KGF, BMP4, EGF, and heparin to the culture.[49]

Mou *et al.* used a similar stepwise approach to differentiate lung progenitors in both murine ESCs and human iPSCs.[50] By addition of BMP4, FGF2, FGF10, Wnt ligands, and retinoic acid they further differentiated multipotent lung progenitors from anterior foregut endodermal cells.[50] The differentiation strategies from Longmire *et al.* and Mou *et al.* reported relatively low efficiencies with no more than 21% of the desired cell type obtained.[49,50] This leaves room for future work to refine and improve lung progenitor and lung epithelial cell differentiation. One interesting factor in the work conducted by Mou *et al.* was the use of an iPSC line derived from a CF patient; this highlighted the potential clinical application of a differentiation protocol for lung progenitors and lung epithelial cells for patient specific drug testing and human disease modeling.[50]

In 2012, Wong *et al.* provided another successful differentiation protocol producing airway epithelial cells from a CF patient derived iPSCs and hESCs.[51] This protocol further differentiated lung progenitors and implemented an air liquid interface *in vitro* culture technique to further mature the lung epithelial cells.[51] They used the CF trans-membrane conductance regulator (CFTR) gene to functionally assay their resulting airway epithelial cells, and achieved approximately 30% efficiency in deriving CFTR functional epithelial cells.[51] However, the lack of functionality tests represents a key barrier in the development of lung epithelial differentiation methods.[47] While CFTR function provides an interesting application of a disease-specific manifestation to determine lung epithelial function, additional functional assays will be necessary to further characterize this cell type.

3.3. Mesoderm

After the invagination of the blastula forms the basis of the endoderm and ectoderm, mesenchymal cells begin to populate the space in between, slowly forming a third germ layer in the center, the mesoderm. The mesoderm gives rise to all muscle types, connective tissues, kidneys, vascular endothelium, and hematopoietic cells. Much like endoderm, the mesoderm also has a unique pattern of signaling required to initiate its specification. The progress of directed differentiation techniques for cardiomyocytes, hematopoietic cells, and vascular endothelial cells derived from this embryonic germ layer is outlined below.

3.3.1. Cardiomyocytes

Cardiomyocytes are the terminally differentiated heart muscle cells. These cells are characterized by spontaneous contraction *in vitro* and syncytium behavior. Cardiomyocytes provide a desirable cell type for clinical applications due to their inability to appreciably replicate and replace lost tissue. Stem cell derived cardiomyocytes could provide a cellular base for cardiovascular disease therapies, in particular for myocardial infarction or heart failure.

In 1985, Doetschman *et al.* described the spontaneous differentiation of mouse embryoid bodies into cardiomyocytes, albeit with very low efficiency, demonstrating the potential to derive cardiomyocytes by *in vitro* culture techniques.[52] In 2002, Xu *et al.* published a feeder cell free method to differentiate cardiomyocytes characterized by spontaneous contraction as well as cellular markers including α-myosin heavy chain, cardiac troponin I and T, and cardiac transcription factor Nkx2.5.[53] This protocol requires the formation of embryoid bodies and used Percoll density centrifugation to isolate a population with 70% purity, but very low (~1%) efficiency.[53] They further reported supplemental factors tested to try to improve differentiation efficiency and showed that DMSO and retinoic acid did not increase cardiomyocyte differentiation while decitabine, a hypomethylating agent, provided efficiency enhancements when added in later stages of differentiation.[53]

Laflamme *et al.* used a similar differentiation strategy in 2005 to derive cardiomyocytes to graft into rat hearts.[54] They showed that human myocardial tissue developed in healthy rat hearts from the hESC derived cardiomyocytes.[54] In 2007, Laflamme *et al.* attempted to graft these cardiomyocytes into infarcted rat hearts with minimal success.[55] As a result they tried using a novel monolayer culture technique with sequential additions of Activin A and BMP4 resulting in a 30% efficient differentiation to cardiomyocytes.[55] In 2008, Yang *et al.* used a similar differentiation technique to derive cardiomyocytes and identify an early cardiac progenitor cell type.[56] By selecting for KDR positive and C-KIT negative cells, they achieved 50% efficiency for cardiomyocyte differentiation.[56] While these strategies are more efficient and use monolayer culture techniques, they rely on centrifugation or gene expression based purification methods to isolate pure populations of cardiomyocytes, which is not ideal for clinical use.

In 2010, Paige *et al.* described the role of Wnt/β-catenin signaling in cardiac differentiation.[57] They showed inhibition of Wnt/β-catenin in early stages of differentiation prevented the formation of cardiomyocytes while late stage inhibition did not prevent cardiomyocyte formation, illustrating a biphasic influence and the critical role of Wnt/β-catenin signaling for mesoderm induction.[57] Willems *et al.* tested the efficacy of several small molecule inhibitors of the Wnt/β-catenin pathway to promote cardiomyocyte differentiation post mesoderm induction.[58] Using Activin A, BMP4, and bFGF to promote mesoderm specification, they performed extensive small molecule screening and identified several small molecules (IWR-1, 53AH, IWP-3, and XAV939) whose addition was sufficient to generate cardiomyocytes from hESC derived mesodermal cells.[58]

In 2012, Lian *et al.* developed a fully defined protocol for highly efficient differentiation of cardiomyocytes using chemical manipulation of the Wnt/β-catenin pathway (Fig. 2).[59,60] This protocol used dense monolayer culture with mesodermal induction by activation of the Wnt/β-catenin pathway via small molecule Gsk3 inhibitor, CHIR99021.[59,60] Inhibition of the Wnt/β-catenin pathway was accomplished by addition of a small molecule porcupine inhibitor: IWP-2, IWP-4 or C59.[59,60] This differentiation strategy yielded 98% cardiac troponin T positive cells, illustrating high efficiency without an extra purification step, and its efficacy was demonstrated with both hESCs and iPSCs.[60] This protocol provided a defined and simple strategy to generate cardiomyocytes; however scale-up challenges and further delineation of the molecular underpinnings of this process would benefit future applications of this technology.

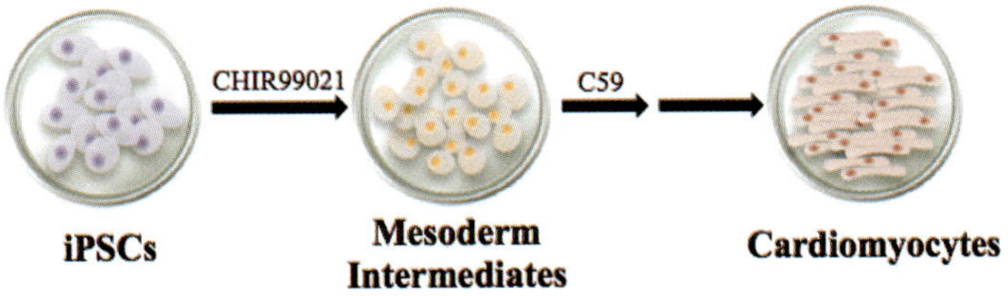

Fig. 2. Schematic of the chemically defined directed differentiation protocol developed by Lian *et al.* using chemical modulators of Wnt/β-catenin pathway activity.[59,60]

3.3.2 Hematopoietic cells

Hematopoietic stem cells are multipotent cells comprising blood components from both myeloid and lymphoid lineages. Autologous and allogeneic donor harvested hematopoietic stem cells are currently used clinically to treat some forms of cancer and immune system disorders (hematopoietic stem cell transplantation). One issue facing this therapy is the limited availability of human leukocyte antigen compatible hematopoietic stem cells. These cells are difficult to harvest and expand and in the case of allogeneic transplant, there is risk of tissue rejection or graft-versus-host disease. Generation of hematopoietic cells from hPSCs could provide an infinite source of these cells. In the case of iPSCs, a hematopoietic stem cell transplant could serve as an alternative autograph cell source.

In 2001, Kaufman *et al.* differentiated hESCs to hematopoietic progenitors expressing cell surface marker CD34 by coculture with murine bone marrow cells or yolk sac endothelial cells.[61] This method did not require any exogenous growth factors, but did use fetal bovine serum.[61] For clinical applications this method is not practical due to its lack of definition and reliance on murine coculture. In 2009, Choi *et al.* published a differentiation protocol using OP9 coculture to derive CD34 and CD43 positive hematopoietic progenitors from hESCs and iPSCs.[62] Again, this technique was not ideal as it involved a coculture with stromal cells reducing the clinical applications of this technique and the resulting cells.

In 2003, Chadwick *et al.* published the first hematopoietic differentiation protocol demonstrating the role of cytokines in promoting hematopoietic lineage definition as marked by the expression of CD45.[63] They used hESCs in an embryoid body culture and supplemented with stem cell factor (SCF), Flt-3 ligand, interleukin-3, interleukin-6, granulocyte colony stimulating factor, and BMP4.[63] Although they identified BMP4 did not play a significant role in hematopoietic differentiation, BMP4 treatment without cytokines enhanced progenitor proliferation.[63] Interestingly, in 2007, Pick *et al.* explored the roles of several different cytokines (VEGF, SCF, BMP4, and FGF2) in inducing hematopoietic differentiation in human embryoid body culture conditions without serum or coculture.[64]

Of the cytokines tested, BMP4 was the only one that promoted development of a primitive streak population, one of the first differentiation steps towards a hematopoietic mesoderm, and BMP4 also promoted expression of the generation of hematopoietic progenitors at low efficiency.[64] Addition of VEGF to the culture significantly improved the yield of hematopoietic progenitors, and SCF and FGF2 addition further increased efficiency.[64]

In 2011, Woods *et al.* outlined a protocol with a completely defined media for hematopoietic induction.[65] They used this media, containing fetal bovine serum, BMP4, TGF-β, VEGF, TPO, EPO, SCF, Flt-3 ligand, holotransferrin, PGE2, and ascorbic acid, to culture human embryoid bodies created from embryonic and iPSCs.[65] While the initial publication required a second step where embryoid bodies were cocultured with OP9 stromal cells, recent advances have eliminated the need for a coculture.[65,66] The commercially available media, MesoTotal, provided efficacy in producing CD34/CD45 positive and CD38 negative hematopoietic cells in a two-step embryoid body system without coculture.[66] Additionally, Ronn *et al.* showed that inhibition of retinoic acid signaling enhanced hematopoietic lineage commitment.[66]

In 2016, Kitajima *et al.* published a successful 2D culture technique to generate hematopoietic progenitors without a coculture.[67] Their protocol initiated the differentiation by activating the Wnt/β-catenin pathway via GSK3 inhibitor and was followed by a series of cytokine additions to the culture.[67] They found that GSK3 inhibition improved the efficiency of deriving cells with CD34 and CD45 markers.[67] Further refinements are still needed to clarify the essential factors for *in vitro* human hematopoietic differentiation. This is further complicated by current difficulties characterizing the true hematopoietic stem cell state.

3.3.3 Endothelial cells

Endothelial cells derived from the mesoderm are found in the heart, lining blood, and lymph vessels. These cells are involved in fluid filtration, recruitment of specialized blood cells, hormone trafficking, and maintaining homeostasis. These cells are known to malfunction in a variety of conditions including atherosclerosis, stroke, and Alzheimer's. As these cells are important components of blood vessel and heart tissue, stem cell derived endothelial cells hold interest for engineering new blood vessels and cell based myocardial repair.

In 2004, Wang *et al.* used hESCs to form embryoid bodies and identified endothelial-like progenitor cells, or hemangioblasts, that were multipotent for endothelial and hematopoietic lineages.[68] These cells were marked by expressions of PECAM-1, Flk-1, and VE-cadherin and lacked CD45 expression.[68] In 2007, Lu *et al.* published a two-step protocol to culture human embryoid bodies and achieve hemangioblasts.[69] Their serum free technique used cytokine additions of BMP4, VEGF, SCF, TPO, and Flt-3 ligand; however their efficiency was very low with less than 1% of the cells actually differentiating to a hemangioblast state.[69] In 2007, Wang *et al.* showed a 2D coculture system with hESCs, MEF cells, and fetal bovine serum supplement that was able to produce hematopoietic progenitors and endothelial progenitors marked by the expression of PECAM-1.[70] While this technique did not rely on spontaneous differentiation of embryoid bodies, which has very low efficiency, it relied on murine coculture minimizing therapeutic application and lacking definition. Interestingly, they demonstrated that when these cells were transplanted into SCID mice, they contributed to blood vessel development and integrated into the circulatory system.[70]

In 2010, James *et al.* initially stimulated human embryoid bodies with BMP4, Activin A, FGF-2, and VEGF and then dissociated the embryoid bodies for further culture.[71] While these conditions produced endothelial progenitors as marked by VE-cadherin expression, the method lacked efficiency, and the endothelial cells were not readily expanded.[71] They found by adding a small molecule inhibitor of TGF-β, SB431542, they were able to increase their yield of VE-cadherin positive cells; however SB431542 addition was only effective when added after the cells differentiated to the point of vascular commitment.[71] In 2013, Kusuma *et al.* developed a 2D culture technique which used serum supplemented media in the initial stage and then treated the cells with VEGF and SB431542.[72] They were able to generate VE-cadherin and PECAM-1 positive cells from both hESCs and iPSCs and showed that these cells would form microcapillary structures in hydrogels.[72]

In 2014, Orlova *et al.* used a defined protocol to generate endothelial cells from pluripotent stem cells.[73] They illustrated that by culturing hPSCs with sequential additions of CHIR99021 and Activin A followed by VEGF and SB431542, they were able to generate VE-cadherin positive endothelial progenitors which later expressed CD34 and KDR.[73] Lian *et al.* published a simplified protocol using only CHIR99021 in the initial differentiation step and a serum and growth factor free media in later steps.[74,75] This protocol produced endothelial progenitors marked by CD34 and VE-cadherin expression within five days; however, interestingly the protocol was only effective with male pluripotent stem cells.[74,75] Future research is necessary to optimize a differentiation technique that works for both male and female cell lines with improved efficiency.

References

1. Nichols J, *et al.* Formation of pluripotent stem cells in the mammalian embryo dependes on the POU transcription factor Oct4. *Cell* **95**, 379–391 (1998).
2. Niwa H, Miyazaki J, & Smith AG. Quantitative expression of Oct-3/4 defines differentiation, dedifferentiation or self-renewal of ES cells. *Nat. Genet.* **24**, 372–376 (2000).
3. Avilion AA, *et al.* Multipotent cell lineages in early mouse development on SOX2 function. *Genes Dev.* **17**, 126–140 (2003).
4. Rodda DJ, *et al.* Transcriptional regulation of Nanog by OCT4 and SOX2. *J. Biol. Chem.* **280**, 24731–24737 (2005).
5. Wang ZX, *et al.* Oct4 and Sox2 directly regulate expression of another pluripotency transcription factor, Zfp206, in embryonic stem cells. *J. Biol. Chem.* **282**, 12822–12830 (2007).
6. Niwa H, Burdon T, Chambers I, & Smith A. Self-renewal of pluripotent embryonic stem cells is mediated via activation of STAT3. *Genes Dev.* **12**, 2048–60 (1998).
7. Matsuda T, *et al.* STAT3 activation is sufficient to maintain an undifferentiated state of mouse embryonic stem cells. *EMBO J.* **18**, 4261–4269 (1999).

8. Sato N, Meijer L, Skaltsounis L, Greengard P, Brivanlou AH. Maintenance of pluripotency in human and mouse embryonic stem cells through activation of Wnt signaling by a pharmacological GSK-3-specific inhibitor. *Nat. Med.* **10**, 55–63 (2004).
9. Cartwright P, *et al.* LIF/STAT3 controls ES cell self-renewal and pluripotency by a Myc-dependent mechanism. *Development* **132**, 885–896 (2005).
10. Li Y, *et al.* Brief report Murine embryonic stem cell differentiation is promoted by SOCS-3 and inhibited by the zinc finger transcription factor Klf4. *Evaluation* **105**, 635–637 (2005).
11. Nakagawa M, *et al.* Generation of induced pluripotent stem cells without Myc from mouse and human fibroblasts. *Nat. Biotechnol.* **26**, 101–106 (2008).
12. Price PJ, Goldsborough MD, Tilkins ML. (1998). Embryonic stem cell serum replacement. International Patent Application WO98/30679. (1998).
13. Amit M, *et al.* Clonally derived human embryonic stem cell lines maintain pluripotency and proliferative potential for prolonged periods of culture. *Dev. Biol.* **227**, 271–278 (2000).
14. Ding VMY, *et al.* FGF-2 modulates Wnt signaling in undifferentiated hESC and iPS cells through activated PI3-K/GSK3?? signaling. *J. Cell. Physiol.* **225**, 417–428 (2010).
15. Thomson J, *et al.* Embryonic stem cell lines derived from human blastocysts. *Science* **282**, 1145–1147 (1998).
16. Williams RL, *et al.* Myeloid leukaemia inhibitory factor maintains the developmental potential of embryonic stem cells. *Nature* **336**, 684–687 (1988).
17. Xu C, *et al.* Feeder-free growth of undifferentiated human embryonic stem cells. *Nat. Biotechnol.* **19**, 971–974 (2001).
18. Chelluri LK. *Stem Cells and Extracellular Matrices. Colloquium Series on Stem Cell Biology* **1** (2012).
19. Amit M, Shariki C, Margulets V, Itskovitz-Eldor J. Feeder Layer- and Serum-free culture of human embryonic stem cells. *Biol. Reprod.* **70**, 837–845 (2003).
20. Ying QL, Nichols J, Chambers I, Smith A. BMP induction of Id proteins suppresses differentiation and sustains embryonic stem cell self-renewal in collaboration with STAT3. *Cell* **115**, 281–292 (2003).
21. Xu RH, *et al.* BMP4 initiates human embryonic stem cell differentiation to trophoblast. *Nat. Biotechnol.* **20**, 1261–1264 (2002).

22. Xu R-H, *et al.* Basic FGF and suppression of BMP signaling sustain undifferentiated proliferation of human ES cells. *Nat. Methods* **2**, 185–190 (2005).
23. Wang G, *et al.* Noggin and bFGF cooperate to maintain the pluripotency of human embryonic stem cells in the absence of feeder layers. *Biochem. Biophys. Res. Commun.* **330**, 934–942 (2005).
24. Beattie GM, *et al.* Activin A maintains pluripotency of human embryonic stem cells in the absence of feeder layers. *Stem Cells* **23**, 489–495 (2005).
25. Vallier L, Alexander M, Pedersen R. A. Activin/Nodal and FGF pathways cooperate to maintain pluripotency of human embryonic stem cells. *J. Cell Sci.* **118**, 4495–4509 (2005).
26. Takahashi K, Yamanaka S. Induction of pluripotent stem cells from mouse embryonic and adult fibroblast cultures by defined factors. *Cell* **126**, 663–676 (2006).
27. Takahashi K, *et al.* Induction of pluripotent stem cells from adult human fibroblasts by defined factors. *Cell* **131**, 861–872 (2007).
28. Yu, J. *et al.* Induced pluripotent stem cell lines derived from human somatic cells. *Science* **318**, 1917–1920 (2007).
29. Okita K, Ichisaka T, Yamanaka S. Generation of germline-competent induced pluripotent stem cells. *Nature* **448**, 313–317 (2007).
30. D'Amour KA *et al.* Efficient differentiation of human embryonic stem cells to definitive endoderm. *Nat. Biotechnol.* **23**, 1534–1541 (2005).
31. Phillips BW, *et al.* Directed differentiation of human embryonic stem cells into the pancreatic endocrine lineage. *Stem Cells Dev.* **16**, 561–578 (2007).
32. Shi Y, *et al.* Inducing embryonic stem cells to differentiate into pancreatic beta cells by a novel three-step approach with activin A and all-trans retinoic acid. *Stem Cells* **23**, 656–662 (2005).
33. D'Amour KA, *et al.* Production of pancreatic hormone–expressing endocrine cells from human embryonic stem cells. *Nat. Biotechnol.* **24**, 1392–1401 (2006).
34. Kroon E, *et al.* Pancreatic endoderm derived from human embryonic stem cells generates glucose-responsive insulin-secreting cells *in vivo. Nat. Biotechnol.* **26**, 443–452 (2008).
35. Shim JH, *et al.* Directed differentiation of human embryonic stem cells towards a pancreatic cell fate. *Diabetologia* **50**, 1228–1238 (2007).
36. Jiang W, *et al. In vitro* derivation of functional insulin-producing cells from human embryonic stem cells. *Cell Res.* **17**, 333–344 (2007).
37. Zhang D, *et al.* Highly efficient differentiation of human ES cells and iPS cells into mature pancreatic insulin-producing cells. *Cell Res.* **19**, 429–438 (2009).

38. Rezania A, *et al.* Maturation of human embryonic stem cell-derived pancreatic progenitors into functional islets capable of treating pre-existing diabetes in mice. *Diabetes* **61**, 2016–2029 (2012).
39. Pagliuca FW, *et al.* Generation of functional human pancreatic beta cells *in vitro*. *Cell* **159**, 428–439 (2014).
40. Schwartz RE, *et al.* Defined conditions for development of functional hepatic cells from human embryonic stem cells. *Stem Cells Dev.* **14**, 643–655 (2005).
41. Shirahashi H, *et al.* Differentiation of human and mouse embryonic stem cells along a hepatocyte lineage. *Cell Transpl.* **13**, 197–211 (2004).
42. Cai J, *et al.* Directed differentiation of human embryonic stem cells into functional hepatic cells. *Hepatology* **45**, 1229–1239 (2007).
43. Song Z, *et al.* Efficient generation of hepatocyte-like cells from human induced pluripotent stem cells. *Cell Res.* **19**, 1233–42 (2009).
44. Duan Y, *et al.* Differentiation and characterization of metabolically functioning hepatocytes from human embryonic stem cells. *Stem Cells* **28**, 674–686 (2010).
45. Basma H, *et al.* Differentiation and Transplantation of Human Embryonic Stem Cell-Derived Hepatocytes. *Gastroenterology* **136**, 990–999.e4 (2009).
46. Hannan NRF, Segeritz C-P, Touboul T, Vallier L. Production of hepatocyte-like cells from human pluripotent stem cells. *Nat. Protoc.* **8**, 430–437 (2013).
47. Kadzik RS, Morrisey EE. Directing lung endoderm differentiation in pluripotent stem cells. *Cell Stem Cell* **10**, 355–361 (2012).
48. Green MD, *et al.* Generation of anterior foregut endoderm from human embryonic and induced pluripotent stem cells. *Nat. Biotechnol.* **29**, 267–272 (2011).
49. Longmire TA, *et al.* Efficient derivation of purified lung and thyroid progenitors from embryonic stem cells. *Cell Stem Cell* **10**, 398–411 (2012).
50. Mou H, *et al.* Generation of multipotent lung and airway progenitors from mouse ESCs and patient-specific cystic fibrosis iPSCs. *Cell Stem Cell* **10**, 385–397 (2012).
51. Wong AP, *et al.* Directed differentiation of human pluripotent stem cells into mature airway epithelia expressing functional CFTR protein. *Nat. Biotechnol.* **30**, 876–882 (2012).
52. Doetschman TC, Eistetter H, Katz M, Schmidt W, Kemler R. The *in vitro* development of blastocyst-derived embryonic stem cell lines:

formation of visceral yolk sac, blood islands and myocardium. *J. Embryol. Exp. Morphol.* **87**, 27–45 (1985).
53. Xu C, Police S, Rao N, Carpenter MK. Characterization and enrichment of cardiomyocytes derived from human embryonic stem cells. *Circ. Res.* **91**, 501–508 (2002).
54. Laflamme MA, *et al.* Formation of human myocardium in the rat heart from human embryonic stem cells. *Am. J. Pathol.* **167**, 663–671 (2005).
55. Laflamme MA, *et al.* Cardiomyocytes derived from human embryonic stem cells in pro-survival factors enhance function of infarcted rat hearts. *Nat. Biotechnol.* **25**, 1015–1024 (2007).
56. Yang L, *et al.* Human cardiovascular progenitor cells develop from a KDR+ embryonic-stem-cell-derived population. *Nature* **453**, 524–528 (2008).
57. Paige SL, *et al.* Endogenous Wnt/Beta-Catenin signaling is required for cardiac differentiation in human embryonic stem cells. *PLoS One* **5**, (2010).
58. Willems E, *et al.* Small-molecule inhibitors of the Wnt pathway potently promote cardiomyocytes from human embryonic stem cell-derived mesoderm. *Circ. Res.* **109**, 360–364 (2011).
59. Lian X, *et al.* Directed cardiomyocyte differentiation from human pluripotent stem cells by modulating Wnt/β-catenin signaling under fully defined conditions. *Nat. Protoc.* **8**, 162–75 (2013).
60. Lian X, *et al.* Robust cardiomyocyte differentiation from human pluripotent stem cells via temporal modulation of canonical Wnt signaling. *Proc. Natl. Acad. Sci.* **109**, E1848–E1857 (2012).
61. Kaufman DS, Hanson ET, Lewis RL, Auerbach R, & Thomson JA. Hematopoietic colony-forming cells derived from human embryonic stem cells. *Proc. Natl. Acad. Sci. USA.* **98**, 10716–10721 (2001).
62. Choi K, *et al.* Hematopoietic and endothelial differentiation of human induced pluripotent stem cells. *Stem Cells* **27**, 559–567 (2009).
63. Chadwick K, *et al.* Cytokines and BMP-4 promote hematopoietic differentiation of human embryonic stem cells. *Blood* **102**, 906–915 (2003).
64. Pick M, Azzola L, Mossman A, Stanley EG, Elefanty AG. Differentiation of human embryonic stem cells in serum-free medium reveals distinct roles for bone morphogenetic protein 4, vascular endothelial growth factor, stem cell factor, and fibroblast growth factor 2 in hematopoiesis. *Stem Cells* **25**, 2206–2214 (2007).
65. Woods N-B, *et al.* Brief report: Efficient generation of hematopoietic precursors and progenitors from human pluripotent stem cell lines. *Stem Cells* **29**, 1158–1164 (2011).

66. Ronn RE, *et al.* Retinoic acid regulates hematopoietic development from human pluripotent stem cells. *Stem Cell Reports* **4**, 269–281 (2015).
67. Kitajima K, Nakajima M, Kanokoda M, Kyba M, & Dandapat A. GSK3 b inhibition activates the CDX/HOX pathway and promotes hemogenic endothelial progenitor differentiation from human pluripotent stem cells. *Exp. Hematol.* **44**, 68–74.e10 (2016).
68. Wang L, *et al.* Endothelial and hematopoietic cell fate of human embryonic stem cells originates from primitive endothelium with hemangioblastic properties. *Immunity* **21**, 31–41 (2004).
69. Lu S, Feng Q, Caballero S, Chen Y. Generation of functional hemangioblasts from human embryonic stem cells. *Nature* **4**, 501–509 (2007).
70. Wang ZZ, *et al.* Endothelial cells derived from human embryonic stem cells form durable blood vessels *in vivo*. *Nat. Biotechnol.* **25**, 317–8 (2007).
71. James D, *et al.* Expansion and maintenance of human embryonic stem cell — derived endothelial cells by TGFb inhibition is Id1 dependent. *Nat Biotechnol* **28**, 161–166 (2010).
72. Kusuma S, *et al.* Self-organized vascular networks from human pluripotent stem cells in a synthetic matrix. *Proc. Natl. Acad. Sci.* **110**, 12601–12606 (2013).
73. Orlova VV, *et al.* Functionality of endothelial cells and pericytes from human pluripotent stem cells demonstrated in cultured vascular plexus and zebrafish xenografts. *Arterioscler. Thromb. Vasc. Biol.* **34**, 177–186 (2014).
74. Lian X, *et al.* Efficient differentiation of human pluripotent stem cells to endothelial progenitors via small-molecule activation of WNT signalling. *Stem Cell Reports* **3**, 804–816 (2014).
75. Bao X, *et al.* Chemically-defined albumin-free differentiation of human pluripotent stem cells to endothelial progenitor cells. *Stem Cell Res.* **15**, 122–129 (2015).

Chapter 5

Magnetic Nanohybrids for Magnetic Resonance Imaging and Phototherapy Applications

Muhammad Zubair Iqbal and Aiguo Wu*

Key Laboratory of Magnetic Materials and Devices
Division of Functional Materials and Nanodevices
Ningbo Institute of Materials Technology and Engineering
Chinese Academy of Sciences
Ningbo 315201, P. R. China
**aiguo@nimte.ac.cn*

Abstract: Cancer, unrestrained development of abnormal cells in the body, is a serious threatening disease which has been defeating many important lives every year. The tumor, a mass of tissues, the clotting of abnormal cells, starts when old cells do not die and grow in as uncontrollable manner. Proper diagnosis at early stage can increase the survival rate. Therefore, scientists and medical experts are focusing their efforts on early detection of premature tumor and proper treatment to improve the life of cancer patients. Nanotechnology has been applied extensively in the field of cancer research, particularly, nanotheranostics for the advancement of effective approaches for diagnosis and treatment. Magnetic nanohybrid particles, combining non-invasive molecular imaging probe with therapeutic particles, have been contributing significantly towards magnetic resonance imaging (MRI) and phototherapy for cancers. Magnetic nanoparticles are successfully applied as MRI contrast agents and excited photosynthesizers are able to produce cytotoxic reactive oxygen species, such as singlet oxygen ($^1O_2$) which can oxidize cellular macromolecules leading to tumor cell ablation. In this chapter, we provide various aspects of magnetic nanohybrid particles, specifically, recent developments in the fabrication of complex hybrid structures, applications of tumor-targeted MRI, photodynamic/photothermal therapy and the outstanding challenges to these approaches.

Keywords: Magnetic nanoparticles, core–shell structures, Janus structures, magnetic resonance imaging, photodynamic and photothermal therapy.

1. Theranostics

Integration of molecular biology with nanotechnology has resulted in a new promising research area called bionanotechnology. "Nanotechnology" is the ability to make efforts at the atomic, molecular and supramolecular stages in order to investigate the system's essential properties and functions. According to biological narration, nanotechnology presents the tools and technical platforms for the exploration and transformation of biological systems, whereas biology offers bioassembled components and inspiration models to nanotechnology.[1–4] The convergence of modern biology with nanoscale science can be understood by graphical representation shown in Fig. 1.

The advanced name of nanobiotechnology is "theranostic nanomedicine". It is the medicinal application of nanobiotechnology which provides highly definite medical interference at the nanoscale level for diagnosis is treatment or prevention of diseases. The term "theranostic" was first coined in 2002 by John Funkhouser, the Chief Executive Officer of PharmaNetics, as a concept of "the ability of the material that combines the diagnostic imaging and modalities of therapy".[6] In other words, the expression "theranostics" was used to describe the ongoing efforts at treatment centers to combine therapeutic and diagnostic abilities into a single platform and to develop more precise, straightforward, individualized therapies for different diseases. Thus, theranostics transports diagnostic imaging agents and

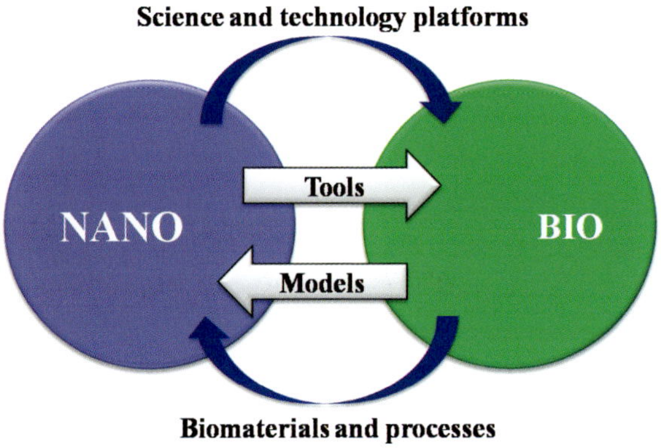

Fig. 1. Main interactions between fields of nanotechnology and biology, Redrawn from Roco (2003), from Ref. 5.

therapeutic drugs simultaneously with same dose for improved treatment. Better molecular understanding is very important to synthesize novel efficient targeted therapy agents and optimize drug selection.[7] Moreover, theranostics endeavors to observe the treatment response, to boost drug efficacy with safety, and to remove the redundant treatment of patients. As a result, theranostic proves to be cost effective for the overall healthcare system. Hence, the promising science of theranostics appears to present a distinctive opportunity to diagnostics and pharmaceutical companies to fulfill the regulatory and financial restrictions imposed by the Food and Drug Administrations (FDA).[8,9] Earlier theranostics, imaging contrast agents were simply mixed to the delivery vehicles loaded with therapy carriers for use in magnetic resonance imaging (MRI), positron emission tomography (PET) and computed tomography (CT). Therefore, the foremost objective of the theranostic research area is to attain the capability to monitor the diseased tissue, drug efficiency, imaging, delivery kinetics, and tune the therapy and dose rate in a controlled manner based on the requirements.[10]

The appearance of nanotechnology has provided the opportunity and made it easier to bring diagnosis and therapy factors closer. Nanostructured materials are assumed to be the fundamental building blocks of practical nanotechnology and thus can be manipulated for particular applications. All man-made and biological systems have been existing at the nanoscale naturally as nanotubes, nanocrystals and nanobiomotors. This natural phenomena has already been existing in the environment and in the living systems, e.g. a DNA molecule is about 2.5 nm wide, while a sodium atom is about 0.2 nm. The promising features of employing nanoparticles as theranostics are their ability to localize the site of disease accurately and reduce the unwanted side effects. Combining imaging, diagnostic and therapeutic functions in single modality or particle has significant influence in biomedical applications.[11]

The dimensions of these nanometric materials prevent them from readily paring through the kidneys. Therefore, circulation in the blood pool is longer, depending on their surface functionalization qualities. Moreover, the novel properties of theranostics nanomaterials observed are attributed to their: (1) shape (spherical versus tube) and morphological structure illustrated by size-induced effects (narrow size distribution, large surface area); (2) chemical composition (high purity and good crystallinity, etc.); (3) solubility in different media; and (4) surface structure (inorganic or organic surface coatings and increased surface reactivity). The above-mentioned properties of nanoparticles are very significant in bionanotechnology. For example, in cancer treatments, if the tumor is of many types, then blood vessels are irregular in shape, dilated, permeable, and can be create fenestrations in endothelial cells. Nanoparticles can easily penetrate into tumor tissues and can be preserved because of weak lymphatic drainage. Thus, the permeability and retention (EPR) effect enhances due to the selective accretion of nanosized particles near tumor tissues.[12–14] In addition, nanoscale particles have high surface area to volume ratios, and the ability of acquiescing high loading capacities. Thus, imaging agents and therapeutic drugs can easily be loaded on nanoparticles because of their surface chemistry. Nanoparticles provide access to the various functional groups on their surface to cloaking agents and targeting ligands which improve the imaging quality, therapeutic function and reduce systemic toxicity. It is noteworthy that nanoparticles with their quantum confinement effects, and small size with large surface area shows unique size-dependent magnetic, optical, and electronic properties.[15–17]

Oberdörster group also proposed the inverse relationship between the particle size and the number of the atoms or molecules on the surface of a spherical nanoparticle.[17] For a nanoparticle of 30 nm diameter, approximately <10% of its constituent atoms/molecules are located on the surface, whereas for a particle of 3 nm diameter about 80% of its atoms/molecules are situated on the surface (2). Therefore, nanoparticles of less than 50 nm have been showing outstanding properties for numerous applications because of the atoms/molecules positioned on their surface.[17] Therefore, small nanosized particles combined with imaging and therapeutic agents have been presenting innovative prospects in current bionanotechnology.

This chapter reviews the magnetic nanoparticles, the properties of magnetic nanoparticles, magnetic nanohybrids, fabrication of magnetic nanohybrids, and theranostic biomedical applications of the magnetic nanohybrid materials. Further, the functionalizations of these magnetic hybrid structures with silica coatings and amphiphilic polymer are illustrated. Finally, the functions of biomolecule conjugates on the surface of magnetic nanohybrids which offer cellular targeting and utilize them as multimodal bioprobes for MRI, drug delivery and phototherapy are highlighted.

2. Magnetic Nanohybrids

Magnetic hybrid nanostructures can be defined as a mixture of two or more components embedded together with magnetic material to form a single system which delivers enhanced multifunctional properties because of its hybridization nature, such as proper arrangement of a magnetic nanoparticle attached on to the tip of a nanorod can be the foundation for various complex heterostructures, e.g. dumbbells, shuttles, chains, microphone, and star shapes. The capability of designing a variety of hybrid nanomaterials is entirely related to the physiochemical, external, and internal interfacing properties of individual components. Besides the primary development in fabrication, curiosity develops to investigate the synergetic properties of HNPs which usually generate at the colloidal boundaries of nanoparticles. Furthermore, the surface of these magnetic nanohybrid structures is modified with biodegradable materials such as silica, targeting ligands and amphiphilic polymers to enhance the biocompatibility.[14,15] Extensively used superparamagnetic iron oxide nanoparticles with various surface functionalities are favorable to carry targeting ligands such as peptides, proteins, DNA, RNA, carbohydrates and enzyme to improve particular tumor targeting.[1,7,16,17] These features and the multifunctionality of magnetic nanohybrids are potentially applicable as contrast agents, smart drug delivery vehicles and in theranostic applications. The delivery via magnetic nanohybrid material to targeted cells is a significant and challenging task in medicinal diagnosis and therapy.

2.1. Preparation of Magnetic Nanohybrids

2.1.1. Seed-mediated Growth Towards Magnetic Nanohybrids

Seed-mediated growth presents an efficient technique for the fabrication of the magnetic hybrid nanostructures with heterogeneous growth. The hybrid part is attached onto the surface of a magnetic core and then sequentially cultivated. In the heterogeneous process, growth via atomic collision and heterogeneous nucleation play a very important role. Excellent progress has been made in the preparation of the core–shell and dumbbell nanostructures by making two different hybrid species. Reverse microemulsion contributes an important character in the synthesis of core–shell magnetic nanohybrids. The individual components which have large lattice space difference produce core–shell structure by growing a uniform layer in the form of shell material around core particles, whereas the individual components which have similar lattice parameters and crystal structures produce dumbbell-shaped structures.[18–20]

Some general synthesis models for the growth of core–shell and dumbbell magnetic hybrid structures are shown in Fig. 2 and can be categorized into synthetic direction. Figures 2(a)–2(c) show an example of direct heterogeneous growth of hybrid species on primary seed. A chemical activation agent is used on the surface of primary core–seed to grow the secondary precursor

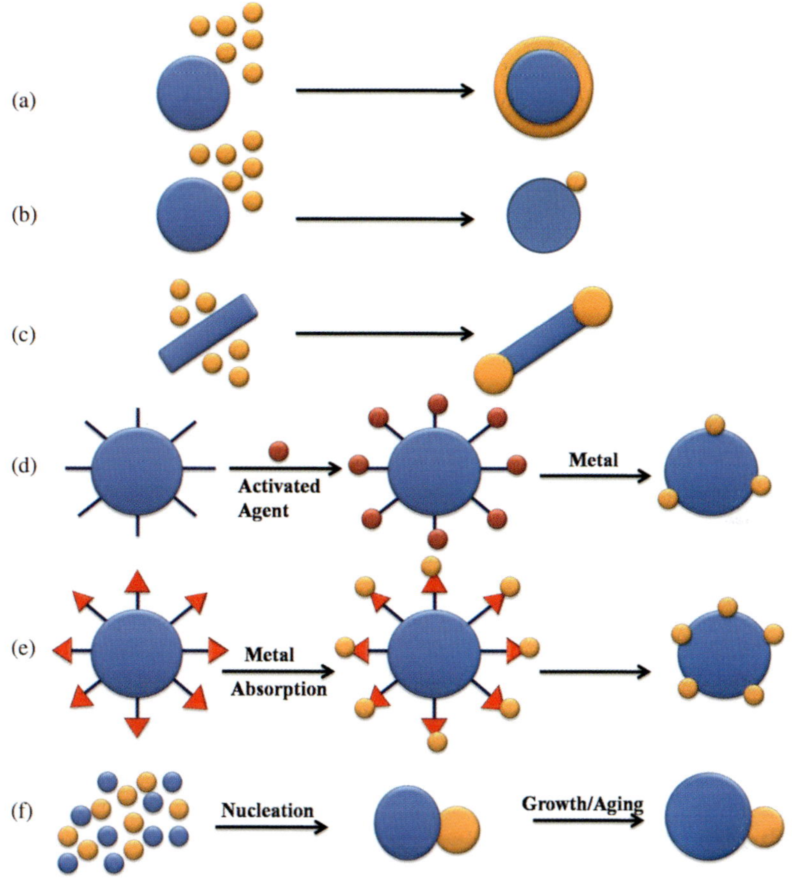

Fig. 2. A general sketch of reaction mechanisms for the formation of the hybrid nanostructures.

(Fig. 2(d)). The primary core is coated with a hydrophilic/hydrophobic polymer and then secondary precursor (Fig. 2(e)) and self-assembled heterogeneous nucleation of both materials (Fig. 2(f)) are introduced. Following there two pathways, a variety of magnetic nanohybrids including metal–oxide oxide–metal, metal–metal, oxide–oxide, and metal–semiconductor structures have been prepared successively.

2.2. Types of Magnetic Nanohybrids

2.2.1. Core–Shell Magnetic Nanohybrids

The past decades, researchers examined single nanoparticles and their properties because such micro/nanoparticles show much better performance than the bulk materials. Afterwards, in the late 1980s, material investigators found that composite or sandwich colloidal particles demonstrate excellent efficiency compared to their corresponding single particles. Thin layers on the surface of nanoparticles produced significant change in their properties and functionalities such as thermal stability, catalytic activity, magnetic–optical–electronic properties and chemical reactivity to develop more layers. In certain prospects, they even exhibit some new characteristics. In recent years, especially during the early 1990s, researchers have put in develop their efforts to synthesized concentric multilayer nanostructured materials with improved properties. Subsequently the terminology, multilayer nanostructure, was recognized by "core/shell" structures and has been receiving attention due to the demand of emerging technologies. Generally, the inner material is called core and outer layered material is refeved to as the shell. These types of architecture can be constructed with inorganic/inorganic and inorganic/organic interactions.[21–25] They may be spherical, eccentric, star-like, centric, or tubular in shape and some are shown in Fig. 3.

The bottom-up approach has been applied more effectively than the top-down approach in the synthesis of core/shell nanoparticles. Moreover, several methods, such as sol–gel, hydrolysis, Stober method, control hydrolysis and microemulsion, have been used to synthesize numerous core–shell materials like Au/SiO_2,[26,27] Ag/SiO_2,[28] ZnO/SiO_2,[29] Au/TiO_2,[30] CdSe/ZnS[31] and also many polymers, for example, PEG, PVP, dextran, chitosan and PVA. However, as we are concerned only with magnetic core–shell nanohybrids in this chapter, the following part will show some recent important examples of complex core–shell nanohybrid structures.

A team from the Department of Chemistry at Massachusetts Institute of Technology synthesized silica magneto-fluorescent nanoparticles complex core–shell hybrid structure.[32] Initially, they prepared

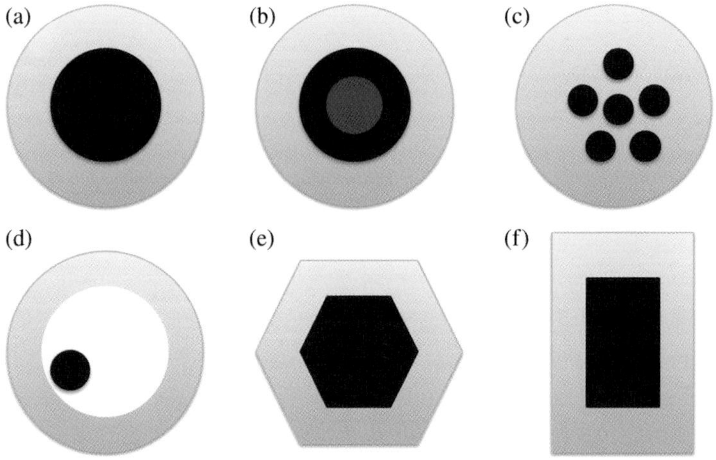

Fig. 3. Various types of core–shell structures: (a) spherical core–shell, (b) nanomatryushka materials, (c) multiple cores in a single shell, (d) moveable core in a hollowshell, (e) hexagonal core–shell and (f) tubular shaped core–shell structure.

CdSe–CdS core–shell nanoparticles and quantum dots (QDs) with Fe_3O_4 magnetic nanoparticles (MNPs), and then fabricated Fe_3O_4–QDs@CdSe–CdS using the coassembling process. Finally, polyethylene–glycol silane conjugated on the surface of nanohybrids to enhance the surface functionality and biocompatibility. In the synthesis process, hydrophobic Fe_3O_4 and QDs was prepared. A surfactant dodecyltrimethylammonium bromide (DTAB) was used to transfer the phase from organic to aqueous solution and then poly(vinylpyrrolidone) (PVP) ethylene glycol (EG) solution was employed to stabilize the nanoparticles as shown in Fig. 4(a). Elemental color mapping results (Fig. 4(b)) demonstrated the evidence of forming core–shell nanohybrids.

Silica shell was used on MFCSNPs to improve the stability and biocompatibility in aqueous media and the size of these particles after coating is about 100 nm as shown in Fig. 4(c) and inset.[33] At room temperature, these silica-coated MFCSNPs reveal superparamagnetism behavior with a saturation magnetization of 15.2 emu/g.

Basically, two methods have been employed extensively for the synthesis of iron oxide nanoparticles: (1) coprecipitation method and (2) thermal decomposition of magnetic oleate in non-ionic solvents.

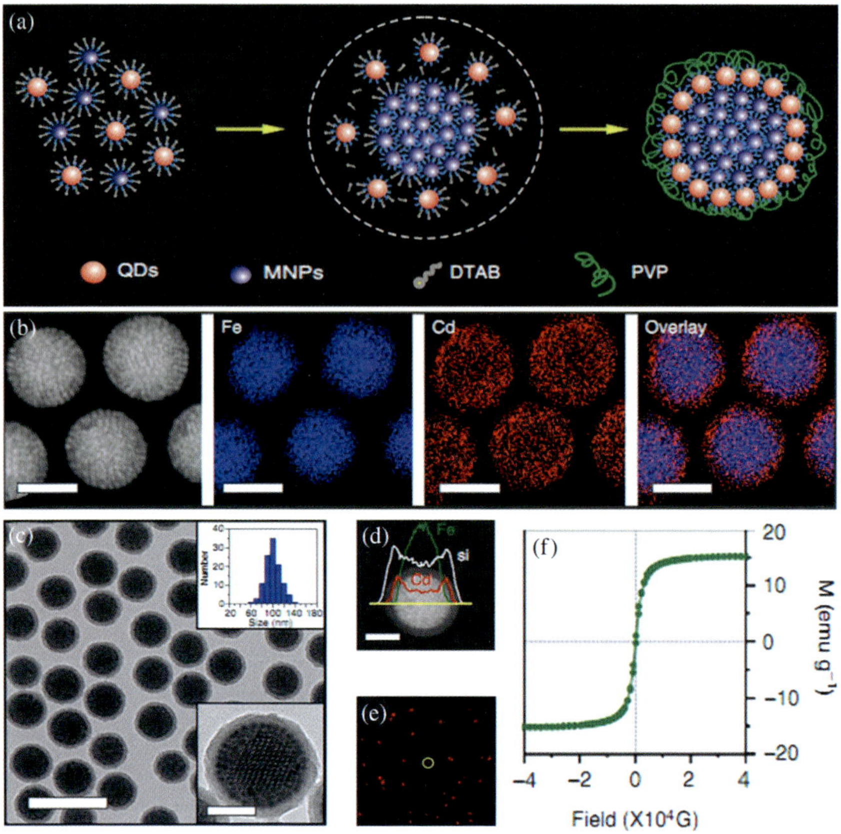

Fig. 4. Fabrication mechanism and characterization of magneto-fluorescent core–shell hybrid nanoparticles (MFCSNPs): (a) synthesis mechanism, (b) TEM and EDS analysis of core–shell nanostructure, scale bars is 80 nm and (c)–(f) TEM, EDS, fluorescence images and magnetic hysteresis of Silica-coated MFCSNPs, respectively. Reprinted with permission from Ref. 32.

Another significant example of magnetic core nanoparticles encapsulated in gold shell is shown in Fig. 5.[34–37]

In the synthesis process, PEI/Ag complex was synthesized by mixing silver nitrate aqueous solution into PEI solution in the presence of cold sodium borohydrate $NaBH_4$. Later, Fe_3O_4@Ag seeds were prepared using hydrothermal method by combining the iron

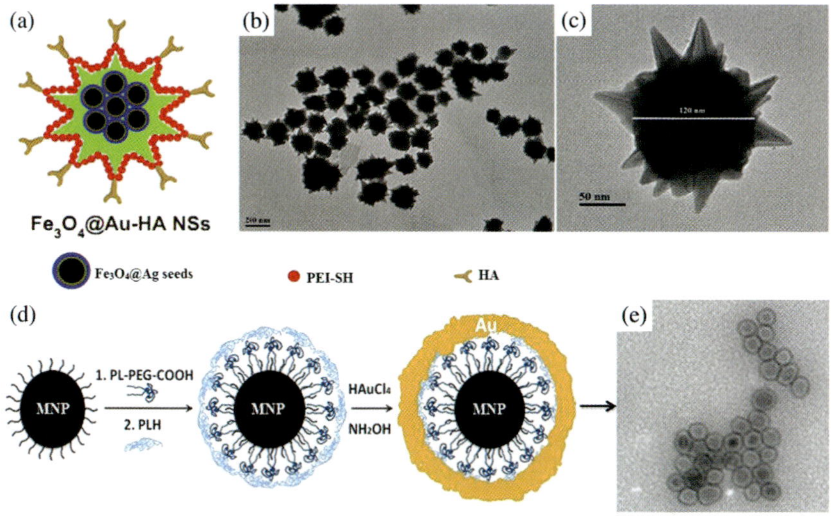

Fig. 5. (a) Schematic representation and (b),(c) TEM picture at low and high magnification of magnetic-gold nanostars (MAuNSs) like nanohybrid structure. (d) Graphic representation of magnetic-gold core–shell Nanoparticles and (e) TEM pictures of monodispersed hydrophobic MNPs coated with gold. Reprinted with permission from Refs. 35 and 37.

salt $FeCl_2 \cdot 4H_2O$ with PEI/Ag complex. Subsequently, gold growth solution was obtained using $HAuCl_4$ in CTAB, $AgNO_3$ (1.1 mg) and ascorbic acid. After Fe_3O_4/Ag solution was introduced to the above gold growth solution to fabricate MAu–NSs, partially thiolated PEI (PEI–SH) was used to enhance the surface properties and biocompatibility as shown in Fig. 5(a). Figures 5(b) and 5(c) illustrates the TEM pictures of MAu–NSs magnetic plasmonic nanohybrid structure at low and high magnification. After the formation of gold leaves onto the surface of magnetic nanoparticles, an obvious plasmon resonance peak of Au appears in the 550–850 nm regions.[38] This method is also applied to fabricate manganese-doped iron oxide@gold core–shell hybrid structure. Gao *et al.* reported an advanced generation of magnetic–plasmonic core–shell hybrid nanocrystals by producing a gap between core and shell as shown in Figs. 5(d) and 5(e).[37] Hydophobic small magnetic particles synthesized using thermal decomposition of iron oleate and then phospholipid–polyethylene glycol (PL–PEG–COOH) was used to transfer the phase from hydrophobic to hydrophilic providing the –COOH functional group on the surface. A layer of poly-L-histidine (PLH), positively charged peptide, is adsorbed onto the external surface of MNPs–PEG via charge interaction and the gold shell is developed by the reduction of gold. The organic molecules around the surface of MNPs present constrain direct gold growth on iron oxide nanoparticles and the gap can be seen in the TEM picture (Fig. 5(e)).

Recently, iron carbides (Fe_3C, Fe_5C_2, and Fe_7C_3) have attained significant attraction in both applied engineering and fundamental science. Iron carbide nanostructures have demonstrated high stability because of carbon atoms on the surface and high saturation magnetization (~140 emu/g) which are promising characteristics for magnetic storage and bioimaging applications.[39] A facile synthesis method provided by Ma *et al.* is shown in Fig. 6.[40,41]

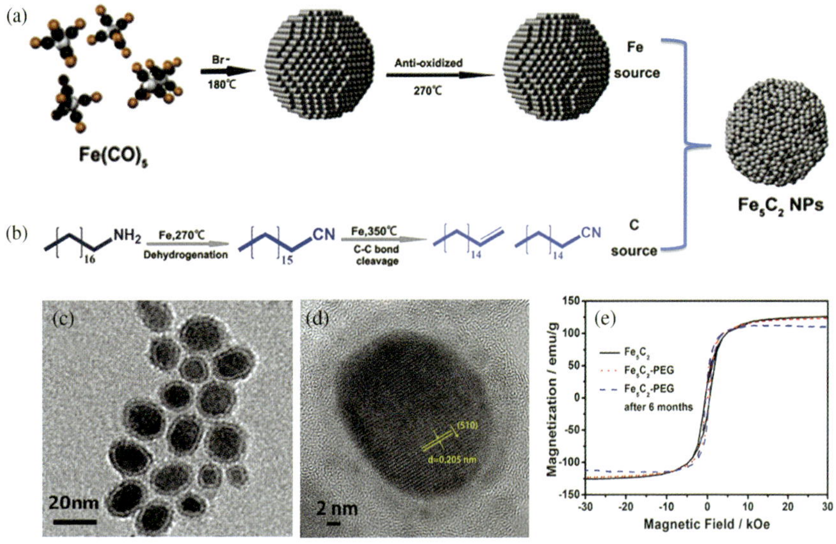

Fig. 6. (a) and (b) Graphical representation of the synthesis mechanism of Fe_5C_2NPs, black spots are carbon, silver are Fe and brown are oxygen, (c) TEM, (d) HRTEM images, and (e) magnetic hysteresis loops for Fe_5C_2NPs after surface modification and (blue line) exposed in water for six months. Reprinted with permission from Refs. 40 and 41.

In the synthesis process, bromide used to control the oxidation of Fe_5C_2 and 1,2-distearoyl-sn-glycero-3-phosphoethanolamine-N-[amino(polyethylene glycol)-2000] (DSPE–PEG–NH_2) was applied on the surface for amine group to link then further. TEM and high resolution TEM results showed that the Fe_5C_2NPs has a core–shell hybrid structure with a diameter of about 20 nm and the amorphous carbon layer is behaving as shell.[40,42] Magnetization hysteresis loop shown in Fig. 6(e) illustrates the superparamagnetic behavior of Fe_5C_2 with a saturation magnetization (Ms) of 125 emu/g at room tempera. Interestingly, carbon layer behave as a protective shell around the magnetic nanoparticles and Ms loss is less than iron oxide nanoparticles (80 emu/g).[43]

Furthermore, Gd-based upconversion paramagnetic core–shell nanohybrids have been attracting great importance towards biomedical applications owing to their distinct properties of exchanging near infrared light (NIR) to ultraviolet (UV) or visible light via an upconversion process.[44] More attractive phenomena occurs when photosynthesizer material layer establishes into upconversion to obtain theranostics application. Many core–shell nanocomposites materials have been studied, such as $NaYF_4$:Yb/Tm@$NaLuF_4$@$NaYF_4$@$NaGdF_4$,[45] $NaYF_4$:Yb/Er@$NaGdF_4$,[46] $NaLuF_4$:Yb,Tm@$NaGdF_4$.[47] Gd^{3+}-doped upconversion magnetic nanoparticles (Gd–UCNPs) can be used for upconversion luminescent (UCL) imaging and as a T_1-weighted contrast agent in MRI, which may enhance the spatial resolution and sensitivity simultaneously.[48–51] Therefore, the fabrication and development of the multifunctional core–shell combining Gd–UCNPs with hollow mesoporous silica nanoparticles are definitely important for the correct diagnosis and effective therapy of cancer.

Recently, Wu *et al.* and Shi *et al.* made great contributions in the synthesis and application of Gd-based upconversion core–shell hybrids prepared by simple hydrothermal and thermal decomposition methods, respectively.[52,53] Figure 7(a) is the schematic illustration of $NaGdF_4$:Yb/Tm@SiO_2@TiO_2 core–shell hybrid structure. Firstly, they synthesized NaGdF4:Yb/Tm nanoparticles and then used versatile Stober method for silica shell. Finally, TiO_2-coated on $NaGdF_4$:Yb/Tm@SiO_2 and employed (3-aminopropyl) triethoxysilane (APTES)

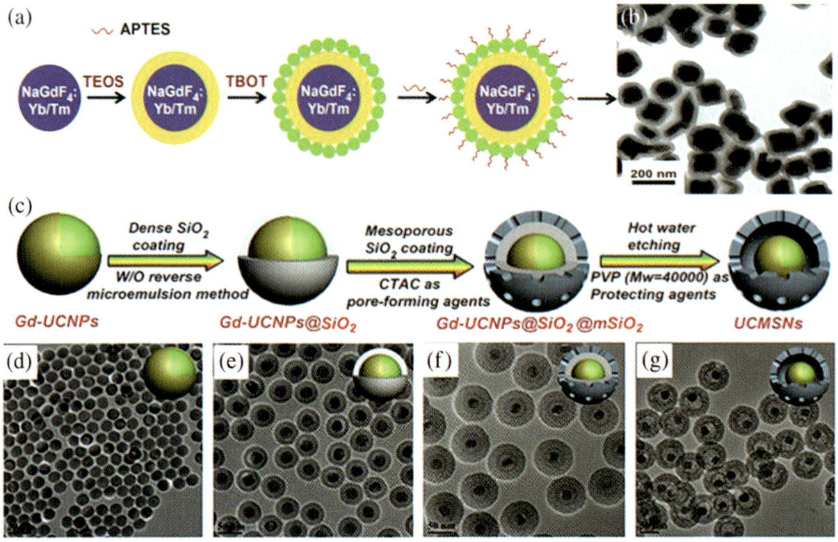

Fig. 7. (a) Synthesis process and (b) TEM picture of FA–Gd–Si–TiNPs. (c) complex fabrication procedure of mesoporus silica-coated Gd–UCNPs–SiO_2 and (d)–(g) corresponding TEM pictures of various growth phases. Reprinted with permission from Refs. 52 and 53.

for amine surface. Fig. 7(b) shows the TEM image of $NaGdF_4$:Yb/Tm@SiO_2@TiO_2. Further, Shi *et al.* designed a $NaYF_4$:Yb/Er/Tm@$NaGdF_4$ UCNPs (Gd–UCNPs) as the core and used reverse microemulsion method for silica coating, and finally utilized CTAC template to deposit mesoporous silica shell on Gd–UCNPs@SiO_2, shown as Gd-UCNPs@SiO_2@mSiO_2. Figure 7(c) showed the schematic diagram and TEM pictures related to the growth process presented in Figs. 7(d)–7(f).

Another complex example of core–shell paramagnetic hybrid nanostructure was presented by Jin *et al.*[54] They demonstrated a novel approach to fabricate multifunctional GdOF:Ln@SiO_2 yolk-like microcapsules. The preparation method (Fig. 8) illustrated the core $Gd(OH)_x(CO_3)_yF_z$ material synthesized using a precipitation method, and then glucose used for carbon coating on $Gd(OH)_x(CO_3)_yF_z$:Ln via hydrothermal process. Later, a popular technique, Stöber method of silica coating was applied on $Gd(OH)_x(CO_3)_yF_z$:Ln@C and finally calcined at 700°C

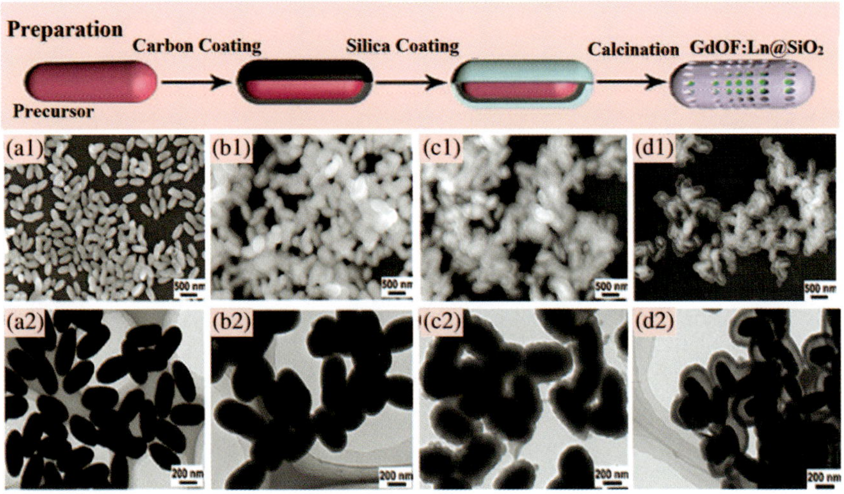

Fig. 8. Preparation of GdOf:Ln@SiO_2 and SEM and TEM images, (a1) and (a2) $Gd(OH)_x(CO_3)_yF_z$:Ln, (b1) and (b2) $Gd(OH)_x(CO_3)_y$Fz:Ln@C, (c1) and (c2) $Gd(OH)_x(CO_3)_yF_z$:Ln@C@mSiO_2, and (d1) and (d2) GdOF:Ln@SiO_2. Reprinted with permission from Ref. 54.

for 3 h to obtain GdOF:Ln@SiO_2 yolk-like core–shell capsules. The core is created because of the shrink of $Gd(OH)_x(CO_3)_yF_z$:Ln to GdOF:Ln through the annealing process and the mesoporous mSiO_2 shell is obtained from the elimination of CTAB. SEM and TEM pictures provide the justification for the proposed growth process shown in Fig. 8.

Additionally, mesoporous structures provide extra pores on the surface which offer to absorb or store materials and functional groups. The obtained GdOF:Ln@SiO_2 nanoparticles were further modified with amine group to conjugate ZnPc photosynthesizer and then decorated carbon dots on the surface of core–shell as shown in Fig. 9. Figures 9(b) and 9(c) are the bright field TEM and STEM images of the GdOF:Ln@mSiO_2ZnPC-CDs. EDS spectra of elements also showed good agreement with synthesized material and purity. The EDS elemental mapping with respect to different materials (Fig. 9(d)) provided uniform dispersion and verification of the yolk-like core–shell nanohybrids structure.

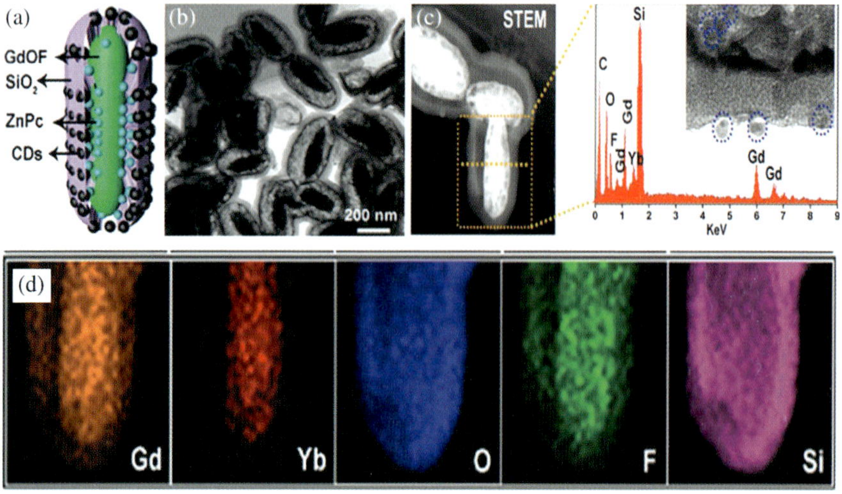

Fig. 9. (a) Physical appearance of prepared GdOF:Ln@mSiO_2ZnPC–CDs hybrids. (b) and (c) TEM and STEM images with EDS compositional analysis, and (d) the elemental mapping images of GdOF:Ln@mSiO_2ZnPC-CDs. Reprinted with permission from Ref. 54.

2.2.2. Janus or heterodimer structures

Another important category of the hybrid structure is the Janus or heterodimer structure. Nowadays, a most interesting and fascinating design of the hybrid structure called "Janus" is attracting much attention owing to its multicompartmental nature. Janus particles,[55–57] named by Casagrande and de Gennes for resembling the two-faced Roman god, are presently composed of a type of soft-matter with different polarity, chemistry, or special physicochemical properties on opposite sides.[55,58,59] Various synthetic approaches have been reported to generate purely inorganic–inorganic, polymer–inorganic heterodimer, droplets, microgels, Janus polymersomes, or fibers and polymer–polymer in different shapes and geometries ranging from nanometers to micrometers. Some beautiful examples of heterostructures other then magnetic hybrids are shown in Fig. 10.[60–62]

All Janus particle synthesis mechanisms are based on a three-step order: masking, phase separation and self-assembly. Self-assembly

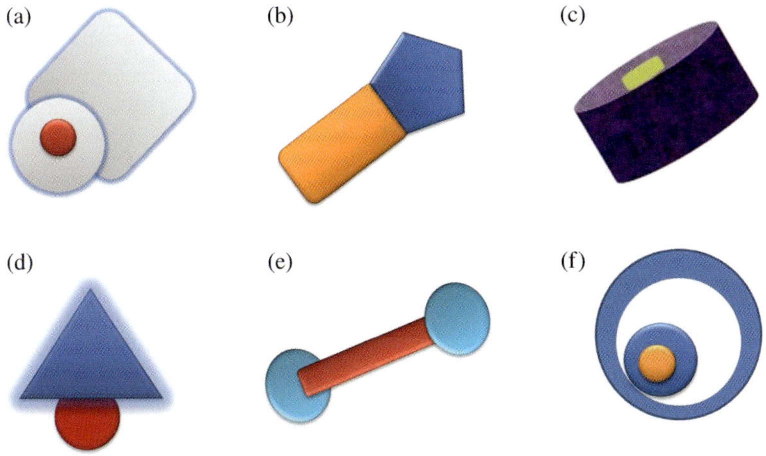

Fig. 10. Graphical representation of various types of complex hybrid nanostructures: (a) UCNP@SiO_2@mSiO_2&PMO composites,[63] (b) microphone-like structure of AgCdSe–Au nanomaterials with average length of Au rod 41 nm and average size of hexagonal AgCdSe being 24 nm,[64] (c) corresponding schematic model of Au–Cu_2S,[65] (d) Au–ZnO hexagonal nanopyramid-like structure,[66] (e) CoO-tipped CdSe/CdS nanorods achieved after oxidation of the Co tips[67] and (f) CdSe/CdS/ZnS–Au@ hollow SiO_2 yolk/shell nanospheres.[68]

plays a vital role in the fabrication of Janus-colloidal nanoparticles, which is a complicated interplay of factors like, entropy, enthalpy and solution dynamics. The general "Janus" self-assembly principle is summarized in Fig. 11, a "freezing-in" of the middle block (B) by covalent or non-covalent cross-linking, while the mutually incompatible end blocks (A and C) point in different directions. The preparation begins in most cases with the synthesis of the copolymers in a common solvent, and upon change of solvent, the self-assembly is induced. In the case of diblock copolymers, two combinations can be chosen for the preparation of Janus particles. Either a combination of AB and BC diblock copolymers, in which the B segments are insoluble in a given solvent and serve as the micellar core, or a mixture of AB and CD diblock copolymers, where B and C have an attractive interaction which forces the two polymers to coassemble is used. In order to use ABC triblock terpolymers for the formation of

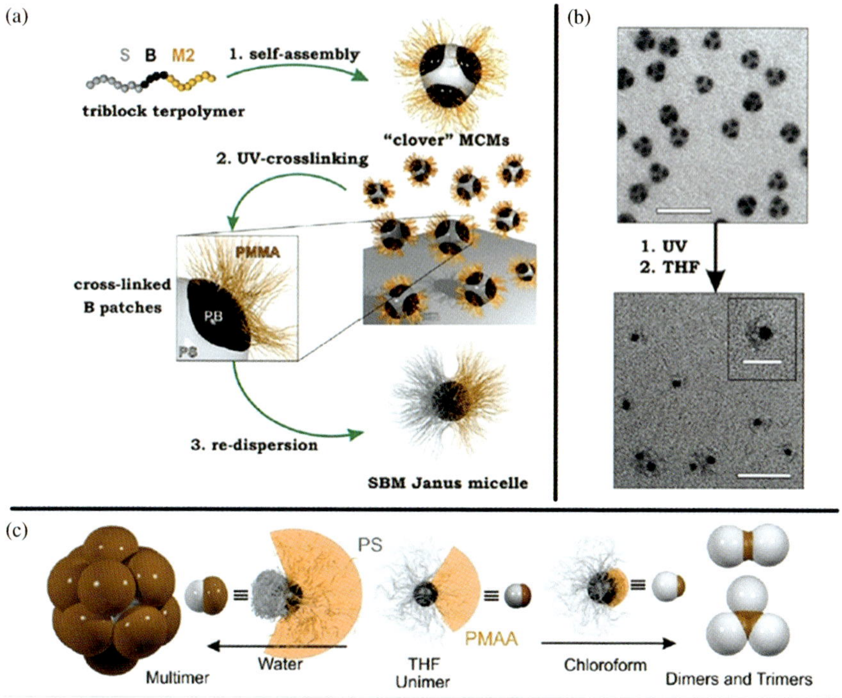

Fig. 11. Schematic illustration of the self-assembly mechanism: (a) Formation of Janus nanostructure by a self-assembly process of multicompartment micelles, (b) TEM images of "clover" multicompartment micelles and corresponding Janus particles (OsO4 staining: PS gray, PB black, and PMMA is not visible due to electron beam degradation), and (c) schematic clustering of asymmetric JPs in dependence of the corona size in water and chloroform (PS is gray and PB is black). Reprinted with permission from Ref. 19.

Janus particles, the terpolymers need to have two outer soluble end blocks, A and C, and an inner insoluble block, B. The simplest method to understand the self-assembly processes is given below in Fig. 11.[19,20,69]

Though the fabrication of Janus-colloidal particles has been evolving, the applications are still in infancy. Janus particles exhibit a number of distinctive properties in comparison to single domain particles due to their asymmetric structure and functionalization. Here, it is possible to alter their magnetic, catalytic, amphiphilic, electrical, optical, and surface active properties. Therefore, Janus particles offer a versatile platform of applications ranging from material science to biology and catalysis.

For instance, very recently, Lucia Curri *et al.* prepared γ–Fe_2O_3–TiO_2 Janus nanoparticles using the concept of hydrophobic interaction between binary nanocrystals.[70,71] Seeded colloidal technique was used to join spherical superparamagnetic iron oxide and photoactive TiO_2 rods in the form of Janus as shown in Fig. 12(a). The size distribution (inset of Fig. 12(a)) showed that the average length of TiO_2 nanorod was about 18 nm and mean diameter of γ–Fe_2O_3 was 8 nm. They also observed the absorption alteration between TiO_2 and Janus γ–Fe_2O_3–TiO_2, and absorption spectra shifted from UV region to visible region because of narrow band gap of maghemite (2.3 eV) as shown in Fig. 12(b).[72–74] Similar kind of Janus nanoparticles can be seen in reported Ref. 71 Further, Wu *et al.* synthesized spherical-shaped Fe_3O_4–TiO_2 Janus nanocomposites by solvo-hydrothermal method.[75] Figure 12(c) demonstrated the various magnification of TEM and HRTEM image of single hetrojunction of Fe_3O_4–TiO_2 that can be seen in Fig. 12(d). The small particle corresponding to TiO_2 with d-spacing of 0.3517 nm, and big particle related to Fe_3O_4 with d-spacing 0.2518 nm are clearly joined in the form of Janus. The crystal structure of prepared multifunctional Fe_3O_4–TiO_2 nanocomposites is shown in Fig. 12(e) and plane[440] and[422] indicated that both materials are hybrid. Moreover, superparamagnetism nature of Fe_3O_4 nanoparticles and Fe_3O_4–TiO_2 Janus nanostructure with saturation magnetization 58.95 and 14.90 emu/g are shown respectively, in Fig. 12(f).

Mesoporous silica shell has obtained significant interest owing to the unique and special properties such as uniform pore size, large pore volume, high surface area, low cytotoxicity and multifunctionality which are very beneficial in biomedical and energy application.[76,77] Therefore, another example of Janus nanohybrid was chose-from polymeric–inorganic material. Shi *et al.* fabricated superparamagnetic Fe_3O_4 nanoparticles coated by SiO_2 combined with polystyrene sphere to form a Janus structure using sol–gel method.[78,79] The schematic illustration is shown in Fig. 13(a). Miniemulsion polymerization has been revealed to be a significant synthetic route to prepare the polymeric–inorganic nanohybrid

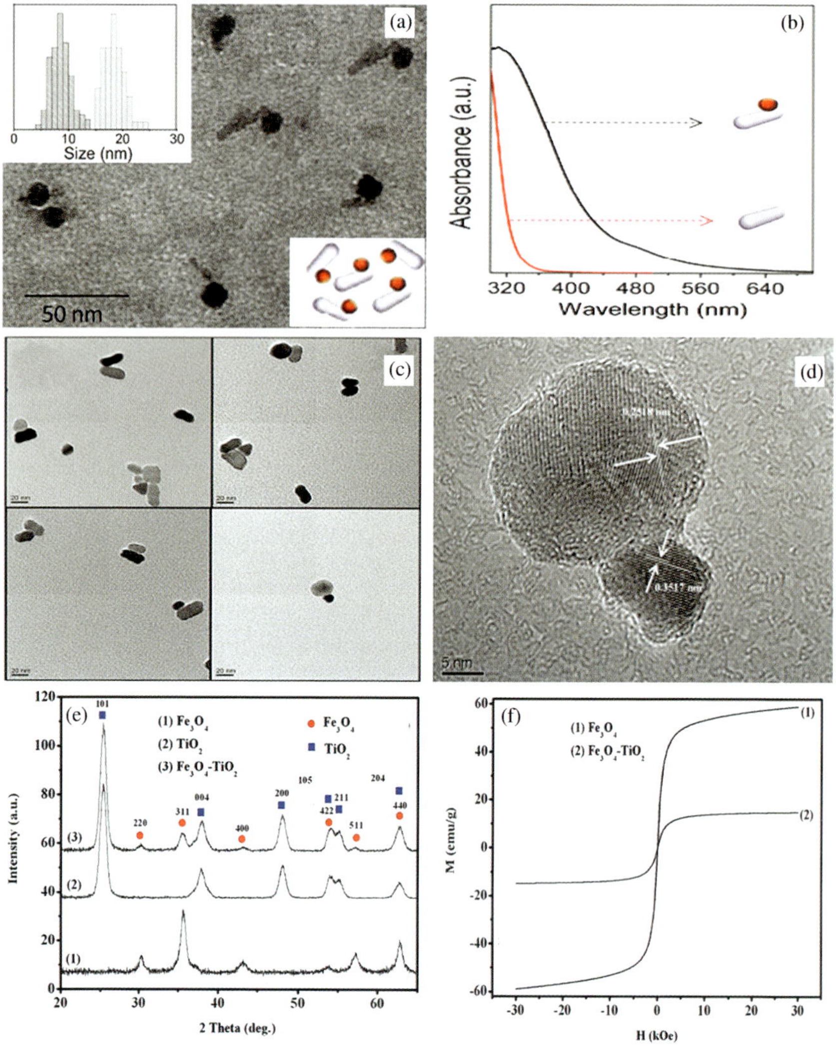

Fig. 12. (a) and (b) TEM picture of chloroform-dispersed γ–Fe_2O_3–TiO_2 hybrids, inset is particle size distribution of (black) Fe_2O_3NCs and (gray) TiO2 NRs and absorption spectra of organically chloroform-dispersed TiO_2NRs (red line) and magnetic hybrids (black line). (c) and (d) TEM, HRTEM images, (e) XRD spectra and (f) magnetic measurements of Fe_3O_4–TiO_2 nanocomposites. Reprinted with permission from Refs. 70 and 75.

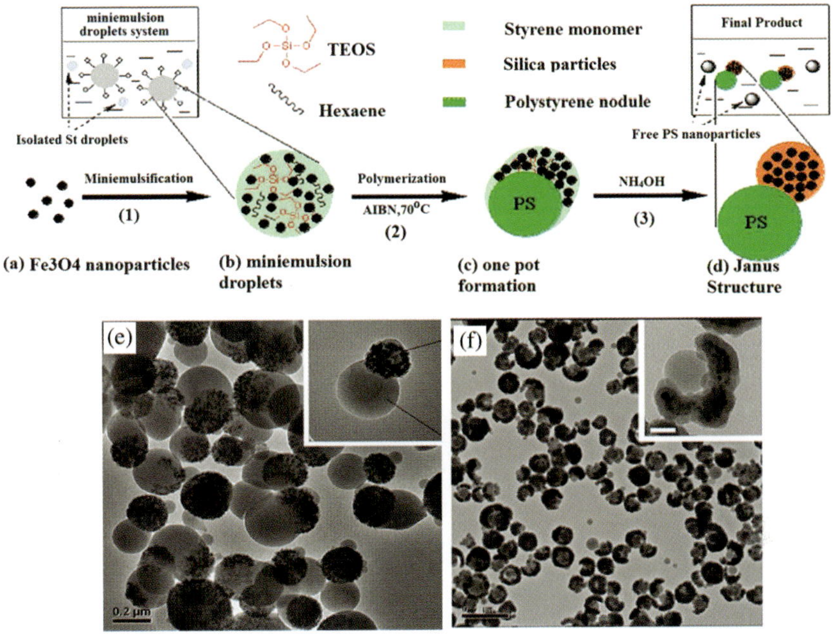

Fig. 13. (a)–(d) Schematic mechanism of polystyrene/Fe_3O_4@SiO_2 magnetic Janus nanohybrids (MJNHs), (e) TEM image of asymmetric Fe_3O_4/silica/polystyrene nanocomposites particles, and (f) TEM images of MJNHs obtained without γ-Meth acryloxypropyltrimethoxysilane molecules (scale bar is 500 nm and inset is 50 nm). Reprinted with permission from Ref. 78.

structure.[80] In the preparation method, they made hydrophobic iron oxide nanoparticles about 10 nm in size and then asymmetric magnetic composite particles were synthesized by mixing of the sol–gel reaction with miniemulsion polymerization. Initially, a little fraction of styrene (St) is changed into polystyrene chain and started to become heterogeneous droplets at stage 3 by accumulating Fe_3O_4@TEOS on one side and polystyrene on an opposite side. Finally, ammonia solution was added to hydrolyze the TEOS to form SiO_2 shell as shown in TEM image (Fig. 13(b)). Later, they repeated the same experiment without γ-Methacryloxypropyltri-methoxysilane (MPS) and the result is shown in Fig. 13(c). Therefore, MPS played an important character in the synthesis of MNHs. Furthermore, these Janus hybrid particles showed superpara-magnetic behavior due to their small sized iron oxide particles.

Lee *et al.* presented a seed-mediated growth mechanism to combine inorganic–inorganic oxide materials to form a Janus-shaped structure.[81,82] Firstly, thermal decomposition method was employed for the preparation of hydrophobic Fe_3O_4 with variable sizes and then these Fe_3O_4 nanoparticles injected into a trioctylamine solution of $Mn(CH_3CO_2)_2$ and oleic acid were used to start the growth process.[83,84] The obtained morphologies at different steps are shown in Fig. 14. Ultra-small iron oxide provided the opportunity for the formation of a uniform thick layer of a MnO in the form of shell around Fe_3O_4 and heterogeneous nucleation occurred when the Fe_3O_4 nanoparticles were

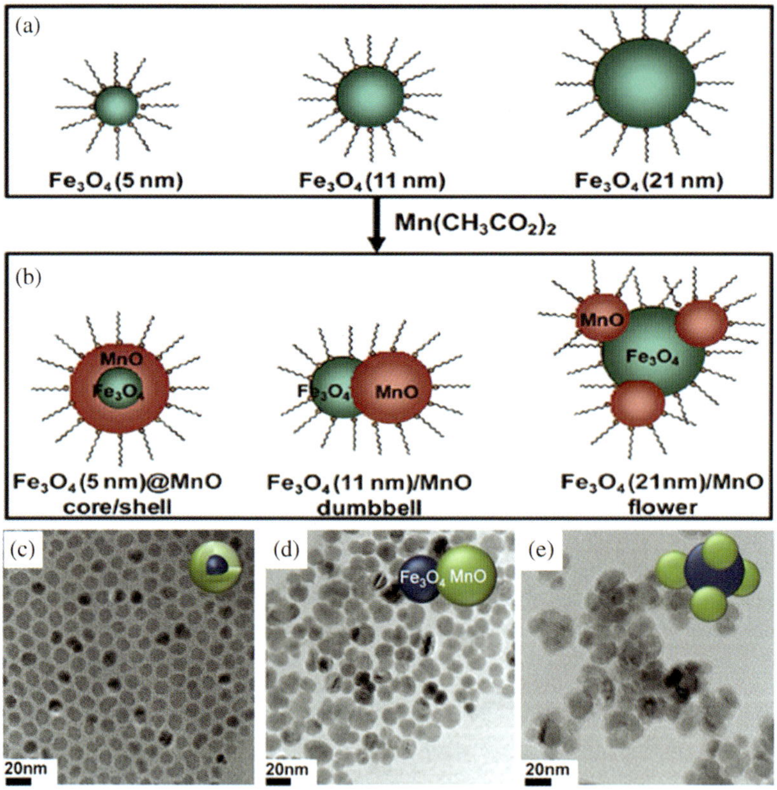

Fig. 14. (a) Seeds of iron oxide with different sizes, (b) seed-dependent growth of Fe_3O_4–MnO magnetic hybrid nanoparticles, (c)–(e) TEM images of Fe_3O_4 (5 nm)@MnO, Fe_3O_4 (11 nm)/MnO dumbbell-like, and Fe_3O_4 (21 nm)/MnO flower, respectively. Reprinted with permission from Refs. 81 and 82.

11 nm in size. The TEM pictures showed the core–shell, dumbbell and flower-shaped morphology of Fe_3O_4–MnO magnetic nanohybrid structure. Later, they used PEG to modify the surface of Janus nanoparticles to enhance the biocompatibility.[85]

In 2006, Chung-Yuan Mou developed a facile strategy to fabricate uniform magnetic-luminescent multifunctional well-ordered mesoporous structure which has good application in the biomedical field.[86] Similar decomposition methods were employed to prepare Fe_3O_4 nanoparticles and were dispersed in cyclohaxene. Subsequently, reverse microemulsion technique was used to synthesize homogeneous silica shell using TEOS and was hydrolyzed by NH_4OH at 50°C. Finally, fluorescent dye N-1-(3–trimethoxy-silylpropyl)-N-fluoresceyl thiourea (FITC–APT–MS) was introduced to synthesize Janus magnetic mesoporous silica/luminescent structure. The self-assembly process is illustrated in Fig. 15(a). The corresponding TEM of the core–shell and Janus-shaped structure are shown in Figs. 15(b) and 15(c). Further confirmation was

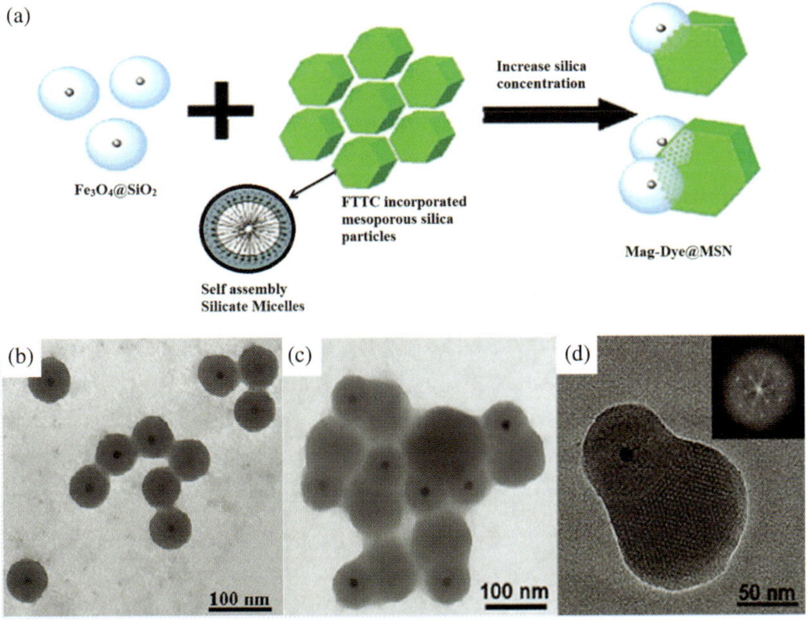

Fig. 15. (a) Synthetic illustration of tumbler-like mesoporous silica nanocomposites (TMSNs). (b) TEM image of Fe3O4@SiO2 core–shell nanoparticles, (c) TEM and (d) HRTEM pictures of TMSNs. Reprinted with permission from Ref. 86.

attained by taking HRTEM images (Fig. 15(d)), which exhibited a well-ordered porous Janus-shaped morphology.

In 2005, a research group from Brown University with a collaboration of Stanford University synthesized magnetic hybrid dumbbell-like Au/Fe_3O_4 nanoparticles using thermal decomposition method.[87] Recently, almost the same method has been employed for Au–Fe_3O_4 with little modification. Firstly, the seeds of AuNPs were grown and then they were injected into the solution of iron pentacarbonyl, $Fe(CO)_5$, with different solvents at high temperature.[88]

Firstly, chloroauric acid ($HAuCl_4$) solution was injected into the surfactant such as oleylamine in the presence of organic solution to prepare AuNPs. The temperature at which the gold precursor was injected into the solution was important to control the size of the particles. Afterwards, the prepared AuNPs seeds were mixed with $Fe(CO)_5$ in 1-octadecene solvent in the presence of oleylamine and oleic acid, heated at 300°C. Figures 16(a)–16(d) was prepared with the similar method. The lattice constant for Fe_3O_4 is 8.394 Å for inverse spinel and gold in the FCC phase is 4.080 Å. The size of the magnetic lattice is twice that of gold with a mismatch of about 3% which facilitates epitaxial growth.[87] The crystal plane {111} was shown direction in epitaxial growth as shown in Fig. 16(b). Similar kind of work was reported by Manna *et al.* and is illustrated in Figs. 16(c)–16(f). They obtained the dumbbell-shaped morphology in the absence of dodecyl dimethyl ammonium bromide and got the heterostructures in the presence of DDAB as shown in Figs. 16(e) and 16(f). DDAB has significant effect on structure and produced urchin-like morphology at higher concentration of ammonium bromide.

Later, Nanfeng Zhang also reported the preparation method of Au–Fe_3O_4 dumbbell nanoparticles via conventional heating process.[89] Finally, they modified the surface of magnetic-plasmonic nanohybrids using silica shell. The simple procedure is illustrated in Fig. 17(a). In detail, the asymmetric Au–Fe_3O_4 magnetic-nanohybrids were dissolved in cyclohexane solution of dodecanethiol to obtain a homogeneous dispersion and then the well-known reverse micro-emulsion method with Brij C10 and NH_3H_2O was used for coating. Finally, tetraethylorthosilicate (TEOS) was added into the above solution for

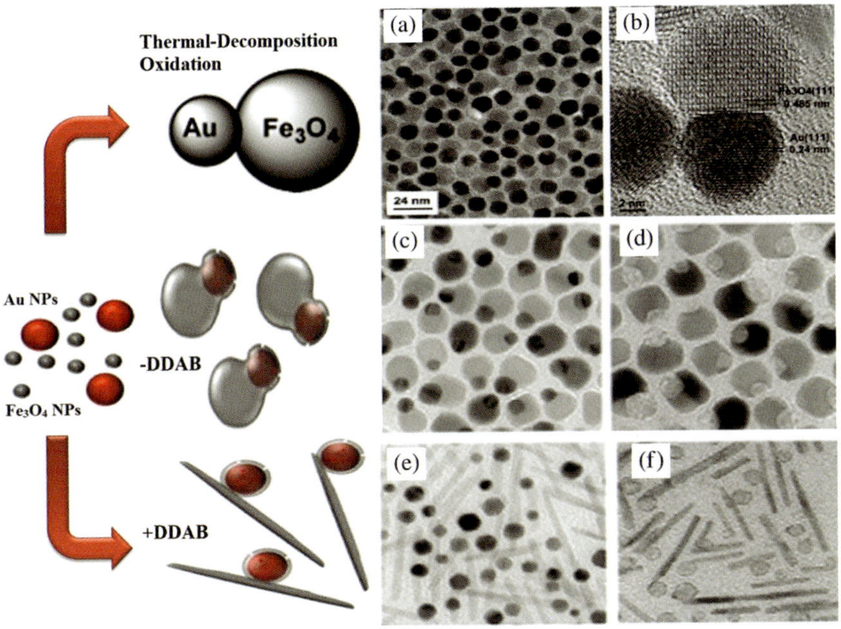

Fig. 16. Schematic presentation for the synthesis of Au–Fe_3O_4 plasmonic–magnetic hybrid nanoparticles. (a) and (b) TEM and HRTEM images of heterodimer or dumbbell-like nanoparticles, (c) and (d) TEM image of Au–Fe_xO_y HSs and after leaching gold from the Au–Fe_xO_y dumbbells, (e) and (f) TEM image of the Au–Fe_xO_y axe-like morphology and corresponding image after leaching Au from the Au–Fe_xO_y hybrid structures. Reprinted with permission from Ref. 87.

SiO_2 coating (Fig. 17(d)). The obtained product was well dispersed in aqueous medium for biological applications.

Another example of magneto–plasmonic hybrid structure is Ag–Fe_3O_4NPs which can be synthesized by controlled nucleation of Ag on the surface of Fe_3O_4NPs (Fig. 18(a)).[90] In facile synthesis, Fe_3O_4NPs dispersed in hexane were mixed with aqueous dispersed $AgNO_3$. The water and hexane are immiscible and the sonication process destroys the hexane layer into tiny droplets that produced microemulsion in the $AgNO_3$ solution. Similar types of structures were also observed in the literature.[91,92]

The Fe_3O_4NPs accumulated at the liquid/liquid border and acted as catalytic centers which were favorable for the growth/nucleation of

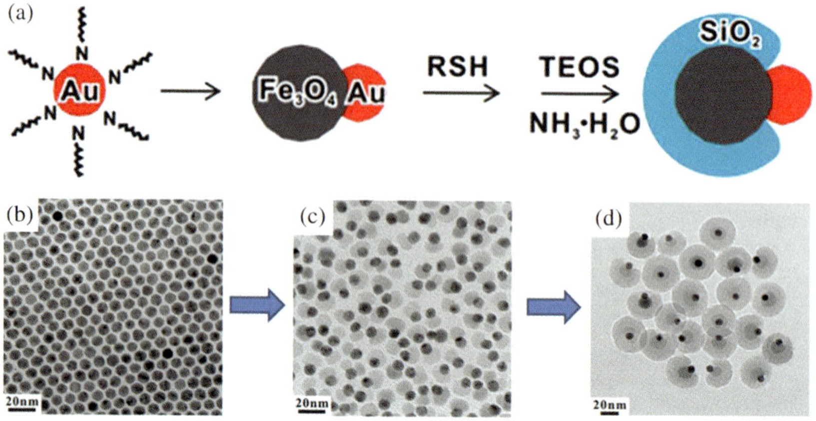

Fig. 17. (a) Schematic illustration of the fabrication of Au–(Fe_3O_4@SiO_2) Janus structure in the presence of alkanethiol and ammonia solution. TEM pictures of (b) OAm-cotaed AuNPs, (c) hydrophobic Au–Fe_3O_4 dumbbell-like structure and (d) Au-(Fe_3O_4@SiO_2) nanoparticles. Reprinted with permission from Ref. 89.

AgNPs. This technique has been utilized to synthesize Ag–FePtNPs and other Ag-based hybrid structures.[93] Some of them are demonstrated in Fig. 18.

Another interesting and complex category of Janus nanoparticles are the ternary dumbbell nanostructures which have been fabricated by seed-mediated growth. In this kind of particles, the third domain was attached by over growth of the desired particle, e.g. Sun *et al.* presented a concept of "tug-of-War" in nanoparticles.[94] In synthetic processes, Au_2NPs are additionally grown-onto Au_1–Fe_3O_4 dumbbell hybrid structures for the synthesis of ternary Au–Au–Fe_3O_4 or Au–Fe_3O_4–Au NPs. The viable growth of Au_2NPs on Au1 surface in the Au–Fe_3O_4 seeds produced a "tug of war" between Fe_3O_4 and Au_2 in the Fe_3O_4–Au_1–Au_2 ternary nanostructures resulting Au_2 extorted Au_1 from the Au_1–Fe_3O_4 structure to developed Au_2–Au_1–Fe_3O_4 ternary binary magnetic nanohybrids. In another study, Fe_3O_4–Au–Fe_3O_4 ternary magnetic–plasmonic nanohybrids were attained by fusing and absorption of the AuNPs from two Au–Fe_3O_4 nanocomposites.[95] The graphical mechanism is given in Fig. 19(a). The corresponding TEM images are also shown in Fig. 19(b).

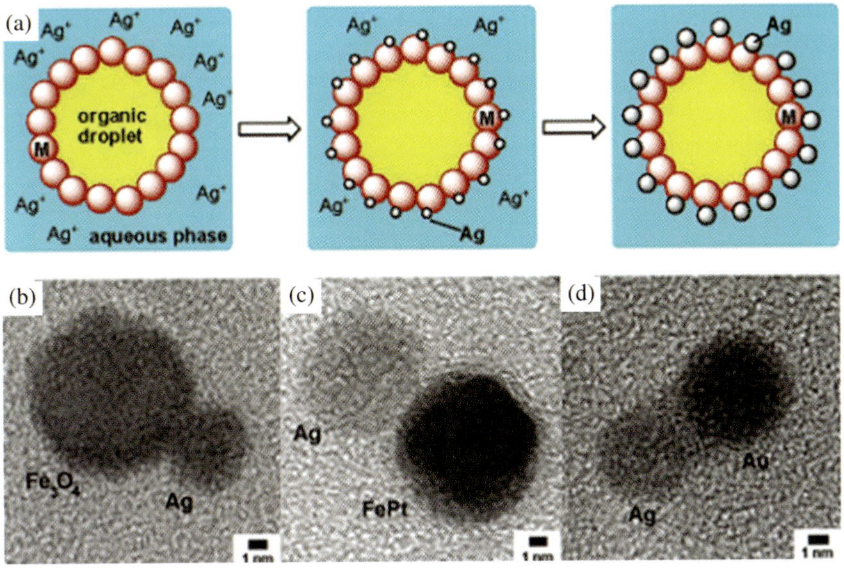

Fig. 18. (a) Schematic description of the formation of Ag-based dumbbell nanoparticles at the liquid/liquid interface. HRTEM images of (b) the Ag–Fe_3O_4, (c) Ag–FePt and (d) Ag–Au plasmonic–magnetic and plasmonic–plasmonic hybrid nanoparticles. Reprinted with permission from Ref. 90.

In the synthesis of Fe_3O_4–Au–Fe_3O_4 structure, Au–Fe_3O_4 solution was heated in the existence of sulfur. Sulfur played a very important role in the fabrication and acted as a glue to combine two AuNPs. It is fascinating that in the preparation of triple-domain nanoparticles, the combination or "tug-of-war" phenomena is observed through the metal particles. Apart from lattice mismatching property, it is also related to special characteristics of metals as compared to oxides and semiconductor chalcogenides materials. Similar kind of phenomena was observed by Schaak *et al.* during the synthesis of Fe_3O_4–Pt–Ag and Ag–Fe_3O_4–Pt as shown in Fig. 19(c).[96]

3. Applications of Magnetic Nanohybrids

Recently, cancer therapy has been improved by various scientific disciplines to facilitate the control on cancer diseases. Several

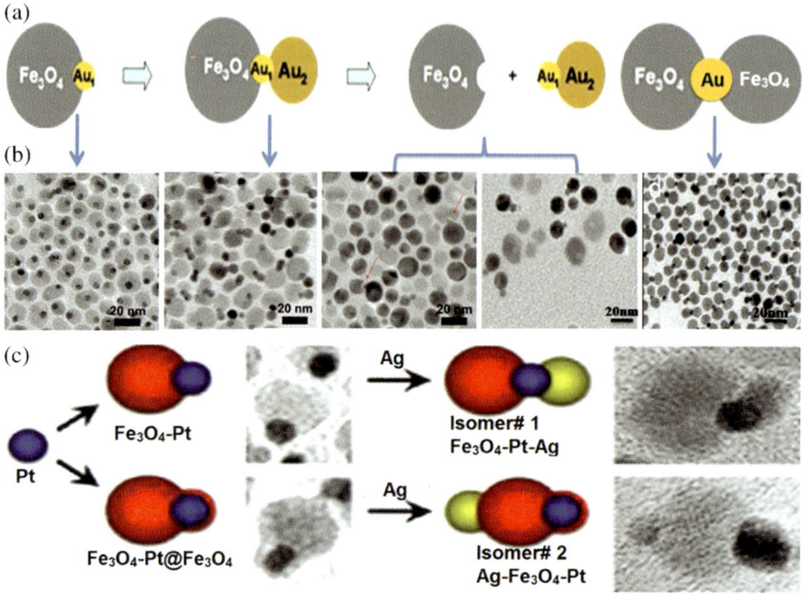

Fig. 19. "Tug-of-war" behavior in Au_2–Au_1–Fe_3O_4 dumbbell nanoparticles. (a) Graphical presentation of the Au_2 nanomaterial overgrowth on Au_1NP and Au_1NP separation from the Fe_3O_4NP, shaping a new dumbbell-like structure of Au_1–Au_2 and the scratched Fe_3O_4 NPs. Also, fusion of the Au component of two Au–Fe_3O_4NPs leads to the synthesis of Fe_3O_4–Au–Fe_3O_4 Janus nanohybrids. (b) TEM pictures of the different growth stages of magnetic–plasmonic nanostructures. (c) Various growth levels of Fe_3O_4–Ag–Pt corresponding with TEM images. Reprinted with permission from Refs. 94–96.

research techniques have been employed in the diagnosis of cancer, such as, positron emission tomography (PET), computed tomography (CT), MRI and optical microscopy in the bioimaging fields. Among them, MRI is a more advanced and promising non-invasive diagnostic technique owing to its sensitivity, selectivity, undisruptive, multiplanar capabilities, and tissue characterization potential and particularly, MRI does not contain ionizing radiation (high-energy radiation that can potentially cause damage to DNA, like the X-rays used for CT scans). It is the innovative technique which is employed mutually in research and clinical applications to map the tissues and organs in the body, for example, in injuries and abnormalities of the brain and spinal cord, heart problems, fibroids, endometriosis, uterine abnormalities and cancerous tumors [97–100]

In 1946, the magnetic resonance (MR) phenomenon was firstly discovered by Purcell and Bloch, but the first image was produced in 1973.[99] Meanwhile, Damadian confirmed in 1971 that certain pathogenic and natural tissues had different proton relaxation times. Finally, in 1977, Damadian, Minkoff and Goldsmith produced the first human scan using this technology. Initially, the scan took 5 h to complete, but the imaging time has been reduced in the portable MRI.

MR is a basic phenomena in MR imaging, in which nuclei of the atom like ^{1}H, ^{13}C and ^{31}P are emitting or absorbing some particular amount of energy from external magnetic fields. Then, magnetic moment will be produced by these specific nuclei which makes them MR active. Hydrogen nuclei ^{1}H is the most frequently used nuclei for imaging in biological atmospheres because human body has water and fat containing hydrogen atoms.

Initially, MRI was developed to present molecular imaging non-invasively. The major problem of MRI in definitive diagnoses is its lower sensitivity compared with nuclear imaging. However, it was observed that the development of contrast agents has facilitated MRI to become the most sensitive imaging technique not only in clinical diagnosis but also a useful instrument for biomedical research.[101,102] Contrast agents enhance sensitivity and/or specificity due to the shortened relaxation time of protons. Therefore, MR signal intensity is based on the local value of transverse ($1/T_2$) or longitudinal ($1/T_1$) relaxation rate of water protons. The relaxivities of $r1$ and $r2$, which are generally measured in $mM^{-1}s^{-1}$, demonstrate the increase in $1/T_1$ and $1/T_2$ per concentration (M) of contrast agent, respectively. Therefore, researchers have been getting attracted to this area and trying to develop promising clinically applicable contrast agents. This part will focus on some recent important examples of magnetic nanohybrids for multifunctionality such as dual MR-imaging, MRI with photodynamic therapy and MRI contract agents coupled with photothermal therapy.[103]

3.1. Multifunctional MRI MNHs

In recent years, the synthesis trend of superparamagnetic nanoparticles (Fe_3O_4) hybrid with paramagnetic nanoparticles or ion (Gd, Mn) has been rising to obtain dual modality MR imaging. Lou *et al.* fabricated the core of iron oxide for T_2 MR imaging and Gd_2O_3 shell for T_1-weighted-images in comparison to water as shown in Fig. 20.[104]

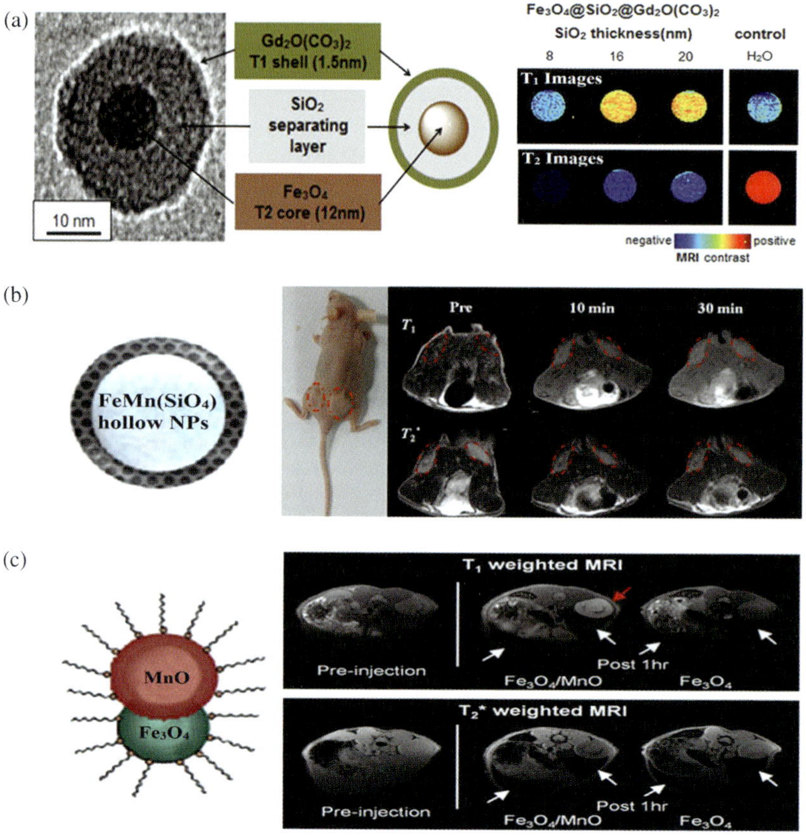

Fig. 20. (a) Graphical, TEM image, T_1- and T_2-weighted MRI color coded images of $Fe_3O_4/SiO_2/Gd_2O(CO_3)_2$ core–shell NPs: 200 μM (Gd) for T_1 image, 100 μM (Fe) for T_2 images as compare to water. (b) T_1- and T_2*-weighted *in vivo* MR images of tumors before and after 10 min and 30 min intravenous injection of $FeMn(SiO_4)$. (c) T_1– and T_2*-weighted MR imaging of abdomen of nude mouse pre and 1 h post-intravenous injection of Fe_3O_4NPs (right) and Fe_3O_4–MnO MHNPs (middle). Reprinted with permission from Refs. 82, 104 and 105.

Table 1. Relaxation properties of different hybrid structures.

Sample name	Size (nm)	*r*1 mM^{-1} S^{-1}	*r*2 mM^{-1} S^{-1}	*r*2/*r*1	References
Fe_3O_4 $SiO_2/Gd_2O(CO_3)_2$ NPs	21.5	3.7	312	84.32	104
$Fe_3O_4/SiO_2/Gd_2O(CO_3)_2$ NPs	29.5	32.9	269.6	8.19	104
$Fe_3O_4/SiO_2/Gd_2O(CO_3)_2$ NPs	32.5	32.2	208	6.45	104
$FeMn(SiO_4)$ hollow nanospheres (pH = 7.4)	80	1.92	92.39	48.11	105
$FeMn(SiO_4)$ hollow nanospheres (pH = 5)	80	0.6	49.43	82.38	105
Fe_3O_4–MnO core–shell	~10	1.3	35.8	28	82
Fe_3O_4–MnO dumbbell	~25	1.4	78.9	56	82
Fe_3O_4–MnO flower	~40	0.6	141	235	82

The signal intensities and relaxivities values (Table 1) of MRI altered with the size of the thickness of SiO_2.

Chen *et al.* synthesized pH-responsive T_1–T_2* dual-modal MRI contrast agents based on iron manganese silicate ($FeMn(SiO_4)$) hollow nanospheres using hydrothermal method.[105] Different pH systems contribute a significant function not only in the ion release from the nanoparticles but also in drug delivery. In this investigation, about 32% Mn^{2+} ions from $FeMn(SiO_4)$ released in acidic media and showed T_1 MR imaging in kidney as well as in tumor area. T_1 and T_2* MR images of the tumor can be shown in Fig. 20(b). The circulation of magnetic nanohybrids such as in liver, lungs and kidneys was also observed. These MR imaging results manifested the excellent enhancement in T_1 due to Mn and T_2 because of Fe. Moreover, the cell viability and histological examinations showed that $FeMn(SiO_4)$) hollow nanospheres are not toxic and developed obvious distinction between the normal and tumors.

Similar release of MnO nanoparticles from Fe_3O_4–MnO Janus nanoparticles was observed under different biological atmospheres and produced T_1 and T_2* MR imaging *in vivo*. Figure 20(c) revealed clearly the difference between MR imaging of liver and kidneys using single Fe_3O_4 and Janus Fe_3O_4–MnO nanoparticles (red arrow indicated kidney and white is used for liver). In Janus nanoparticles, T_1 single was observed in the kidney because of MnO and T_2 single was found in the liver after post injection of the magnetic nanohybrids. The corresponding *r*1 and *r*2 relaxation times can be seen in Table 1. In summary, the superparamagnetic nanoparticles joined with paramagnetic are excellent candidates for dual T_1 and T_2 MR imaging with enhancement in signal strength.

3.2. Multifunctional MRI and PDT MNHs

Later, researchers combined magnetic particles and paramagnetic Gd-based upconversion nanoparticles with photosynthesizers to obtain MRI and photodynamic therapy. In fact, rare-earth doped UCNPs can exchange NIR excitation to ultraviolet or light visible to stimulate photosensitizers which will help produce cytotoxic $^1O_2$ for killing cancerous cells. In this way, we can enhance PDT efficiency and be able to treat deep-seated tumors.[106,107]

Folic acid triggered Gd–Si–Ti hybrid nanocrystals were prepared by Wu *et al.* and obtained MRI and PDT response. The MR relaxation time *r*1 was observed to be 4.53 mM^{-}1s^{-}1, and the ratio of *r*2/*r*1 was about 1.92, demonstrated good T_1 contrast agent. These particles exhibited good performance and low toxicity *in vitro*. Figure 21(a) demonstrated the MR images and normalized tumor volume *in vivo*. The obvious brightness enhancement observed in T_1-weighted MR images before and after intratumoral injection which was not observed in T_2 MRI, indicated good T_1 performance *in vivo* MR imaging. Later, they performed PDT effect using NIR light 980 nm at the tumor site by injecting FA–Gd–Si–TiNPs and obtained excellent treatment effect to inhibit tumor growth as shown in Fig. 21(a3). Shi *et al.* reported the first example of trimodal synergetic therapy using UCNPs coupled with hematoporphyrin photosynthesizer. Nanotheranostic system for cancer therapy was achieved by Gd–UCNPs as the core, HP and Dtxl capsulated inside the mesoporous silica. The obtained relaxation time *r*1 = 5.62 mM^{-1}s^{-1} is greater than Gd–UCNPs@SiO$_2$@mSiO$_2$ 3.85 mM^{-1}s^{-1}. Later, MRI was performed by the injected the nanoparticles at tumor site and

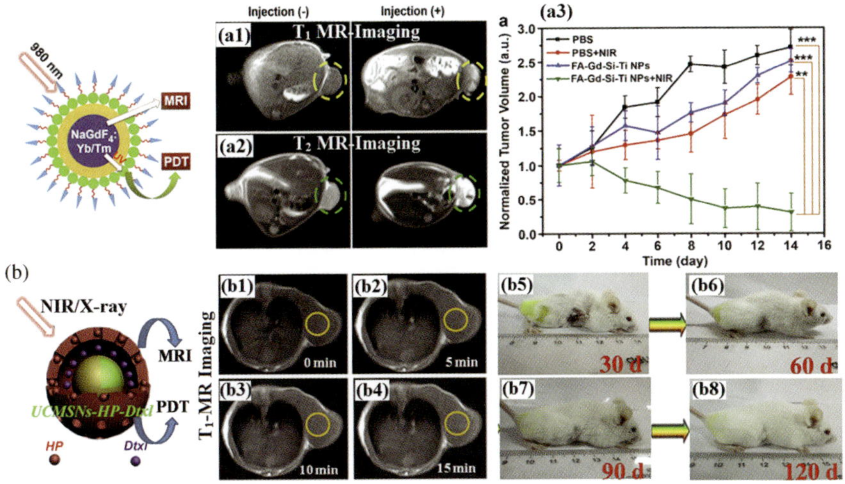

Fig. 21. (a1), (a2) T_1 and T_2-weighted *in vivo* MR images of MCF-7 tumor-bearing nude mice before and after 2 h of injection (100 mL FA–Gd–Si–TiNPs (500 mg/mL) in PBS by intratumoral injection) and (a3) photodynamic therapy effect of various nanoparticles on tumor size. (b) *In vivo* T_1-weighted MR images of a 4*T1*-tumor bearing mouse after intravenous injection of UCMSNs at selected time points: (a) 0 min; (b) 5 min; (c) 10 min; (d) 15 min. Reprinted with permission from Refs. 52 and 53.

images (Figs. 21(b1)–21(b4) taken at different intervals of time. The MRI results showed that signal intensity in tumor was increased by time period and accumulated at tumor area due to enhanced permeability and retention effect (EPR) of tumors. Moreover, *in vivo* therapeutic effect of UCMSNs–HP on mice 4T_1 tumor is shown in Figs. 21(b5)–21(b8). UCMSNs–HP–Dtxl + RT + NIR showed significant influence to kill the tumor which they also observed in histological examination after sacrifice into the mice. Hence, these MRI and PDT results showed that a UCMSNs–HP material has potential against cancer activity.

Another excellent and first report was published by Wu *et al.*[75] They synthesized Fe_3O_4–TiO_2 Janus hybrid nanoparticles for T_2-weighted MRI contrast agent and PDT applications. The longitudinal and transverse relaxivities ($r1$ and $r2$) of Fe_3O_4–TiO_2 magnetic nanohybrids with various concentrations were 1.444 mM^1s^1 and

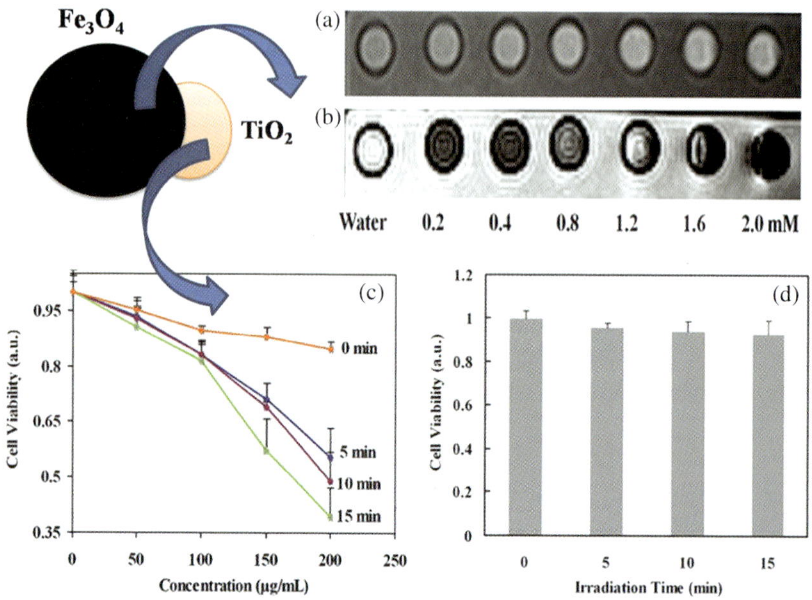

Fig. 22. (a) and (b) T_1 and T_2 MR images of Fe_3O_4–TiO_2 magnetic nanocomposites, (c) Cell viability after UV light irradiation on MCF-cells with Fe_3O_4–TiO_2 nanocomposites preferred time and (d) MCF-7 cells without Fe_3O_4–TiO_2 nanocomposites. Reprinted with permission from Ref. 75.

54.691 mM^1s^1, respectively, resulting in these magnetic composites having high T_2 effect. The prepared Fe_3O_4–TiO_2NPs showed very good T_2-weighted MRI performance as compared to water and T_1 images shown in Figs. 22(a) and 22(b). They also observed the PDT effect using the UV irradiation light on MCF-7 cell line with and without nanocomposites. The cell viability was examined to be more than 90% in the absence of nanoparticles. However, viability reduced to 35% after 15 min of UV-irradiation in the presence of Nanoparticles (Fig. 22(c)). As a result, Fe_3O_4–TiO_2 magnetic nanohybrids are significant agents for MRI diagnosis and PDT of cancer.

3.3. Multifunctional MRI and PTT MNHs

The complex MHNPs were reported by Lin *et al.* to obtain the trimodality anticancer therapy function such as chemotherapy, PDT and PTT.

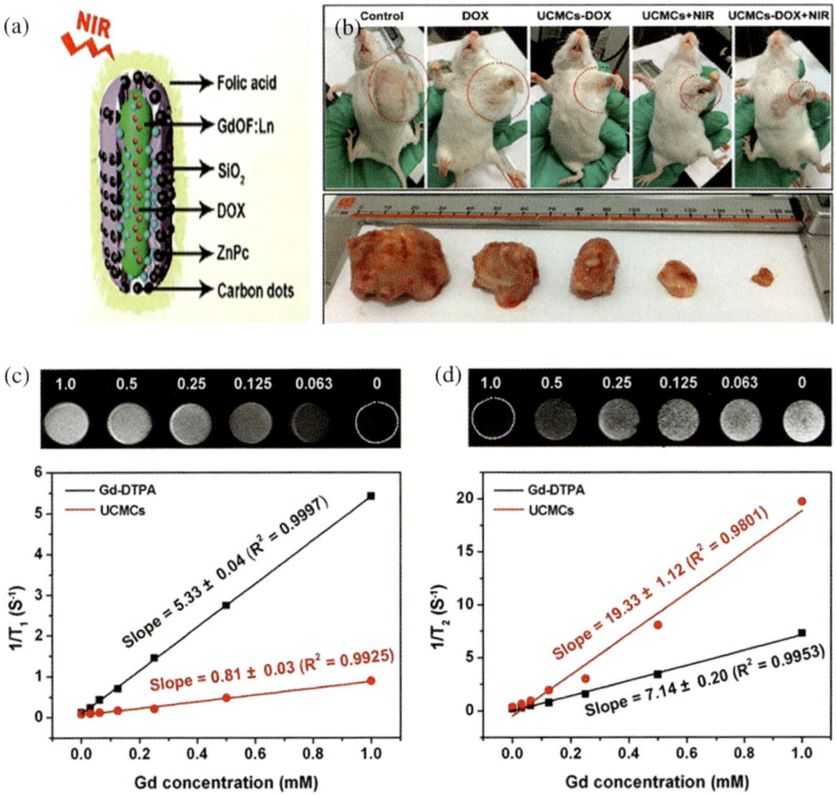

Fig. 23. (a) Representative pictures of mice after various intratumoral treatments: control, DOX, UCMCs–DOX, UCMCs+NIR irradiation, and UCMCs–DOX+NIR irradiation. (b) Snaps of tumor tissue attained after 14 days. (c) and (d) *In vitro*, T_1 and T_2 MRI and corresponding relaxation rate ($1/T_1$, $1/T_2$) with different concentrations of Gd. Reprinted with permission from Ref. 54.

The structure is given in Fig. 23(a).[54] Here, we are considering MRI and PDT part. GdOF:Ln@SiO_2–ZnPc–CDs microcapsules were injected to the H22 tumor-bearing mice at the tumor place and the results of different groups are shown in Fig. 23(b). Remarkable change in the tumor volume was observed between control and UCMCs–DOX under NIR irradiation a is shown in Fig. 23(b). After 14 days of treatment, for mice treated with UCMCs–DOX+NIR, that tumor size did not only stop growing but also decreased from original volume as can be seen in individual tumor size comparison pictures. Further, they conducted T_1 and T_2-weighted images of MHNPs. Both imaging intensities were enhanced by increasing the concentration of Gd as shown in Figs. 23(c) and 23(d). Afterwards, the corresponding relaxivities *r*1 and *r*2 were calculated to be 0.81 $mM^{-1}s^{-1}$ and 19.33 $mM^{-1}s^{-1}$. Interestingly, this composite demonstrated better T_2 performance as compared to common Gd–DTPA complex. Therefore, these complex hybrid structures have potential use in imaging and therapeutic applications.

Schematic illustration (Fig. 24) of Fe_5C_2 theranostic nanoparticles is reported by Hou *et al.* The prepared nanocrystals were successfully

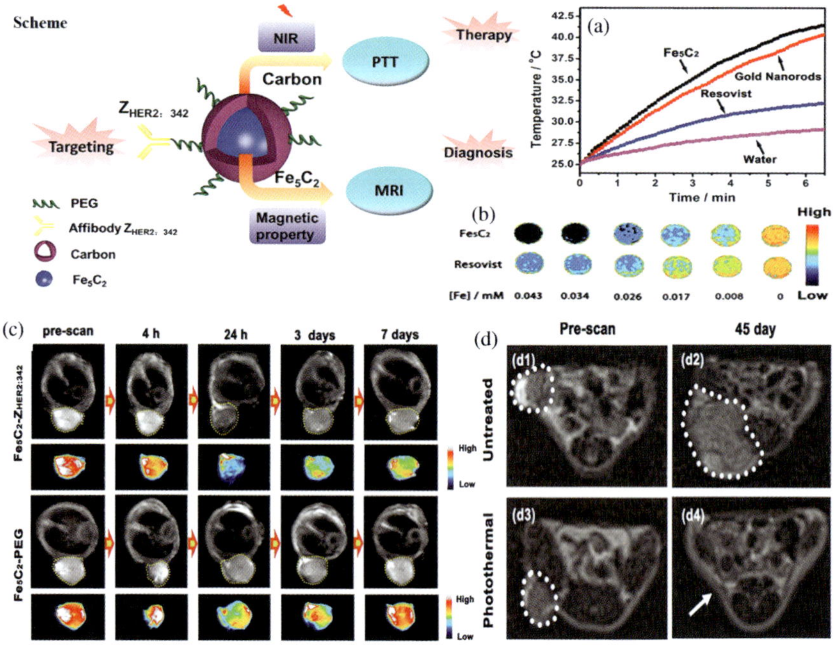

Fig. 24. Schematic illustration of a complex structure of Fe_5C_2 MHNPs as a targeted theranostic candidate. (a) Temperature analysis curves for water and the prepared aqueous dispersions Fe_5C_2 MHNPs/Resovist/AuNRs with similar concentration (0.5 mM Fe or Au) under continuous radiation of an 808 nm NIR laser at 2 W cm^{-2}. (b) T_2 MR images of Fe_5C_2 MHNPs (upper) in comparison to Resovist (down) at different concentrations of iron. (c) *In vivo* T_2-weighted MRI of SKOV-3 tumor-bearing mice pre and postintravenous injection of Fe_5C_2–$Z_{HER2:342}$ and Fe_5C_2–PEG. (d) MR images of untreated and treated with NIR+MHNPs. Reprinted with permission from Ref. 41.

employed in multifunctional systems such as PTT and MRI. The designed probe Fe_5C_2 coated by (DSPE–PEG–NH_2) conjugated with affinity proteins $Z_{HER2:342}$ for ovarian-cancer targeting. The affinity proteins such as $Z_{HER2:342}$ are very effective for ovarian-cancer targeting.[42] The temperature increment graph of Fe_5C_2 using NIR laser radiation (λ = 808 nm, 2 W cm^{-2}) exhibited heat which is higher than AuNRs (aspect ratio of 3.5) and Resovist (Fig. 24(a)) as shown in Fig. 24.[108] Also, MRI relaxation time *r*2 (311.94 $mM^{-1}s^{-1}$) was greater than Resovist 173.95 $mM^{-1}s^{-1}$. MR images (Fig. 24(b)) of Fe_5C_2 also showed excellent imaging enhancement as compared to Resovist.

Further, Fe_5C_2 nanoparticles conjugated with PEG and $Z_{HER2:342}$ were examined for *in vivo* T_2–MRI contrast agent. The nanoparticles were injected into SK–OV–3 tumor-bearing mice and performed T_2 weighted MR images shown in Fig. 24(c) before and after intravenous injection. In both groups, imaging intensities were dropped at the tumor sites which can be seen in the pseudocolor pictures. The MRI results demonstrated that the Fe_5C_2 coupled with $Z_{HER2:342}$ probe attained efficient *in vivo* tumor ablation and excellent MRI without any other effect. Later, they employed prepared magnetic hybrids in photothermal therapy and the results are shown in Fig. 24(d). MRI results reveal obvious tumor ablation between the untreated mice and mice with treated Fe_5C_2–$Z_{HER2:342}$ nanoparticles. This research highlights the significant potential of Fe_5C_2 MHNPs as multifunctional probes for MR imaging and PTT of cancer.

Xiangyang Shi *et al.* took one step forward in the synthesis of magnetic gold nanostars Mau–NSs to obtain trimodality purposes.[34,35] Fe_3O_4@Au hybrid nanostars functionalized using polyethyleneimine (PEI) were successfully used in T_2 MRI contrast agent and photothermal therapy of cancer. They investigated the *r*2 MR transverse relaxation time of Mau–NSs which is about 144.39 $mM^{-1}s^{-1}$ (Fig. 25) greater than many reported values. The signal intensity enhancement in the MRI can be seen in the inset of Fig. 25(a) which might be related to the aggregation of iron oxide nanoparticles.

Next, *in vivo* MR imaging was analyzed using the BALB/c mice bearing HeLa tumor with respect to time. The prepared nanoflowers ([Fe] = 5.0 mM, 0.1 mL) in aqueous solution were injection

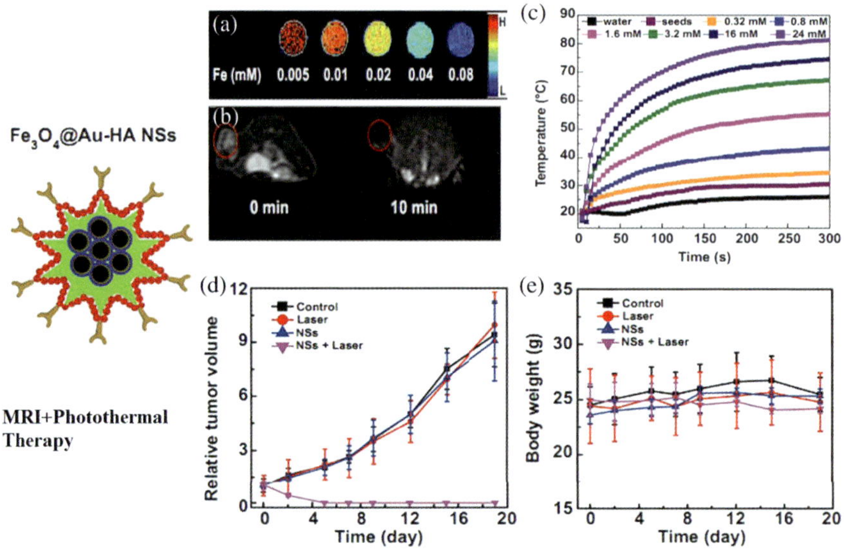

Fig. 25. The illustration of Fe_3O_4@Au core/shell magnetic gold nanostars Mau–NSs. (a) T_2-weighted MR images of Mau–NFs at different concentrations of Fe. (b) *In vivo* T_2-weighted MR images of HeLa tumor-bearing mice pre and post-intravenous injection of Mau–NFs at different times. (c) Photothermal effect of Mau–NSs under 5 min irradiation (915 nm laser (1.2 W/cm^2) for 300 s). (d) and (e) Relative tumor size and body weight analysis after the treatment for 19 days. Reprinted with permission from Ref. 35.

via intravenously. T_2-weighted MR images shown in Fig. 25(b) at preinjection and after 10 mins. The intensity (641 before injection) at the tumor site that decreased from 41% after 10 mins post injection, might be related to the accumulation of MAu–NSs at the tumor area. This MR imaging of Mau–NSs provided an evidence of good biodistribution behavior. Furthermore, *in vitro* photothermal influence of Mau–NSs using NIR laser (irradiated under a 915 nm laser (1.2 W/cm^2) for 300 s) was studied and temperature was increased from about 23°C to 81.2°C at 24 mM gold concentration as mentioned in Fig. 25(c). After they obtained good results *in vitro*, nanoflowers were further applied *in vivo* for potential effect of temperature on tumor cells HeLa. Mau–NSs were injected intravenously into the mice with a dose rate of 40.0 mg/kg of gold and then NIR irradiation at the tumor site for 5 min at 10 min post injection. There was very minute differences observed in the tumor growth when they used NSs and NIR separately. However, Mau–NSs+NIR produced significant heat to kill the tumor (Fig. 25(d)). Moreover, mice with various treatment groups maintained their body weights throughout the experimental time period which confirmed that NSs are unable to generate toxicity to the mice as shown in Fig. 25(e). This study suggested that MAu–NSs have the ability to ablate and stop the tumor regrowth.

4. Conclusion and Prospective

Magnetic hybrid nanoparticles can be considered as the most problematic or complex colloidal particles in existence. In this chapter, we have highlighted the most recent strategies focused on the complex design, synthesis and theranostic applications of these multifunctional magnetic nanohybrids materials. MRI contrast probes integrated with therapeutic agents have demonstrated excellent ability in MR signal enhancement and also the capability for tumor cell ablation via reactive oxygen species (ROS). These combined features of magnetic hybrid compounds present a novel platform for multifunctional purposes such as diagnosis and therapeutic functions. At present, most researches focused on the fabrication part, but the applications are still in infancy. Although some MNHPs have revealed excellent tools for extensive applications, a number of challenges have to be investigated deeply. Therefore, the following characteristics might need better concentration in the future. In the fabrication part, the interface between two compounds and their morphology have not been controlled very well to provide stability. Therefore, this issue still needs to be addressed by developing new synthesis techniques or improving the existing ones. Moreover, demand for non-toxic novel MRI contrast agents with ultrasensitive imaging and advanced functionality is very high. Magnemite or magnetite particles may impose additional danger because the pH in the endolysosomes is low. The acidic environment produced by ferric ions can destroy many biochemical processes which are very critical to the process. Some stable coating may decrease this effect. Magnetic hybrids based nanotheranostic systems can directly relate to the biological atmosphere *in vivo* and to the therapeutic outcomes such as biodistribution, biodegradability and accumulation in tumor areas. However, the studies on MNHs for tumor diagnosis, targeting and therapy are still in the labs and have not been implemented clinically because of some inherent problems, such as the stability issues, biosafety and biodegradability during scale-up production which have not been addressed. PDT and PTT showed significant response for skin diseases but still failed at deep tissue penetrations. Much effort is essential for efficient research of the correct diagnosis, biosafety issues and the efficiency of drug delivery through MNHs before these theranostic nanoparticles enable clinical trials. In general, the multifunctional MNHs are promising candidates of MRI diagnosis and phototherapy for cancers in future.

Acknowledgments

This work was supported by the National Natural Science Foundation of China (U1432114, 81550110258), NSFC–Guangdong Province Joint Project on National Supercomputer Centre in Guangzhou (NSCC-gZ) (Aiguo Wu), Key Breakthrough Program of Chinese Academy of Sciences (KGZD-EW-T06) and the Postdoctoral Science Foundation of China (2014M561799). Furthermore, the authors also acknowledge Shanghai Synchrotron Radiation Facility at Line BL15U (No. h15sr0021) used for X-ray fluorescence imaging.

References

1. Kharisov BI. A review for synthesis of nanoflowers. Recent Patents on Nanotechnology; 2.
2. Jin R, Cao Y, Mirkin CA, Kelly KL, Schatz GC, Zheng JG. Photoinduced conversion of silver nanospheres to nanoprisms. *Science* **294**, 1901–1903 (2001).
3. Li G, Peng C, Zhang C, Xu Z, Shang M, Yang D, *et al.* Eu3+/Tb3+-Doped La2O2CO3/La2O3 Nano/microcrystals with multiform morphologies: Facile synthesis, growth mechanism, and luminescence properties. *Inorg. Chem.* **49**, 10522–10525 (2010).

4. Yang H, Xia Y. Bionanotechnology: Enabling biomedical research with nanomaterials. *Adv. Mater.* **19**, 3085–3087 (2007).
5. Roco MC. Nanotechnology: Convergence with modern biology and medicine. *Curr. Opin. Biotechnol.* **14**, 337–346 (2003).
6. Funkhouser J Rp. The theranostic revolution. *Curr. Drug. Discov.* **2** (2002).
7. McCarthy JR. Multifunctional agents for concurrent imaging and therapy in cardiovascular disease. *Adv. Drug Delivery Rev.* **62**, 1023–1030 (2010).
8. Pene F, Courtine E, Cariou A, Mira J-P. Toward theragnostics. *Crit. Care Med.* **37**, S50–S58 (2009).
9. Hong-Guang Xie FWF. Pharmacogenomics steps toward personalized medicine. *Pers. Med.* **2**, 13 (2005).
10. Reineke SSKaTM. *Theranostics: Combining Imaging and Therapy* **22**, 25 (2011).
11. Hu S-H, Gao X. Nanocomposites with spatially separated functionalities for combined imaging and magnetolytic therapy. *J. Am. Chem. Soci.* **132**, 7234–7237 (2010).
12. Maeda H, Greish K, Fang J. The EPR effect and polymeric drugs: A paradigm shift for cancer chemotherapy in the 21st Century. In: Satchi-Fainaro R, Duncan R, editors. Polymer Therapeutics II. Berlin, Heidelberg: Springer Berlin Heidelberg, pp. 103–121 (2006).
13. Alexis F, Pridgen E, Molnar LK, Farokhzad OC. Factors affecting the clearance and biodistribution of polymeric nanoparticles. *Mol. Pharm.* **5**, 505–515 (2008).
14. Iyer AK, Khaled G, Fang J, Maeda H. Exploiting the enhanced permeability and retention effect for tumor targeting. *Drug Discov. Today* **11**, 812–818 (2006).
15. Somerville MH, Greenberg DR, del Alamo JA. Temperature and carrier density dependence of mobility in a heavily doped quantum well. *Appl. Phys. Lett.* **64**, 3276–3278 (1994).
16. de Picciotto R, Stormer HL, Pfeiffer LN, Baldwin KW, West KW. Four-terminal resistance of a ballistic quantum wire. *Nature* **411**, 51–54 (2001).
17. Oberdörster G, Oberdörster E, Oberdörster J. Nanotoxicology: An emerging discipline evolving from studies of ultrafine particles. *Environ. Health Perspectives* **113**, 823–839 (2005).
18. Jana NR, Gearheart L, Murphy CJ. Seed-mediated growth approach for shape-controlled synthesis of spheroidal and rod-like gold

nanoparticles using a surfactant template. *Adv. Mater.* **13**, 1389–1393 (2001).
19. Gröschel AH, Walther A, Löbling TI, Schmelz J, Hanisch A, Schmalz H, *et al.* Facile, Solution-based synthesis of soft, nanoscale janus particles with tunable Janus balance. *J. Am. Chem. Soc.* **134**, 13850–13860 (2012).
20. Gröschel AH, Schacher FH, Schmalz H, Borisov OV, Zhulina EB, Walther A, *et al.* Precise hierarchical self-assembly of multicompartment micelle. *Nat. commun.* **3**, 710 (2012).
21. Ghosh CR, Paria S. Core/Shell nanoparticles: Classes, properties, synthesis mechanisms, characterization, and applications. *Chem. Rev.* **112**, 2373–2433 (2012).
22. Ali Z, Abbasi AZ, Zhang F, Arosio P, Lascialfari A, Casula MF, *et al.* Multifunctional nanoparticles for dual imaging. *Anal. Chem.* **83**, 2877–2882 (2011).
23. Levin CS, Hofmann C, Ali TA, Kelly AT, Morosan E, Nordlander P, *et al.* Magnetic–plasmonic core–shell nanoparticles. *ACS Nano.* **3**, 1379–1388 (2009).
24. Reddy LH, Arias JL, Nicolas J, Couvreur P. Magnetic nanoparticles: design and characterization, toxicity and biocompatibility, pharmaceutical and biomedical applications. *Chem. Rev.* **112**, 5818–5878 (2012).
25. Lee H, Shin T-H, Cheon J, Weissleder R. Recent developments in magnetic diagnostic systems. *Chem. Rev.* **115**, 10690–10724 (2015).
26. Solovyeva VA, Vu KB, Merican Z, Sougrat R, Rodionov VO. One-Pot Synthesis of Au@SiO2 Catalysts: A click chemistry approach. *ACS Comb. Sci.* **16**, 513–517 (2014).
27. Huang J, Zhang M, Wang J, Hu X, Luque R, Lam FLY. A comprehensive study on the effect of preparation methods for Au-core@shell silica materials in room temperature oxidative amide formation. *J. Mater. Chem. A* **3**, 789–796 (2015).
28. Tang F, He F, Cheng H, Li L. Self-assembly of conjugated polymer-Ag@SiO2 hybrid fluorescent nanoparticles for application to cellular imaging. *Langmuir*: **26**, 11774–11778 (2010).
29. Raevskaya AE, Panasiuk YV, Stroyuk OL, Kuchmiy SY, Dzhagan VM, Milekhin AG, *et al.* Spectral and luminescent properties of ZnO-SiO2 core-shell nanoparticles with size-selected ZnO cores. *RSC Adv.* **4**, 63393–63401 (2014).

30. Ruan Q, Fang C, Jiang R, Jia H, Lai Y, Wang J, *et al.* Highly enhanced transverse plasmon resonance and tunable double Fano resonances in gold@titania nanorods. *Nanoscale* **8**, 6514–6526 (2016).
31. Baek S-W, Shim J-H, Seung H-M, Lee G-S, Hong J-P, Lee K-S, *et al.* Effect of core quantum-dot size on power-conversion-efficiency for silicon solar-cells implementing energy-down-shift using CdSe/ZnS core/shell quantum dots. *Nanoscale* **6**, 12524–12531 (2014).
32. Chen O, Riedemann L, Etoc F, Herrmann H, Coppey M, Barch M, *et al.* Magneto-fluorescent core-shell supernanoparticles. *Nat. comm.* **5**, 5093 (2014).
33. Zhang J, Thurber A, Hanna C, Punnoose A. Highly shape-selective synthesis, silica coating, self-assembly, and magnetic hydrogen sensing of hematite nanoparticles. *Langmuir* **26**, 5273–5278 (2010).
34. Li J, Zheng L, Cai H, Sun W, Shen M, Zhang G, *et al.* Facile one-pot synthesis of Fe3O4@Au composite nanoparticles for dual-mode MR/CT imaging applications. *ACS Appl. Mater. Interfaces.* **5**, 10357–10366 (2013).
35. Li J, Hu Y, Yang J, Wei P, Sun W, Shen M, *et al.* Hyaluronic acid-modified Fe3O4@Au core/shell nanostars for multimodal imaging and photothermal therapy of tumors. *Biomaterials* **38**, 10–21 (2015).
36. Huang J, Guo M, Ke H, Zong C, Ren B, Liu G, *et al.* Rational design and synthesis of gammaFe2 O3 @Au magnetic gold nanoflowers for efficient cancer theranostics. *Adv. Mater.* **27**, 5049–5056 (2015).
37. Jin Y, Jia C, Huang SW, O'Donnell M, Gao X. Multifunctional nanoparticles as coupled contrast agents. *Nat. Comm.* **1**, 41 (2010).
38. Huang J, Zong C, Shen H, Liu M, Chen B, Ren B, *et al.* Mechanism of cellular uptake of graphene oxide studied by surface-enhanced raman spectroscopy. *Small* **8**, 2577–2584 (2012).
39. Schnepp Z, Wimbush SC, Antonietti M, Giordano C. Synthesis of highly magnetic iron carbide nanoparticles via a biopolymer route. *Chem. of Mater.* **22**, 5340–5344 (2010).
40. Yang C, Zhao H, Hou Y, Ma D. Fe5C2 nanoparticles: A facile bromide-induced synthesis and as an active phase for Fischer-Tropsch synthesis. *J. Am. Chem. Soc.* **134**, 15814–15821 (2012).
41. Yu J, Yang C, Li J, Ding Y, Zhang L, Yousaf MZ, *et al.* Multifunctional Fe5 C2 nanoparticles: A targeted theranostic platform for magnetic resonance imaging and photoacoustic tomography-guided photothermal therapy. *Adv. Mater.* **26**, 4114–4120 (2014).

42. Orlova A, Magnusson M, Eriksson TLJ, Nilsson M, Larsson B, Höidén-Guthenberg I, *et al.* Tumor imaging using a picomolar affinity HER2 binding affibody molecule. *Cancer Res.* **66**, 4339–4348 (2006).
43. Sun S, Zeng H, Robinson DB, Raoux S, Rice PM, Wang SX, *et al.* Monodisperse MFe2O4 (M = Fe, Co, Mn) Nanoparticles. *J. Am. Chem. Soc.* **126**, 273–279 (2004).
44. Liu Q, Sun Y, Li C, Zhou J, Li C, Yang T, *et al.* 18F-Labeled magnetic-upconversion nanophosphors via rare-earth cation-assisted ligand assembly. *ACS Nano.* **5**, 3146–3157 (2011).
45. Shen J-W, Yang C-X, Dong L-X, Sun H-R, Gao K, Yan X-P. Incorporation of computed tomography and magnetic resonance imaging function into NaYF4:Yb/Tm upconversion nanoparticles for *in vivo* trimodal bioimaging. *Anal. Chem.* **85**, 12166–12172 (2013).
46. Qiao R, Liu C, Liu M, Hu H, Liu C, Hou Y, *et al.* Ultrasensitive *in vivo* detection of primary gastric tumor and lymphatic metastasis using upconversion nanoparticles. *ACS Nano.* **9**, 2120–2129 (2015).
47. Sun Y, Zhu X, Peng J, Li F. Core–Shell lanthanide upconversion nanophosphors as four-modal probes for tumor angiogenesis imaging. *ACS Nano.* 7, 11290–11300 (2013).
48. Wang L, Liu J, Dai Y, Yang Q, Zhang Y, Yang P, *et al.* Efficient gene delivery and multimodal imaging by lanthanide-based upconversion nanoparticles. *Langmuir.* **30**, 13042–13051 (2014).
49. Gnanasammandhan MK, Idris NM, Bansal A, Huang K, Zhang Y. Near-IR photoactivation using mesoporous silica-coated NaYF4:Yb,Er/Tm upconversion nanoparticles. *Nat. Protocols.* **11**, 688–713 (2016).
50. Liu C, Gao Z, Zeng J, Hou Y, Fang F, Li Y, *et al.* Magnetic/upconversion fluorescent NaGdF4:Yb,Er nanoparticle-based dual-modal molecular probes for imaging tiny tumors *in vivo*. *ACS Nano.* **7**, 7227–7240 (2013).
51. Ni D, Zhang J, Bu W, Xing H, Han F, Xiao Q, *et al.* Dual-targeting upconversion nanoprobes across the blood–brain barrier for magnetic resonance/fluorescence imaging of intracranial glioblastoma. *ACS Nano.* **8**, 1231–1242 (2014).
52. Zhang L, Zeng L, Pan Y, Luo S, Ren W, Gong A, *et al.* Inorganic photosensitizer coupled Gd-based upconversion luminescent nanocomposites for *in vivo* magnetic resonance imaging and near-infrared-responsive photodynamic therapy in cancers. *Biomaterials* **44**, 82–90 (2015).

53. Fan W, Shen B, Bu W, Chen F, He Q, Zhao K, *et al.* A smart upconversion-based mesoporous silica nanotheranostic system for synergetic chemo-/radio-/photodynamic therapy and simultaneous MR/UCL imaging. *Biomaterials* **35**, 8992–9002 (2014).
54. Lv R, Yang P, He F, Gai S, Li C, Dai Y, *et al.* A yolk-like multifunctional platform for multimodal imaging and synergistic therapy triggered by a single near-infrared light. *ACS Nano.* **9**, 1630–1647 (2015).
55. Hu J, Zhou S, Sun Y, Fang X, Wu L. Fabrication, properties and applications of Janus particles. *Chem. Soc. Rev.* **41**, 4356–4378 (2012).
56. Song Y, Liu X, Chen S. CHAPTER 14 Janus nanoparticles by interfacial engineering. *Functional Nanometer-Sized Clusters of Transition Metals: Synthesis, Properties and Applications*: The Royal Society of Chemistry, Thomas Graham House, Science Park, Milton Road, Cambridge CB4 0WF, UK, pp. 407–433 (2014).
57. Yi Y, Sanchez L, Gao Y, Yu Y. Janus particles for biological imaging and sensing. *Analyst* (2016).
58. Tang C, Zhang C, Liu J, Qu X, Li J, Yang Z. Large scale synthesis of Janus submicrometer sized colloids by seeded emulsion polymerization. *Macromolecules* **43**, 5114–5120 (2010).
59. Glotzer SC, Solomon MJ. Anisotropy of building blocks and their assembly into complex structures. *Nat. Mater.* **6**, 557–562 (2007).
60. Tian L, Zhang B, Li W, Li X, Fan X, Jia X, *et al.* Facile fabrication of Fe3O4@PS/PGMA magnetic Janus particles via organic-inorganic dual phase separation. *RSC Advances* **4**, 27152–27158 (2014).
61. Schick I, Lorenz S, Gehrig D, Tenzer S, Storck W, Fischer K, *et al.* Inorganic Janus particles for biomedical applications. *Beilstein J. Nanotechnol.* **5**, 2346–2362 (2014).
62. Kaewsaneha C, Tangboriboonrat P, Polpanich D, Eissa M, Elaissari A. Janus colloidal particles: Preparation, properties, and biomedical applications. *ACS Appl. Mater. Interfaces.* **5**, 1857–1869 (2013).
63. Li X, Zhou L, Wei Y, El-Toni AM, Zhang F, Zhao D. Anisotropic growth-induced synthesis of dual-compartment Janus mesoporous silica nanoparticles for bimodal triggered drugs delivery. *J. Am. Chem. Soc.* **136**, 15086–15092 (2014).
64. Liang S, Liu X-L, Yang Y-Z, Wang Y-L, Wang J-H, Yang Z-J, *et al.* Symmetric and asymmetric Au–AgCdSe hybrid nanorods. *Nano Lett.* **12**, 5281–5286 (2012).

65. Kim Y, Park K-Y, Jang DM, Song YM, Kim HS, *et al.* Synthesis of AuCu2S Ćore shell nanocrystals and their photocatalytic and electrocatalytic activity. *J. Phys. Chem. C.* **114**, 22141–22146 (2010).
66. Li P, Wei Z, Wu T, Peng Q, Li Y. Au–ZnO hybrid nanopyramids and their photocatalytic properties. *J. Am. Chem. Soc.* **133**, 5660–5663 (2011).
67. Lawrence JH, Mathew MB, Younghum S, Adam GS, Philip TD, Nathaniel ER, *et al.* Directing the deposition of ferromagnetic Ćobalt onto Pt-tipped CdSe@CdS Ńanorods: Synthetic and mechanistic insights. *ACS Nano.* **6**, 8632–8645 (2012).
68. Huang C-M, Cheng S-H, Jeng U-S, Yang C-S, Lo L-W. Formation of CdSe/CdS/ZnS-Au/SiO2 dual-yolk/shell nanostructures through a Trojan-type inside-out etching strategy. *Nano Res.* **5**, 654–666 (2012).
69. Lattuada M, Hatton TA. Preparation and controlled self-assembly of Janus magnetic nanoparticles. *J. Am. Chem. Soc.* **129**, 12878–12889 (2007).
70. Valente G, Depalo N, de Paola I, Iacobazzi RM, Denora N, Laquintana V, *et al.* Integrin-targeting with peptide-bioconjugated semiconductor-magnetic nanocrystalline heterostructures. *Nano Res.* **9**, 644–662 (2015).
71. Buonsanti R, Grillo V, Carlino E, Giannini C, Curri ML, Innocenti C, *et al.* Seeded growth of asymmetric binary nanocrystals made of a semiconductor TiO2 rodlike section and a magnetic γ-Fe2O3 spherical domain. *J. Am. Chem. Soc.* **128**, 16953–16970 (2006).
72. Nolan M. Electronic coupling in iron oxide-modified TiO2 leads to a reduced band gap and charge separation for visible light active photocatalysis. *Phys. Chem. Chem. Phys.* **13**, 18194–18199 (2011).
73. Mallik A, Bryan S, Puukila S, Chen A, Khaper N. Efficacy of Pt-modified TiO(2) nanoparticles in cardiac cells. *Exp. Clin. Cardiol.* **16**, 6–10 (2011).
74. Thurn KT, Arora H, Paunesku T, Wu A, Brown EMB, Doty C, *et al.* Endocytosis of titanium dioxide nanoparticles in prostate cancer PC-3M cells. *Nanomedicine* 7, 123–130 (2011).
75. Zeng L, Ren W, Xiang L, Zheng J, Chen B, Wu A. Multifunctional Fe3O4-TiO2 nanocomposites for magnetic resonance imaging and potential photodynamic therapy. *Nanoscale* **5**, 2107–2113 (2013).

76. Iqbal MZ, Ma X, Chen T, Zhang Le, Ren W, Xiang L, *et al.* Silica-coated super-paramagnetic iron oxide nanoparticles (SPIONPs): a new type contrast agent of T1 magnetic resonance imaging (MRI). *J. Mater. Chem. B.* **3**, 5172–5181 (2015).
77. Gomes MC, Cunha Â, Trindade T, Tomé JPC. The role of surface functionalization of silica nanoparticles for bioimaging. *J. Innov. Opt. Heal. Sci.* 0:1630005.
78. Wang Y, Xu H, Ma Y, Guo F, Wang F, Shi D. Facile one-pot synthesis and morphological control of asymmetric superparamagnetic composite nanoparticles. *Langmuir* **27**, 7207–7212 (2011).
79. Wang F, Pauletti GM, Wang J, Zhang J, Ewing RC, Wang Y, *et al.* Dual surface-functionalized Janus nanocomposites of polystyrene/Fe(3)O(4)@SiO(2) for simultaneous tumor cell targeting and stimulus-induced drug release. *Adv. Mater.* **25**, 3485–3489 (2013).
80. Antonietti M, Landfester K. Polyreactions in miniemulsions. *Prog. Polym. Sci.* **27**, 689–757.
81. Lee KS, Anisur RM, Kim KW, Kim WS, Park T-J, Kang EJ, *et al.* Seed size-dependent formation of Fe3O4/MnO hybrid nanocrystals: Selective, magnetically recyclable catalyst systems. *Chem. Mater.* 24, 682–687 (2012).
82. Im GH, Kim SM, Lee D-G, Lee WJ, Lee JH, Lee IS. Fe3O4/MnO hybrid nanocrystals as a dual contrast agent for both T1- and T2-weighted liver MRI. *Biomaterials* **34**, 2069–2076 (2013).
83. Masala O, Seshadri R. Spinel ferrite/MnO core/shell nanoparticles: chemical synthesis of all-oxide exchange biased architectures. *J. Am. Chem. Soc.* **127**, 9354–9355 (2005).
84. Yin M, O'Brien S. Synthesis of monodisperse nanocrystals of manganese oxides. *J. Am. Chem. Soc.* **125**, 10180–10181 (2003).
85. Shtykova EV, Huang X, Remmes N, Baxter D, Stein B, Dragnea B, *et al.* Structure and properties of iron oxide nanoparticles encapsulated by phospholipids with poly(ethylene glycol) tails. *J. Phys. Chem. C.* **111**, 18078–18086 (2007).
86. Lin Y-S, Wu S-H, Hung Y, Chou Y-H, Chang C, Lin M-L, *et al.* Multifunctional composite nanoparticles: Magnetic, luminescent, and mesoporous. *Chem. Mater.* **18**, 5170–5172 (2006).
87. Yu H, Chen M, Rice PM, Wang SX, White RL, Sun S. Dumbbell-like bifunctional Au–Fe3O4 nanoparticles. *Nano Lett.* **5**, 379–382 (2005).

88. George C, Genovese A, Qiao F, Korobchevskaya K, Comin A, Falqui A, *et al.* Optical and electrical properties of colloidal (spherical Au)-(spinel ferrite nanorod) heterostructures. *Nanoscale* **3**, 4647–4654 (2011).
89. Wu B, Tang S, Chen M, Zheng N. Amphiphilic modification and asymmetric silica encapsulation of hydrophobic Au-Fe3O4 dumbbell nanoparticles. *Chem. Comm.* **50**, 174–176 (2014).
90. Hongwei Gu ZY, Jinhao Gao, C. K. Chang, and Bing Xu. Heterodimers of nanoparticles: Formation at a liquid liquid interface and particle-specific surface modification by functional molecules. *J. Am. Chem. Soc.* **127** (2005).
91. Sotiriou GA, Hirt AM, Lozach PY, Teleki A, Krumeich F, Pratsinis SE. Hybrid, silica-coated, Janus-like plasmonic-magnetic nanoparticles. *Chem. Mater.* **23**, 1985–1992 (2011).
92. Sotiriou GA, Visbal-Onufrak MA, Teleki A, Juan EJ, Hirt AM, Pratsinis SE, *et al.* Thermal energy dissipation by SiO2-coated plasmonic-superparamagnetic nanoparticles in alternating magnetic fields. *Chem. Mater.* **25**, 4603–4612 (2013).
93. Wang H, Shang P, Zhang J, Guo M, Mu Y, Li Q, *et al.* One-step synthesis of high-coercivity L10-FePtAg nanoparticles: Effects of Ag on the morphology and chemical ordering of FePt nanoparticles. *Chem. Mater.* **25**, 2450–2454 (2013).
94. Wang C, Wei Y, Jiang H, Sun S. Tug-of-War in nanoparticles: competitive growth of Au on Au–Fe3O4 nanoparticles. *Nano Lett.* **9**, 4544–4547 (2009).
95. Shi W, Zeng H, Sahoo Y, Ohulchanskyy TY, Ding Y, Wang ZL, *et al.* A general approach to binary and ternary hybrid nanocrystals. *Nano Lett.* **6**, 875–8781 (2006).
96. Hodges JM, Biacchi AJ, Schaak RE. Ternary hybrid nanoparticle isomers: Directing the nucleation of Ag on Pt–Fe3O4 using a solid-state protecting group. *ACS Nano.* **8**, 1047–1055 (2014).
97. Gillies RJ, Bhujwalla ZM, Evelhoch J, Garwood M, Neema M, Robinson SP, *et al.* Applications of magnetic resonance in model systems: Tumor biology and physiology. *Neoplasia* **2**, 139–151.
98. Degani H, Chetrit-Dadiani M, Bogin L, Furman-Haran E. Magnetic resonance imaging of tumor vasculature. *Throm. Haemostasis* **89**, 25–33 (2003).
99. Lauterbur PC. Image formation by induced local interactions: examples employing nuclear magnetic resonance. *Nature* **242**, 190–191 (1973).

100. Katti G, Ara SA, Ayesha Shireen. Magnetic resonance imaging (MRI) — A review. *Int. J. Dent. Clin.* **3** (2011).
101. Bowtell R. Medical imaging: Colourful future for MRI. *Nature* **453**, 993–994 (2008).
102. Issa B, Obaidat IM, Albiss BA, Haik Y. Magnetic nanoparticles: Surface effects and properties related to biomedicine applications. *Int. J. Mol. Sci.* **14**, 21266–21305 (2013).
103. Na HB, Song IC, Hyeon T. Inorganic nanoparticles for MRI contrast agents. *Adv. Mater.* **21**, 2133–2148 (2009).
104. Yang M, Gao L, Liu K, Luo C, Wang Y, Yu L, *et al.* Characterization of Fe3O4/SiO2/Gd2O(CO3)2 core/shell/shell nanoparticles as T1 and T2 dual mode MRI contrast agent. *Talanta* **131**, 661–615 (2015).
105. Chen J, Zhang WJ, Guo Z, Wang HB, Wang DD, Zhou JJ, *et al.* pH-responsive iron manganese silicate nanoparticles as T1-T2* dual-modal imaging probes for tumor diagnosis. *ACS Appl. Mater. Int.* 7, 5373–5383 (2015).
106. Guan M, Dong H, Ge J, Chen D, Sun L, Li S, *et al.* Multifunctional upconversion-nanoparticles-trismethylpyridylporphyrin-fullerene nanocomposite: A near-infrared light-triggered theranostic platform for imaging-guided photodynamic therapy. *NPG Asia Mater* 7, e205 (2015).
107. Tian G, Zhang X, Gu Z, Zhao Y. Recent advances in upconversion nanoparticles-based multifunctional nanocomposites for combined cancer therapy. *Adv. Mater.* **27**, 7692–7712 (2015).
108. Nikoobakht B, El-Sayed MA. Preparation and growth mechanism of gold nanorods (NRs) using seed-mediated growth method. *Chem. Mater.* **15**, 1957–1962 (2003).

Chapter 6

Plasmonic Nanoparticles Application in Biosensor and Bioimaging

Mengmeng Liu

The Institute of Nanomedicine East Hospital
Tongji University School of Medicine
Shanghai, P. R. China

Abstract: Plasmonic nanoparticles hold great promising potential in applications for biosensor and bioimaging due to their unique optical property, tunable physical morphology, and surface characteristics. They extinct incident light with a certain wavelength efficiency, caused by coherent oscillation of the conduction electrons on their surface at nanoscale, and their strong interaction is known as localized surface plasmon resonance (LSPR). Meanwhile, their LSPR spectral, highly associated with their size, shape, and microenvironment surrounding them, provide us a conventional way to investigate the biomolecules' interaction, such as cell protein expression and DNA hybridization, and map the LSPR probe at the single molecule level. Additionally, plasmonic nanoparticle-based energy transfer, surface-enhanced Raman scattering, and photothermal therapy further enlarge its application in analyzing biomolecule and killing tumor *in vivo*.

Keywords: Plasmonic nanoparticles, localized surface plasmon resonance (LSPR), DFM, energy transfer, surface-enhanced raman scattering (SERs).

1. Introduction

Plasmonic nanoparticles have been widely applied as gene-regulating agents, imaging agents, and in drug delivery and photoresponsive therapeutics.[1] They exhibit intense absorption and scattering a cross sections that are several orders of magnitude higher than fluorescence without photobleaching and photoblinking. This excellent optical property originates from localized surface plasmon resonance (LSPR). LSPR represents the strong oscillation of plasmonic nanoparticle surface electrons when their electric field resonance with a certain wavelength incident light is at a confined nanoscale. The LSPR has two important effects.[2] Firstly, it gives a huge local electric field enhancement at the metal nanoparticle surface. Secondly, it creates a strong enhanced light absorption and scattering by the nanoparticle at the LSPR frequency in the visible to near-UV region of spectrum, and the shift of the plasmon resonance displays high sensitivity to the surrounding environment even at single-molecule level. And in this chapter, we will focus on explanation of LSPR theory, fabrication of LSPR structures, LSPR detection technology and plasmonic application-based on LSPR observation.

Besides, plasmonic nanoparticles could participate in energy transfer, and act as donor in plasmon resonance energy transfer (PRET) form and as acceptor in nanometal surface energy transfer (NEST). Due to the enhanced surrounding electric field, plasmonic nanoparticles could enhance enormously the Raman intensity of molecule, which adsorb on their surface. Moreover, plasmonic nanoparticle conjugation, such as cancer therapeutic agents, also absorbs near-infrared (NIR) light and generates heat photothermal therapy (PTT) or reactive oxygen species to kill cancer cells. In this chapter, we will introduce the theory and application of plasmonic nanoparticle-based energy transfer and, surface-enhanced Raman scattering (SERs) and therapy briefly.

2. Theory of Localized Surface Plasmon Resonance

Surface plasmons (SPs) are coherent oscillation of conduction electrons on a metal surface excited by electromagnetic radiation at a metal-dielectric interface. When it confined itself to a particle of a size comparable to the wavelength of light, that is, the particle's free electrons participate in the collective oscillation, surface plasmon takes the form of LSPR.[2]

LSPR significantly enhances the local electric field at the metal nanoparticle surface, which is the basis for the SERs, and increases light absorption and scattering by the nanoparticle at the SPR frequency. For noble metal nanoparticles, with a negative real and small positive imaginary dielectric constant over a range of wavelengths, the SPR frequency occurs at visible wavelengths. As the scattering wavelength of nanoparticle depends on its size and shape, the refractive index of the surrounding medium and internanoparticle coupling interactions, addition of biomolecule, that would change the electric field of nanoparticle, results in scattering shift. Due to this effect, plasmonic nanoparticle has been engineered to be a sensor with high specificity and sensitivity. To fully understand LSPR, would like to briefly introduce the scattering theory.

2.1. Mie Theory

Gustav Mie proposed a theory to calculate the scattering and absorption cross-section of a small spherical particle in 1908.[3] Assuming the small nanoparticles ($d<<\lambda$) interact with a plane-wave incident, the total scattering, extinction, and absorption cross-sections can be expressed as follows:

$$\sigma_{\text{sca}} = \frac{2\pi}{|k|^2} \sum_{L=1}^{\infty} (2L+1)\left(a_L^2 + b_L^2\right), \tag{1}$$

$$\sigma_{\mathrm{ext}} = \frac{2\pi}{|k|^2}\sum_{L=1}^{\infty}(2L+1)\left[\mathrm{Re}\left(a_L + b_L\right)\right], \tag{2}$$

$$\sigma_{\mathrm{abs}} = \sigma_{\mathrm{ext}} - \sigma_{\mathrm{sca}}, \tag{3}$$

where k is the incoming wave vector and L are integers representing the dipole, quadrupole, and higher multipoles of the scattering. The scattering coefficient can be simplified somewhat by introducing the Riccati–Bessel functions:

$$\psi_n(\rho) = \rho j_n(\rho), \xi_n(\rho) = \rho h_n^{(1)}(\rho). \tag{4}$$

Then, a_L and b_L are the following parameters:

$$a_L = \frac{m\psi_L(mx)\psi_L'(x) - \psi_L'(mx)\psi_L(x)}{m\psi_L(mx)\xi_L'(x) - \psi_L'(mx)\xi_L(x)}, \tag{5}$$

$$b_L = \frac{\psi_L(mx)\psi_L'(x) - m\psi_L'(mx)\psi_L(x)}{\psi_L(mx)\xi_L'(x) - m\psi_L'(mx)\xi_L(x)}. \tag{6}$$

Here, n is the complex refractive index of the metal, and n_m is the refractive index of the surrounding medium. Also, r is the radius of the particle, and $k_m = 2\pi/\lambda_m$ is defined as the wave number in the medium rather than the vacuum.

If the nanoparticle is assumed to be very small as compared to the wavelength, the Riccati–Bessel function can be approximate by power series. Following Bohren and Huffman, if one and keeps only terms to order, and switches the refractive index to dielectric function, then Equations (1) and (2) simplify to

$$\sigma_{\mathrm{ext}} = \frac{18\pi\varepsilon_m^{\frac{3}{2}}V}{\lambda}\frac{\varepsilon_2(\lambda)}{\left[\varepsilon_1(\lambda) + 2\varepsilon_m\right]^2 + \varepsilon_2(\lambda)^2}, \tag{7}$$

$$\sigma_{sca} = \frac{32\pi^4 \varepsilon_m^2 V^2}{\lambda^4} \frac{(\varepsilon_1 - \varepsilon_m)^2 + (\varepsilon_2)^2}{[\varepsilon_1 + 2\varepsilon_m]^2 + (\varepsilon_2)^2}, \tag{8}$$

where V is the particle volume, and $\varepsilon_m = n_m^2$ is the medium's dielectric function. Also n_R and n_I are the real part and imaginary part of metal dielectric index, respectively.

Else, the particles' size increase beyond the Rayleigh size limit (particle diameter >1/20 of incident wavelength), many important changes will occur in the scattering and absorption properties. These include new resonance bands that appear in the absorption and scattering spectra profile and shift toward longer wavelength, angular distribution of scattered light intensity behavior maxima and minima, the scattering cross-section increases slower than r^6 and degree of polarization decreases, and the scattered light intensity becomes concentrated in the forward direction. These changes may arise from the increasing contributions from electric and magnetic multipoles in addition to the usual electric dipole contribution.

2.2. Gan Theory

Limitations of Mie theory is it that is applicable only to spherical particles. Richard Gans extended Mie's result to spheroidal particles of any aspect ratio in the small particle approximation in 1912.[4] He conclude that the absorption cross-section for a prolate spheroid is very similar to a sphere, which is shown in Equation (7). This is as follows:

$$\sigma_{abs} = \frac{\omega}{3c}\varepsilon_m^{\frac{3}{2}} V \sum_j \frac{\left(\frac{1}{P_j^2}\right)\varepsilon_2}{\left\{\varepsilon_1 + \left[\frac{1-P_j}{P_j}\right]\varepsilon_m\right\}^2 + \varepsilon_2^2} \tag{9}$$

Here, for example, a 3D prolate spheroid, P_j includes P_a, P_b, P_c termed depolarization factors, meanwhile, for each axis of the particle,

$a > b = c$. The depolarization factors isotropically alter the values of ε1 and ε2 and the resulting LSPR peak frequencies. Explicitly, they are as follows:

$$P_A = \frac{1-e^2}{e^2}\left[\frac{1}{2e}\ln\left(\frac{1+e}{1-e}\right)-1\right]$$

$$P_B = P_C = \frac{1-P_A}{2}, \tag{10}$$

where e is the following factor, which includes the particle aspect ratio R:

$$e = \left[1-\left(\frac{B}{A}\right)^2\right]^{\frac{1}{2}} = \left(1-\frac{1}{R^2}\right)^{\frac{1}{2}}. \tag{11}$$

The resulting extinction spectrum display non-sphere nanoparticles have two peaks: one result from the transverse plasmon mode from the x and y contributions to the sum, and the other come form the longitudinal plasmon mode from the z contribution. Equation (9) could be used to analyze the effect of the aspect ratio on the LSPR peak wavelength. Increase in the aspect ratio of non-sphere nanoparticles could result in an increase in weighting $\varepsilon m\ [(1-P_j)\ /P_j\ ]$, which is 2 for spherical particles. This leads to a red-shift of the plasmon peak with increasing aspect ratio, as well as increased sensitivity to the dielectric constant of the surrounding medium.

2.3. Large Particles Theory

In order to obtain quantitative results for larger particles, the logarithmic derivative was introduced in the context of computing scattering coefficient for a sphere.[5]

$$D_n(\rho) = \frac{d}{d\rho}\ln\psi_n(\rho). \tag{12}$$

We may therefore recast Equations (1) and (2) in the form:

$$a_n = \frac{\left[D_n(mx)/m + \frac{n}{x}\right]\psi_n(x) - \psi_{n-1}(x)}{\left[D_n(mx)/m + \frac{n}{x}\right]\xi(x) - \xi_{n-1}(x)}, \tag{13}$$

$$b_n = \frac{\left[mD_n(mx) + \frac{n}{x}\right]\psi_n(x) - \psi_{n-1}(x)}{\left[mD_n(mx) + \frac{n}{x}\right]\xi(x) - \xi_{n-1}(x)}, \tag{14}$$

where we have used the recurrence relations:

$$\psi_{n+1}(x) = \frac{2n+1}{x}\psi_n(x) - \psi_{n-1}(x),$$

$$\xi_n = \psi_n - i\chi_n, \tag{15}$$

$$D_{n-1} = \frac{n}{\rho} - \frac{1}{D_n + \frac{n}{\rho}}.$$

These relations are evaluated beginning with:

$$\psi_{-1}(x) = \cos(x), \quad \psi_0(x) = \sin(x),$$

$$\chi_{-1}(x) = \sin(x), \quad \chi_0(x) = \cos(x), \tag{16}$$

where $$D_{\text{Nmax}} = 0.0 + i0.0,$$

$$\text{Nmax} = \text{Max}\left(x + 4x^{\frac{1}{3}} + 2, |mx|\right) + 15.$$

2.4. Nanoparticle Coupled Models

Nobel nanoparticles interact with induced light to produce the LSPR phenomenon. When two or more nanoparticles move close to each other, their enhanced electric field will overlap and induce more complex phenomenon,[6–9] that is hot spot, Fano resonance and so on.

Nordlander *et al.* developed a theory to explain Fano resonance, analogous to hybridization model in atom bonding.[10–13] Take Dimer system as example, briefly, when the polarizer is oriented perfectly along the interparticle axis, only the σ-mode is seen in the scattering spectrum. As σ-mode is formed by head-on hybridizing, charge with the opposite sign is accumulated on opposite sides of the gap to decrease frequency, and its net charge is high to increase intensity, whereas only the π^*-mode contributes when the polarizer is oriented perfectly perpendicular to the interparticle axis. Contrast to σ-mode, π^*-mode is slightly blue-shifted relative to the single particle resonance and is lower in intensity. For complicate systems, it is necessary to reduce the system to a simple system.

In the experiment, Reinhard B.M group found that the ratio of spectral shift of particles (with diameters in the range of 20–70 nm) is a first approximation, independent of the nanoparticles diameter and follows a universal scaling relationship to individual nanoparticles resonance wavelength. For AuNPs, the universal scaling relationship as a function of interparticle separations, S, and particle diameter, D, was determined as follows[14,15]:

$$\frac{\Delta\lambda_{\text{plas}}}{\lambda_0} = 0.18e^{\frac{-\frac{S}{D}}{0.23}} \qquad (17)$$

2.5. Refractive Index Sensitivity

The electric field of nanoparticle associates with its surrounding environment, subtle solvent change and biomolecule binding, which also affects its local electric filed, significantly leads to scattering light shift or change. Generally, the scattering wavelength shift is approximately linearly associated with refractive index of the surrounding medium, for example, the nanorod's longitudinal peak wavelength red-shift from 660 nm to 735 nm when refractive indices of buffer change from 1.000 to 1.428.[16]

The nanoparticle's refractive index sensitivity (S), reported in nanometers of peak shift per refractive index unit (nm/RIU), represents its sensitivity as plasmonic effect and is various for different materials.

$$S = \frac{d\lambda_p}{dn}. \tag{18}$$

For example, the refractive index sensitivity S of gold nanospheres with 50 nm in diameter is 60 nm/RIU,[17] while silver nanosphere with same size is about 160 nm/RIU.[18] As another example, refractive index sensitivity of gold nanocubes with 30–50 nm in size is 83 nm/RIU[19] while 146 nm/RIU[20] for silver nanocubes in the same size. These results indicate that the scattering light of Ag nanosphere is much more sensitive for environment change than gold nanosphere. Meanwhile, nanoparticle's refractive index sensitivity also relates with its shape. It is discovered that silver nanotriangles had a much higher sensitivity (350 nm/RIU) than that of spheres (160 nm/RIU).[18] As the refractive index sensitivity directly determines the sensitivity of sensors to engineer novel sensor with high sensitivity and specificity, much effort has been spent in studying the relationship between refractive index sensitivity and the structure of plasmonic nanoparticles. In briefly, particles with sharp tips (nanotriangles, bipyramids) exhibit especially high refractive index sensitivities. For example, Sun *et al.* demonstrated that gold nanoshell has a much higher S (409 nm/RIU) than that of nanosphere (60 nm/RIU) in the same diameter.[17] For gold nanorods (AuNRs), Lee *et al.* showed that the sensitivity increases from 157 nm/RIU to 497 nm/RIU for aspect ratio increasing from 1.0 (spheres) to 3.4.[21–23]

3. Fabrications of Plasmonic Nanoparticles

LSPR, associated with noble metal nanostructure, results in a strong oscillation of its surface electrons and creates strong electromagnetic near-field enhancement as well as sharp spectral absorption and scattering peaks when its electric field resonates with a certain wavelength incident light at the nanoscale.[2,24,25] As the electric field of nanoparticle depends on its size, shape, structure and surrounding environment,[5,19,26] its plasmon property, corresponding with its LSPR, the wavelength of the exciting light are tunable. Today, advances in colloidal synthesis methods enable nanoparticle with tunable shape,[27,28] such as nanosphere, nanorods, nanostars, varied sizes[29–32] and therefore wide range of applications in catalysis, optics, chemical and biological sensing, and medical therapeutics to be obtained.

During the last two decades, to acquire nanoparticle with certain plasmonic property and strong LSPR effect, considerable effort has been focused on improving the methods of synthesizing and organizing nanoparticles with certain shape, structure, solubility, stability, and functionality. Although numerous preparative methods have been reported, these methods generally can be divided into two approaches, "top-down"[33–37] (physical manipulation) and "bottom-up" (chemical transformation). In the following section, I would like to briefly introduce several popular methods of synthesizing and modifying plasmonic nanoparticles from these approaches.

3.1. Bottom-Up Approach (Chemical Methods)

Bottom-up approach, mainly based on chemical methods by reducing metal-salt in the presence of surface stabilizers,[38,39] provides us a serial of metal nanoparticles with controllable size, morphology, coupled structure and plasmonic property. We will briefly introduce some chemical methods to synthesize gold nanoparticle (AuNP) and silver nanoparticle (AgNP), which show strong LSPR effect and are widely applied in biosensor and bioimaging.

3.1.1. Gold nanoparticles

Among the conventional methods of AuNPs synthesis by reduction of gold (III) derivatives, the most popular method is reducing $HAuCl_4$ with citrate in water, which was proposed by Turkevitch in 1951.[40] In this method, citric acid acts as both reducing and stabilizing agents and provides AuNPs with diameters of 20 nm. G. Frens' work in 1973[41] was enabled controlling over AuNPs size (16–147 nm) by varying the feed ratio of gold salt to sodium citrate. Further studies show reduction of $HAuCl_4$ by sodium citrate,[42] following a kinetically controlled seeded growth strategy, resulting in monodisperse citrate-stabilized AuNPs with a uniform quasi-spherical shape of up to ~200 nm and a narrow size distribution. This method proceeds via four overlapping steps: nucleation, aggregation of nuclei, slow isotropic growth via reduction and/or coalescence/Ostwald ripening, and rapid consumption of the gold precursor.[43] Mechanistically, gold reduction occurs through a concerted redox reaction, whereby citrate is both chemically oxidized by chloroauric acid and thermally oxidized by heating to form dicarboxyacetone (DCA).[44–46] Electrons from DCA serve to reduce $AuCl_3$ to AuCl, which is believed to form a bidentate complex with DCA. Particle nuclei forms via a disproportionation reaction, where AuCl complexes combine to form zero-valent gold atoms and an $AuCl_3$ complex. Then, the inhibition of any secondary nucleation during homogeneous growth was complete by adjusting the reaction conditions: temperature, gold precursor to seed particle concentration, and pH.[46,47]

A two-phase synthetic strategy, (the Brust–Schiffrin method), which utilizes strong thiol — gold interactions to protect AuNPs with thiol ligands, was achieved by Brust and Schiffrin.[48] Due to the strong thiol–gold interaction, these AuNPs can be repeatedly isolated and redissolved in common organic solvents without irreversible aggregation or decomposition, and they can be easily handled and functionalized just as stable organic and molecular compounds. In this method, $AuCl_4-$ is transferred from aqueous phase to toluene using the surfactant tetraoctylammonium bromide (TOAB) and reduced by sodium borohydride ($NaBH_4$) in the presence of dodecanethiol. On addition of $NaBH_4$, a quick color change from orange to deep brown takes place in organic phase. The AuNPs are generated in toluene with controlled diameters in the range 1.5–5 nm. Various reaction conditions, such as gold/thiol ratio, temperature, and reduction rate, can be used to tune the particle size.[28] For example, larger thiol/gold mole ratios give smaller average core sizes, and fast reductant addition and cooled solutions produced smaller, more monodisperse particles.

A higher abundance of small core sizes (<2 nm) is obtained by quenching the reaction immediately following reduction or by using sterically bulky ligands.[49–53] Subsequently, Brust–Schiffrin procedure, also called monolayer-protected clusters (MPCs), was applied for the synthesis of other stable AuNPs,[54–59] containing functional thiols groups.

Rod-shaped AuNPs, which transfer the energy of NIR light into heating efficiently, are synthesized by a seed-growth method. Murphy and coworkers[60] firstly used ascorbic acid and silver nitrate to replace electrochemical reduction and produced AuNRs for various aspect ratios, including 4.6 ± 1, 13 ± 2, 18 ± 2.5. However, under the high CTAB concentration and low pH (~2.5) growth condition, the ascorbic acid used in this synthesis reduces gold derivatives to Au (I) state[61] instead of metallic state. But addition of small seed particles with about 3 nm in diameter into the Au(I) solution results in complete reduction to metallic gold, which is catalyzed by the surface of the seeds and leads to the gradual change in shape from quasi-spherical to rod-like crystal. Subsequently El-Sayed[62] improved this method by utilizing CTAB-capped seeds rather than the citrate-capped seed particles to produce rod-like morphology in high yield, as shown in Fig. 1. In brief, small (~1.5 nm diameter) seed nanoparticles are formed via classical borohydride reduction of Au salt in the presence

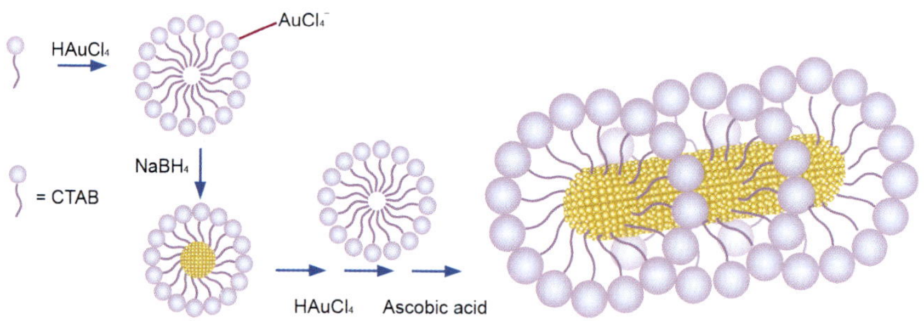

Fig. 1. Gold (III) chloride quantitatively displaces Br^- counter ions in micelles of CTAB and subsequent borohydride reduction produces small (~1.5 nm diameter) seed nanoparticles surface stabilized by a CTAB-bilayer. Then Au (III) bound to CTAB micelles is reduced to Au (I) by ascorbic acid. Directional growth of gold AuNRs occurs via crystallographically preferential reduction of Au (I) onto the seed nanoparticles.

of a cationic surfactant stabilizer. The solution turned pink on addition of the reducing agent, ice-cold sodium borohydrate, indicating the formation of particles. These seeds are then added to a solution containing additional surfactant and Au salt that has been mildly reduced by ascorbic acid, forming dehydroascorbic acid and resulting in an $AuCl_2$ — surfactant complex.[63,64] The color of the solution changed from pink to reddish brown, indicating the formation of AuNRs. Other studies improve controlling of the aspect ratio by varying the amount of silver nitrate, the concentration ratio of Au seed to Au3+ ions, the reducing agent, capping agent, and reaction temperature and time intervals. In seed grown methods, AuNRs grown using CTAB-protected gold seeds in the presence of Ag+ bear Au[65] faces on the sides of the rods and {100} faces on the ends.[66] Silver under potential deposition (UPD) is the reduction of Ag^+ to Ag^0 at a metal substrate with a surface potential less than the standard reduction potential of Ag^+. A greater positive shift in potential for silver UPD on gold surfaces is observed on {110} surfaces compared to the {100} or {111} faces,[67,68] suggesting that deposition of silver on the sides of the rods ({110} facets) should be faster than on the ends ({100} facets). Fast deposition of silver, followed by strong CTAB binding (via bromide) inhibits gold growth on the sides of the rods and leads to preferential growth of gold at the ends. It is notable that the bonding interaction between Au and cetyltrimethylammonium bromide (CTAB) is typical of almost all chemical conjugation schemes involving AuNPs. Br^- anions are known to form atomic surface layers on Au and other metals/semiconductors (i.e. adlayers) and also form electrostatically coordinated CTA^+cations and their micelles. The general consensus for AuNR–CTAB bonding involves the formation of a Br^- adlayer on the AuNR surfaces, which then coordinates N head groups on CTA^+. The aliphatic tail of surface-bound CTA^+ then associates with another CTA^+ cation-oriented with its $N^+ Br^-$ head group facing outwards.[69]

3.1.2. Silver nanoparticles

The most common approach for Ag NPs synthesis is chemical reduction by organic and inorganic reducing agents. In general, silver ions

(Ag^+) in aqueous or non-aqueous solutions were reduced by various reducing agents, such as sodium borohydride ($NaBH_4$), sodium citrate and polyol to formate metallic silver (Ag_0). Following that, Ag_0 agglomerate into oligomeric clusters and eventually formate metallic colloidal silver particles.[70–72] It is important to use protective agents to stabilize dispersive nanoparticles during the course of metal nanoparticle preparation and protect the nanoparticles that can be absorbed on or bind onto nanoparticle surfaces, avoiding their agglomeration. The presence of surfactants comprising functionalities (e.g. thiols, amines, acids, and alcohols) for interactions with particle surfaces can stabilize particle growth and protect particles from sedimentation, agglomeration,[73] or losing their surface properties. One of popular approaches to quickly generate silver colloids was reported by Lee and Meisel in 1982.[74] In their method, boiling $AgNO_3$ was reduced by citrate in an aqueous solution to generate silver nanoparticles.[75,76] During synthesis, the citrate ions serve as a reducing agent and a stabilizer, and meanwhile, they may also bind with Ag^+ ions or Ag_2 dimers in the early stages of the reaction.[29,77–80] Additionally, changes of pH will slightly alter the shape of Ag nanoparticles. Besides, the polyol process represents a robust and versatile method for generating Ag nanostructures with wide different and well-defined shapes.[77–81] Achievement on tunable control over nucleation, growth and final shape were realized by varying the reaction conditions, such as temperature, reagent concentration, and presence of trace ions.[72,82–86] In a typical synthesis as displayed in Fig. 2, ethylene glycol serves as the source of reductant to generate metal atoms from AgO_3 at an elevated temperature, meanwhile, a polymeric capping agent, such as poly (vinyl pyrrolidone) (PVP), was added to facilitate control on nanoparticles' shape.[79,87–90] In this method, seeds with different sizes grow into nanoparticles with different shapes.[79] Small clusters of fluctuating structure aggregates into larger clusters for seeds. There are three predominant seed structures: single crystalline, single twinned, and multiply twinned. Typically, the single crystalline seeds grow into cuboctahedrons or nanocubes, depending on relative growth rates along the {100} and {111} directions.[85] Additionally, through oxidative etching, these cuboctahedrons or nanocubes can be applied as

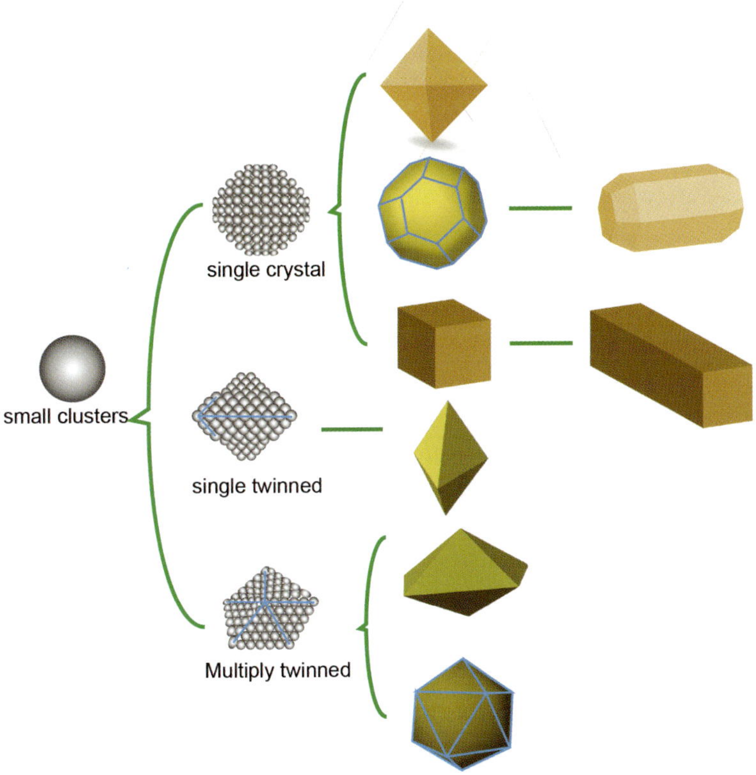

Fig. 2. A schematic illustrating the reaction pathways that lead to AuNPs having different shapes. First, a precursor is reduced or decomposed to form nuclei (small clusters). Once the nuclei have grown past a certain size, they become seeds with a single-crystal, singly twinned, or multiply twinned structures. Twin planes are delineated in the drawing with blue lines. Figure modified with permission from Ref. 90. Copyright 2011, American Chemical Society.

seeds to anisotropically grow into nanorods and nanobars, respectively.[91] The single twinned seeds produce right bipyramids or nanobeams with a truncated and ultimately sharp morphology.[77] Besides, multiply twinned seeds grow into rods or nanowires with a pentagonal cross-section directly. In general, cubes, bipyramids, and pentagonal wires are usually prepared by polyol process. In order to improve the purity of production, a number of ways, such as oxidative etching, have been investigated. Oxidative etching removes multiply twinned seeds and, therefore, influence the relative proportions of different Ag seeds. For example, Cl^- is a stronger ligand, the Cl^-/O_2 pair dissolve both multiply twinned and singly twinned seeds. Close monitoring of the reaction process by UV–Visible spectroscopy and electron microscopy revealed that twinned particles are formed first but later dissolved by oxidative etching. Thus, after a certain reaction time, single-crystal seeds with a spherical profile predominate, and continue growing into nanocubes with all facets being {100}.[72,92] Meanwhile, Br^- is slightly weaker than Cl^-, the Br^-/O_2 pair will only dissolve multiply twinned seeds, resulting in a right bipyramids dominated solution.

3.2. Physical Methods

The major challenge of wet-synthesis lies in controlling the position and orientation of nanoparticles with sufficiently high resolution when nanoparticles placed on substrates. Unlike chemical synthesis, lithography, deposition, and other forms of nanofabrication offer precise control over deliberately varied shapes and configurations and therefore open avenues to both fundamental investigations and applications.[93] Conventional lithographic techniques such as electron beam lithography (EBL) or focused ion beam lithography (FIB) utilize polymeric resists to fabricate masks for deposition or etching of metal with high resolution.[94,95] Take EBL for example, a focused electron beam is scanned over a substrate coated with a thin layer of resist (EBL usually used PMMA as resist), exposing specific regions according to a programmed pattern.[96,97] Halas and Nordlander *et al.* utilize EBL to organize nanostructures with precise size and shape and to generate complex optical phenomenon like Fano resonance.[98–101]

4. Technic to Detect Localized Surface Plasmon Resonance

As a typical plasmon spectroscopy technique, dark-field microscopy (DFM) coupled with spectrometer was developed for the observation of metal nanoparticles. The primary advantage of this technique is that it provides high contrast images and makes it possible to collect the scattering light of plasmon nanoparticles at the single-particle level. To date, DFM is widely utilized to monitor biological events, such as protein association and overexpression. Many modifications to the conventional dark-field system like employing different illumination methods, light sources and various CCDs, have been developed to improve the signal and time resolution of DFM to identify single molecular binding events.

4.1. Conventional Dark-Field Microscopy

In a typical DFM system, the white light of a Halogen lamp pass through a condenser to tilt the incident illumination and are focused at the sample to excite the plasmons. The condenser is equipped with a circular block at the lens to permit light with high angular transmit. Most of the light is directly transmitted through the sample and is not collected by the objective below the stage because the numerical aperture of objective is smaller than the condenser, causing the dark background. Only the scattered light from the sample could be collected to image sample in true-color or to analyze single particle spectra.

4.2. Improvements in Sensitivity and the Speed of Dark-Field Microscopy

The main limitation of conventional DFM is the low signal intensity for nanoparticles smaller than 50 nm. To obtain sufficient SNR of single nanoparticle, an intense light source and long recording time (up to seconds) are both required, which makes DFM in suitable for fast dynamic observation. To improve SNR and the temporal resolution, several optical modifications have been developed. Noji *et al.* developed the objective-type evanescent illumination DFM by simply replacing the dichroic mirror with a perforated mirror in a conventional microscope.[13] In this system with total internal reflection (TIR)-illumination, the incident laser is reflected at the water–glass interface and forms the evanescent field on the coverslip surface. The evanescent field induces the oscillating dipoles within the particle, leading to conversion of the non-propagating near-field light into the conventional propagating scattered light. The scattered light is detected by camera while the returning laser is stopped, inducing dark-field effects. Taking advantage of the focus on a thin layer by TIR illumination, background noise was decreased and images with high SNR were acquired. In a study of the rotary mechanism of F_1–ATPase 1–2 nm, spatial precision and 9.1 μs temporal resolution were demonstrated by the successful application of this home-built TIRDFM system. Later, to achieve higher sensitivity, white light laser with average power output of 2.8 W was reported to supply a strong light source to illuminate the particles in TIR geometry. With the help of the improved system, the dynamic evolution of a single protein binding to a single AuNP was observed with millisecond time resolution.[102]

Although TIR illumination can decrease the background noise, this technique is limited to a thin region adjacent to the interface. Baumberg *et al.* used a supercontinuum laser to perform DF spectroscopy in a focused mode, they equipped a circular block in the illumination beam path to focus the ring-shaped laser on the sample, and the illumination intensities yielded were up to 10 MW/cm^2.[103] Under intense illumination, the growth of a single gold nanorod was observed in real time with 8 ms time resolution. More importantly, this method can be used to probe deeply into a sample.

To reduce the toxicity of the illumination light to living cells, He *et al.* developed quiet a different method. They utilized a new technique that named dual-wavelength difference (DWD) imaging to remove noise in the living cell environment.[27] Briefly, the scattering intensity of nanoparticle probe at resonance frequency is much higher than noise signal from living cells, while at non-resonance frequency scattering intensity of nanoparticle is similar with noise signals. Then, noise free images could be obtained by subtracting the signal of reference beam, of which wavelength is far away from the resonance frequency, from probe poem beam at the resonance. With this novel strategy, the diffusion dynamics of nucleic acids functionalized AuNPs on the cell membranes were explored.

4.3. 3D Orientational Imaging with Dark-Field Microscopy

The orientational dynamics of nanocargos would be a key parameter for revealing the fundamental mechanisms of vital biological processes, such as the interactions between the nanocargo and nearby biomolecules. 3D DFM provides access to solve this problem.[104,105] Recently, He and Zhu successfully recorded the 3D orientation angles of single anisotropic nanoparticle with standard optical DFM by the method of defocused imaging.[29] The field distribution pattern in the defocused DFM images depends on the aspect ratio of the AuNPs, the numerical aperture of the objective, the defocusing distance and the polarization direction of the incident radiation, thus, by comparing the experiment results at fixed condition with the simulated images-3D orientations of nanoparticle was confirmed. They improved this method subsequently by adding another focused laser to monitor transverse oscillation modes simultaneously. The dual-wavelength DFM has been proven to obtain the localization and orientational information at the same time by the combination of focused transverse illumination and defocused longitudinal mode.

Further, He *et al.* presented another method to directly decipher the 3D orientation of individual AuNRs in solution or inside living cells with planar illumination.[104] They illuminated the sample with two orthogonal sheets of light and obtained the longitudinal and transverse scattering spectra simultaneously. DFM was improved by inserting a birefringent crystal in the detection path before CCD camera. They utilized this method to directly monitor the rotational dynamics of individual AuNRs, which modified with different surface ligands and dispersed in glycerol/water mixtures, and with the help of proper surface function, they track the rotational movements of individual AuNR transported by kinesin motor protein on microtubules inside living cells.

Since the 3D orientation of anisotropic metallic nanoparticles was obtained by matching the measured pattern with the simulated images in convention 3D DFM, high degrees of uncertainty are involved. As TIR microscopy reduces the background noise greatly, Fang's group developed TIR scattering microscope to follow 3D rotational movement of anisotropic nanoparticles with a subdegree uncertainty.[106–108] By combination of defocused imaging and the rotation of the polarization direction of light incident upon the sample, the 3D orientation of stationary AuNPs was resolved with a higher precision in TIR scattering microscope. In another work, a dual color TIR scattering microscope was used to track the orientation change of AuNRs when interacting with synthetic lipid membranes by analyzing orientation-dependent intensity fluctuations of surface plasmon resonance enhancement along the long and short axis simultaneously.[108] AuNR orientation changes in the sample plane can be tracked in details using polarized light that propagates parallel to the prism surface, while out-of-plane orientation changes can be tracked utilizing the long axis of the AuNR and its interaction with polarized light propagating perpendicularly to the prism surface.

5. Application based on Localized Surface Plasmon Resonance

The typical LSPR sensing application of plasmonic nanoparticles is to detect changes of their surrounding environment through shifts of scattering profile.[109] This kind of a sensor has been designed to detect bimolecular binding events[110,111] close to the surface and molecular adsorb on the surface,[112] to characterize carbohydrate–protein interactions[113] and measure the binding of a protein to DNA.[114] For instant, a sensor[115] was designed to detect conformational changes in a surface-bound construct of the calcium-sensitive protein calmodulin. Increases in calcium concentration induce a 0.96 nm red-shift in the spectral position of the LSPR extinction maximum (λ max). Addition of a calcium-chelating agent forces the protein to return to its original conformation and is detected as a reversal of the λ max-shift.

5.1. Localized Surface Plasmon Resonance Change to Sensor Catalytic Reaction

Due to the quantum effects and free electrons on their surface, AuNPs of small scale exhibit catalytic activity and attract many scientists to study its mechanism. It's discovered that catalytic reactions occurring on the surface of AuNPs bring electron injection into the nanoparticles and change their local environment. This change surrounding nanoparticles result in a shift in DF spectra, which can be obviously monitored by DFM at the single molecular level. As spectra shift caused by catalytic reactions is small and hardly observed by other technologies, such as UV, AFM, TEM and microscopy, a serial of novel nanosensors system-based DFM has been invented to investigate and monitor catalysis reactions.[116] Fan *et al.* proposed to apply red-shift in DFM by enlargement of AuNPs to sensor catalytic reaction.[117,118] They discovered that AuNPs with 50 nm size, acting like glucose oxidation (GOX), produces H_2O_2, which, in turn, induces the AuNP' seeded growth *in situ* in the presence of $HAuCl_4$. They demonstrated that the surface of AuNPs prefer to adsorb singlestranded (ss) DNA than doublestranded (ds) DNA due to nitrogen containing nucleotides. For example, addition of growing solution (H_2O_2 and $HAuCl_4$) leads to a 100 nm red-shift of scattering peak for bare AuNPs (Fig. 3(a)). The ss-DNA and the ds-DNA suppressed the catalytic activity of AuNPs and, in turn, their growth in varying degrees by adsorbing on its surface. Especially, the scattering profile of nanoparticle exhibited almost identical in the presence of ss-DNA with bare nanoparticle. Therefore, the conclusions that the adsorption ss-DNA on the surface of AuNPs inhibits their GOX-like catalytic activity and finally inhibits AuNPs' further growth were obtained.[117] With this sensor based on DFM, they not only monitor the mechanism of AuNPs growth but also propose a novel method for regulating the GOX-like catalytic activity of AuNPs with the help of DNA nanotechnology.[118]

Furthermore, this nanosensor system of AuNP-catalyzed GOX was extended to detect biomolecules, such as adenosine triphosphate (ATP).[118] Similar to the mechanism mentioned above, an anti-ATP aptamer physically covers the surface of AuNPs and inhibits its catalytic activity. In the presence of ATP, the aptamer forms tertiary structure, resulting in it detaching from the surface of AuNPs. Therefore, AuNPs' catalytic activity recovers and the AuNPs start to grow in the presence of a growing solution, resulting in a visible red-shift in scattering spectra. Experiments indicate that this ATP aptamer-based plasmonic nanoparticle sensor has a detection limit of nM. This system holds great promise for real-time imaging of multiple targets by using

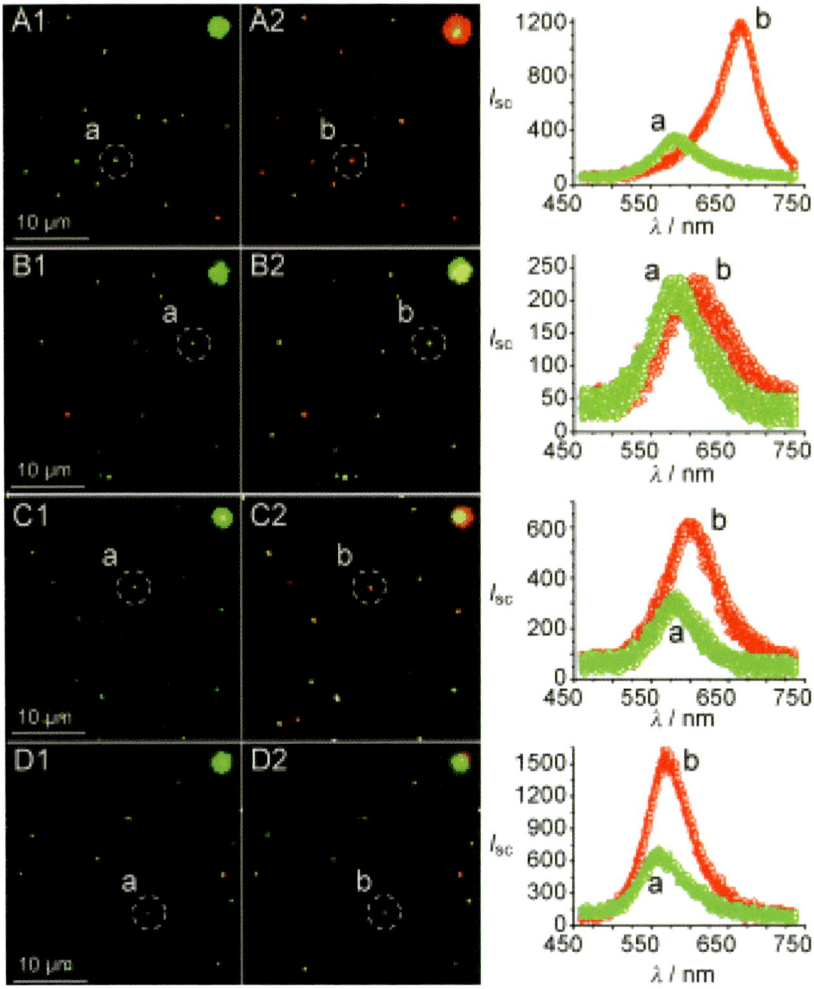

Fig. 3. Left: DFM images of large AuNPs (50 nm in diameter) before (1) and after (2) the 25 min glucose-induced enlargement process. (A) Bare AuNPs, (B) ss-DNA–AuNPs, (C) ds-DNA–AuNPs, (D) ds–DNA with 1-base mismatches interacting with AuNPs. Right: The light scattering spectra corresponding to individual nanoparticle marked with the circle in the DFM images. I_{sc} = scattering intensity. Figure modified with permission from Ref. 109. Copyright 2014, Elsevier B.V.

different aptamers *in vitro* and *in vivo*. It is noted that color change in images and peak shift in spectra can be displayed with a single dot by DFM when catalytic reaction happens, indicating that each individual nanoparticle in this system acts as an individual plasmonic probe.

Besides, AuNPs were also used as contrast and recognition reagents. The catalytic enlargement of AuNPs conjugate associated with biorecognition complexes is used to yield conductive patterns that follow biosensing process.[119–121] Inspired by catalytic deposition of production on AuNPs by catalysis, AuNPs have been applied for the optical and electrochemical detection of biocatalytic processes. Long's group[122] utilized AuNPs (50 nm in diameter) to monitor the gold-catalyzed reduction of Cu_2+ ions on AuNPs forming Au@Cu core–shell nanoparticles by NADH or by NAD+ cofactor-dependent enzyme/substrate system. The plasmon spectra of Au@Cu nanoparticles red-shifted followed the increase in the concentration of NADH. Therefore, it could be applied to map the distribution of NADH in cells, following *in vitro* intracellular metabolic pathways and to screen drugs affecting cell metabolism.[122] Meanwhile, Fan *et al.* applied AuNPs with 50 nm as building blocks for surface attachment of double stranded (ds-) DNA and HRP.[99] As HRP initiates the polymerization of aniline on ds-DNA templates with the addition of H_2O_2,[123] leading to the growth of AuNPs, which changes its true color and shifts plasmon band, which lay the foundation for plasmonic detection of H_2O_2. Another work based on the DA-induced seeded growth of AuNPs was able to monitor the release of DA from living PC12 cells and highlight the concomitant relation of ATP-stimulated release of DA is with the Ca_2^+ influx, and the influx of Ca_2^+ is through ATP-activated channels.[124] In construct, the depression of growth was applied to detect RDX residues in LFP. In the absence of RDX residues, NADH mediated the reduction of Cu_2^+ and deposition of Cu_0 on AuNP seeds (50 nm), leading to the coating of AuNP with a shell of Cu, which results in a red-shift in the LSPR scattering spectrum, while, in the presence of RDX, NADH competes between Cu_2^+ and RDX. As a result, the catalytic growth of AuNPs is inhibited, leading to a weakened nanoplasmonic response.[125]

5.2. Particle Plasmon-Coupled Localized Surface Plasmon Resonance

Advances in DNA nanotechnology allow us to organize nanoparticles with specific distance, and this distance change also affects their plasmonic effect. For example, when two plasmonic particles approach each other within one particles' diameter, these particles' plasmon couples, resulting in enhancing the electromagnetic field between two nanoparticles and a visible red-shift in the scattering spectral. Therefore, DFM and Rayleigh spectroscopy find great applications in determination of the distance between each other.

5.2.1. Applications in assembles and disassembles

Taking advantage of the color change of AuNPs solution from red to blue caused by the aggregation of AuNPs,[126–128] Fan *et al.* developed a series of aptamer sensors to exploit interactions between aptamers and AuNPs by observing its color change, such as potassium ion-specific G-quartet probe, T-rich DNA probe, C-rich DNA probe, anti-ATP aptamer and anticocaine aptamer.[129–131] For example, Fan *et al.* developed a biosensor assay based on cocaine–AuNPs to identify cocaine in latent fingerprints (LFPs).[132] Briefly, two pieces of rationally engineered anticocaine aptamers were attached to different AuNPs. The presence of cocaine in LFPs, working as a molecular linker, induces the aggregation of AuNPs by reassembling the two pieces of aptamers, resulting in green-to-red color change of the scattered light in DFM images (Fig. 4). This strategy provides a quasi-quantitative method to identify cocaine loadings in LFPs. Similarly, Long's group designed a sensor through the status of AuNPs aggregation to monitor Cu^{+}-catalyzed click reaction at the single-nanoparticle level by DFM.[133] By contrast, Lee and the coworkers applied a core-satellite nanoassembled substrate for colorimetric biomolecular detection through red-to-green changes in the DFM images.[134] In the presence of target biomolecules, the core-satellite nanoassembled substrate disassembled and a blue shift in the spectrum was demonstrated by DFM. Furthermore, stepwise peak shifts in LSPR were monitored by

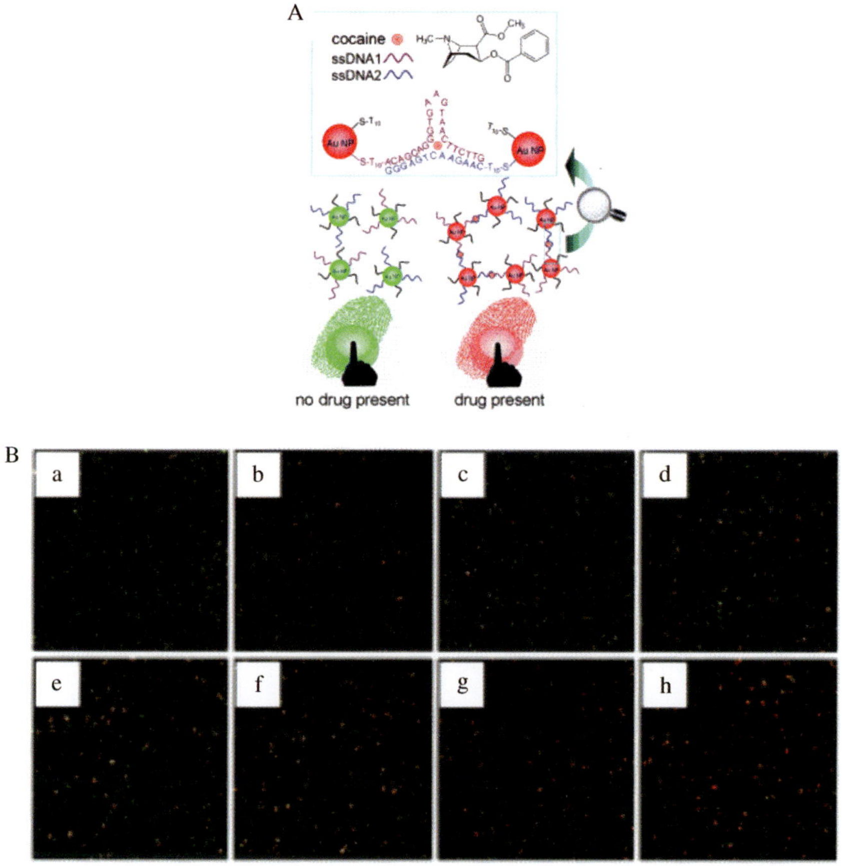

Fig. 4. (A) Principle of nanoplasmonic imaging of LFPs and identification of cocaine in LFPs by DFM. (B) Magnified dark-field images of LFPs containing different loadings of cocaine. (a) 0 g, (b) 1.5×10^{-7} g, (c) 3×10^{-7} g, (d) 7.5×10^{-7} g, (e) 1.5×10^{-6} g, (f) 3×10^{-6} g, (g) 1.5×10^{-5} g, and (h) 3×10^{-5} g. Figure modified with permission from Ref. 109. Copyright 2014, Elsevier B.V.

DFM when the binding event of nanoparticles was carefully investigated, and the steps showed good agreement with a number of the additional particles observed by SEM (Fig. 5). This fact confirmed the capability of single-binding detection by DFM.[135] Similar to Lee's, a dimer sensor system was designed to detect miR210. In this system,

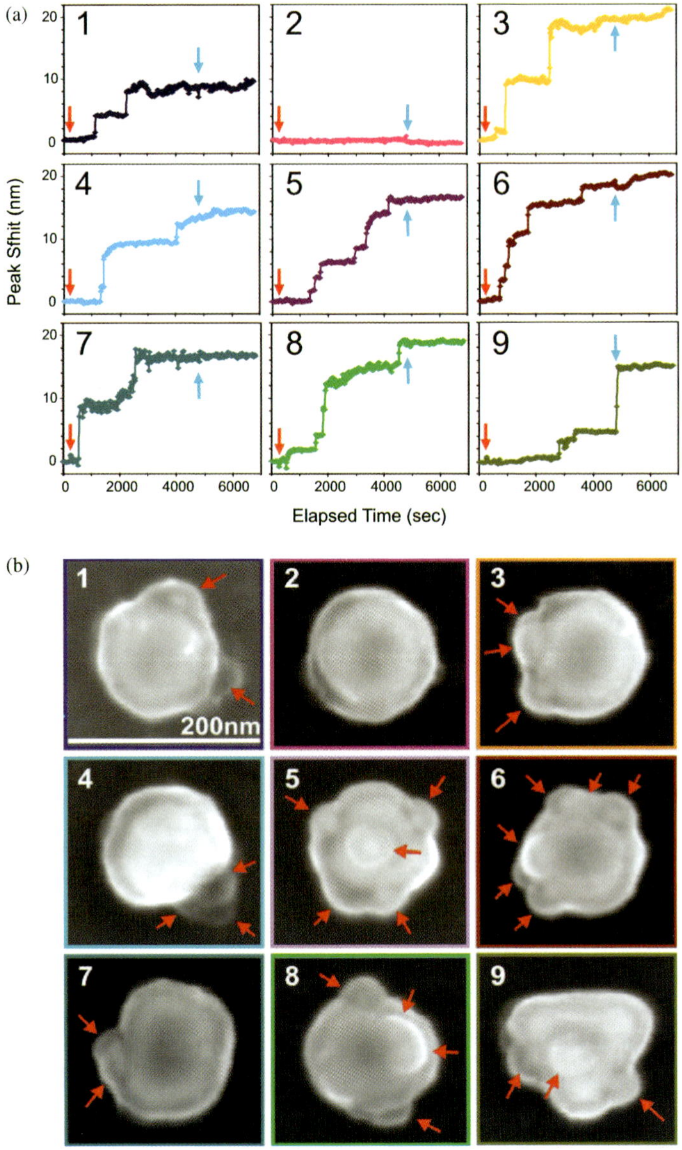

Fig. 5. (a) Adsorption kinetic curves for the nine particles. The peak shift is plotted as a function of elapsed time. At the time of 250 s, the ad-particle solution was injected, and at 4800 s, the flow-cell was rinsed with buffer solution, as shown by red and blue arrows, respectively. (b) SEM pictures of the corresponding particles in (a) after adsorption measurement. The arrows indicate the ad-particle positions. Figure modified with permission from Ref. 109. Copyright 2014, Elsevier B.V.

dimers were modified by stem-loop DNA sequence, and they geometrically extend upon binding to the target resulting in an increase in the energy of the hybridized bonding plasmon mode. Therefore, fractional blue-shift related to the concentration of miR210 could be detected.[136]

5.2.2. Applications in plasmonic ruler

Plasmon ruler, like its name, is capable of determining the distance of different interparticles by measuring their light scattering profile. Preliminary works suggest that plasmonic rulers measure and monitor dynamic distance changes over 1–100 nm length scale in biology.[137–139] Alivisatos[9] and coworkers firstly verified that plasmon coupling could be used to continuously monitor separations of gold–silver dimer up to 70 nm for >3,000 s. In Ginger's work, plasmonic rulers were implemented by using reconfigurable DNA nanostructure, whose length extended when the complementary strands were introduced, and a visible blue-shift displayed in DFM reveal the nanostructure change.[140]

5.2.3. Nanostructures complex than dimer

One promising method for increasing the sensitivity of LSPR sensor is to apply the plasmon hybridization mentioned above. The solution to fabricate plasmon hybridization is to assemble plasmonic nanoparticles into a multicomponent nanoarchitecture, dubbed nanoantenna.[118,141] For example, Larsson and Alivisatos fabricated nanoantenna composed of Au plasmonic nanoparticles and Pd nanocatalysts by using nanolithography, on the basis of which they plasmonically studied gas-sensing and gas-phase catalysis.[142] Kinkhabwala *et al.* showed a 1340-fold fluorescence enhancement for a single molecule in a gold bowtie nanoantenna with 10 nm spacing.[143] Fan *et al.* report a DNA-based bottom-up approach for constructing a programmable nanohalo architecture that indirectly monitor solution-phase catalysis at single-particle level. Especially, the plasmonic L–AuNPs and catalytic S–AuNPs are brought to close proximity through DNA self-assembly. The locally enhanced electromagnetic field in the gap between L– and S–AuNPs allows indirect and sensitive monitoring of the catalytic reaction, which provides ready extraction of physicochemical parameters for quantitative understanding of AuNP-based catalysis.[143,144]

5.3. Localized Surface Plasmon Resonance Imaging in Cells

The distance-dependent electromagnetic coupling of nanoparticles provides a unique approach for real-time monitoring nanoparticles assembly and subsequent trafficking during nanoparticles ternalization. Reinhard *et al.* used this method to investigate the interactions of negatively charged AuNPs with J774A.1 muring macrophages.[14] They demonstrated that AuNPs uptake is mediated by scavenger receptors, and this internalization is associated with a time- and concentration-dependent nanoparticles assembling process. Detailed study showed that both caveolin- and clathrin-dependent endocytosis domain this uptake process. The clathrin-dependent endocytosis leads to non-agglomerated nanoparticles uptake within the cells while caveolin-dependent endocytosis induces the formation of compact nanoparticles clusters. Sokolov *et al.* applied DFM to track anti-EGFR AuNPs, and found a progressive red-shift greater than 100 nm in the nanoparticle plasmon resonance wavelength over a time period of 60 min.[145] They demonstrated that changes in the scattering peak are associated with EGFR trapped by the cell membrane, early endosomes, and late endosomes. Combining these experiment data with statistical analysis of RGB-channel, they create a real-time map of EGFR regulatory states in living cells.

In addition, there are many other works on the determination of noble nanoparticle interacting with cellular, for example, a microscopic method was developed to quantify the number of nanoparticle and the degree of their aggregation in the cells.[146,147] The major advantage of this method is to eliminate the influence of nanoparticles binding to the outer surface of the cells. Fan *et al.* also reported some work on the interaction between AuNPs and cells, such as the uptake methods, the transport pathways and the regulation by DNA nanotechnology.[148]

Immunolabeled nanoparticles are nanoparticles with one or more antibodies on their surface for recognizing specific targets. El-Sayed's *et al.* chose non-malignant epithelial cell line (HaCaT) and two malignant oral epithelial cell line (HOC 313 clone 8 and HSC 3) as model cells, they incubated them with monoclonal antiepidermal growth factor receptor (anti-EGFR) antibodies modified AuNPs, then recorded SPR scattering images and SPR absorption spectra of the cells.[149] Compared with AuNPs control, anti-EGFR antibody modified AuNPs specifically and homogeneously bind to the surface of cancer cells. Furthermore, Wax *et al.* used anti-EGFR antibody modified AuNPs to quantify the EGFR expression in different cell lines with DFM.[5] They demonstrate that this method perform fluorescence assays in quantitatively measuring EGFR expression *in vitro*, providing the information of local refractive index through their scattering spectra. Following that, Reinhard's *et al.* worked onto investigate immunolabeled nanoparticles.[150–152] They designed a secondary antibody labeling strategy that amplifies the number of binding sites per EGFR on the cell surface and introduces additional structure flexibility in the attachment chemistry. The quantitatively obtained experimental relationship between the scattering peak wavelength measured by DFM and the immunolabel density on the cell surface was recorded by SEM. They applied this method to monitor the size and dynamics of ErbB1-enriched membrane domains, quantify differential ErbB1 and ERbB2 cell surface expression and analyze spatial nanoclustering of the lateral organization of Cell-Surface CD24 and CD44.

As noted, AuNPs, nanorods, and nanoshells with unique properties have shown great potential in anticancer treatment. A vast amount of work studying the effect of AuNPs on the cells have been done.[153–157] In order to selectively transport the AuNPs into the cancer cell nucleus, El-Sayed *et al.*[153,158] chose RN AuNPs as research model, RN AuNPs stand for AuNPs bioconjugated with an arginine–glycine–aspartic acid peptide (RGD) and a nuclear localization signal (NLS) peptide. With the help of DFM, cytokinesis arrest was observed in real time. They found that nanomaterials in the cell nucleus could specifically affect the cellular function, and 0.4 nM RN AuNPs could affect the extending of cytoplasmic bridge, and then leads to cell apoptosis. Following that, a detailed work on concentration-dependent has been done. They found that cancer cell communities have cellular attraction, clustering, and bystander killing at low concentrations, while the degree of attraction and clustering is diminished and most cells in community died at high nanoparticle concentrations.

6. Other Application

Besides the nanoparticle's plasmic property, its optical property, such as Förster resonance energy transfer (FRET), surface enhanced Raman scattering (SERs) and nonlinear optical properties, also have been applied to construct biosensors for biosensing and bioimaging. For example, with the help of its high surface area to volume ratio, a serial of biomolecules and biomarkers can be conjugated to a nanoparticle, constructing a multifunction probe for detecting various material, ranging from small organic molecules to proteins.

6.1. Energy Transfer

FRET technology is widely applied in detecting dynamic changes at single molecule level. The non-radiative transfer of excitation energy between donor and acceptor interacts by dipole–dipole when distance between them is smaller than the Förster distance (typically less than 10 nm).[159,160] To trigger FRET, the emission frequency of the donor should match the absorption frequency of the acceptor.[161–163] Due to their large molar extinction coefficients, AuNPs can be used as acceptors to quench the donors.[164] When metal nanoparticles are used as acceptors, the energy transfer phenomenon is named NSET. While AuNPs also can be used as a donor to be quenched by absorbed molecular acceptors this energy transfer form is named as PRET.[165]

6.1.1. Plasmon resonance energy transfer

When the plasmon resonance spectrum of a plasmon sensor intentionally matches the absorption spectrum of chemical or biological molecules adsorbed on their surface, energy transferred by PRET induces a wavelength-dependent quenching in the scattering profile.[166] Due to the strong scattering intensity in the visible region, AuNPs are considered to be perfect material for the studying of PRET. For instance, As the amine complex of $[Cu(TMSen)]^2$ has optical absorption peaks in the visible range (around 550 nm), AuNPs with 50 nm in diameter, of which scattering peak lay around 550 nm, was used to form nanoplasmonic probe, this nanoplasmonic probe is shown in Fig. 6. This probe with ethylenediamine moiety modified

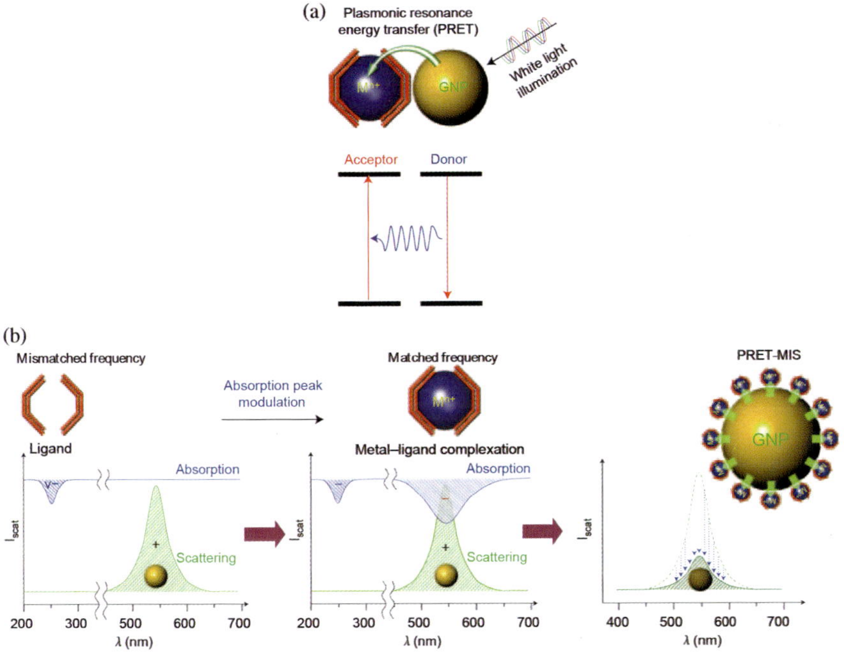

Fig. 6. Plasmonic resonance energy transfer-based metal ion sensing (PRET–MIS). (a) When the transition metal ion (blue) binds with the matching ligand (red), owing to the new absorption band, transferred to the metal–ligand complex. (b) There is no spectral overlap between ligands without the metal ion and the GNP (left). When the electronic absorption frequency of the metal–ligand complex matches with the Rayleigh scattering frequency, the selective energy transfer is induced by this spectral overlap (middle) and the distinguishable resonant quenching on the resonant Rayleigh scattering spectrum is observed (right). Reprinted with permission from Ref. 161 Copyright 2009, NPG.

on its surface was used to selectively recognize Cu^{2+} down to 1 nM over K^+, Na^+, Ca^{2+}, Mg^{2+}, Ni^{2+}, Co^{2+}, in the aqueous solutions.[161] Another example, AuNPs transfer their energy to adsorbed cytochrome c, and "spectral dips" are displayed in the single nanoparticle scattering spectrum, where the positions of spectral dips match the cytochrome c absorption peak positions of about 550 nm.[167] Meanwhile, since nanoplasmonic optical sensor do not photobleach or blink, real time production of cytochrome c in living HepG2 cells has been dynamically imaged for long term using PRET spectroscopy.[162] Highly sensitive and selective metal ion sensing has also been enabled by PRET spectroscopy.[163,168] Besides gold nanosphere, nanorods are also applied as nanoplasmonic sensors to monitor the opening process in molecular beacon.[165]

In addition to offering high spatial resolution owing to the small nanometer-scale size of the biosensor, this method is 100–1,000 times more sensitive and much more faster than organic reporter-based methods.[166]

6.1.2. Nanometal surface energy transfer

In order to describe the rate of energy transfer in NEST from a dipole to a metallic surface interband transition, Chance *et al.* used a formula, which is similar to FRET, however, Persson and Lang extend this formula and create a surface energy transfer rate.[169–172] Thus, energy transfer to a surface follows a very different distance trend and magnitude of interaction. In brief, in NEST, energy transfer distances are much more longer than typical Förster distance (typically double), and quenching of dyes of different emission frequencies forms the visible range to the NIR and could be realized by the same nanoparticles rather than different dyes.[164]

To bring the NEST sensor into the application, metal nanoparticles quench the fluorescence of the fluorophores adsorbing on the surface by chemical binding or electrostatic interaction at first. Then, the target molecules bind selectively to the probe and, in turn, the fluorescence restores. The restored intensity is proportional to the concentration of the target molecular structure. For instance, Ray *et al.*

developed a NEST sensor to monitor the cleavage of ds-DNA by nucleases.[164] A fluorphore-labeled ss-DNA was first modified on the surface of AuNPs. Due to the flexiblity of ss-DNA, it loops back and adsorbs to the surface of the AuNPs forming an arch-like structure, and fluorescence of ss-DNA was quenched by AuNPs. Then, target DNA was added to hybrid with ss-DNA, forming ds-DNA, and the fluorescence signal was enhanced by a factor of 120 after the cleavage of this ds-DNA by S1 nuclease. Meanwhile, Mirkin's *et al.* developed a NEST-based sensor for mRNA detection in living cells, also called nanoflare sensor (Fig. 7(a)). In this system, AuNPs were modified with oligonucleotides, which could hybridize with short dye-terminated fluorescent reporter sequence, and the fluorescence was dark due to AuNP quenching. When target mRNA appears, the recognized oligonucleotides hybridize with the complementary target sequence and

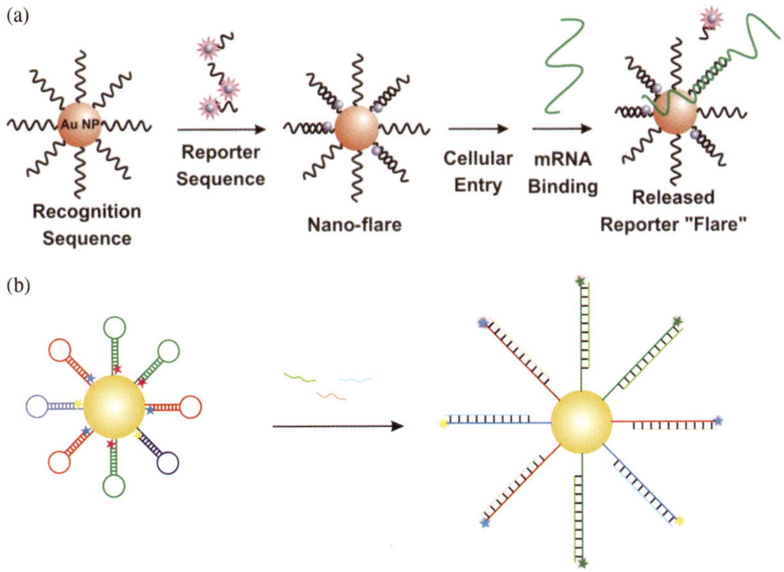

Fig. 7. Several representative AuNP-based activatable assays for DNA detection: (a) Target mRNA is capable of displacing the short Cy5 labeled reporter strand from the hybridized sequence on AuNPs, inducing the fluorescence recovery. Reprinted with permission from Ref. 173. Copyright 2007, American Chemical Society. (b) AuNPs self-assembled with multicolor stem–loop probes.

the fluorescence was bright like a flare.[173] Later, this system was extended for multiplexed detection of different targets simultaneously in living cells.[174–176] Besides nanoflare, NEST sensor based on molecular beacon, composed of a loop structure and a stem structure,[177,178] has been employed. In the hairpin structure, the nucleotides at each terminus, where the fluorophore and quencher are anchored respectively, are complementary to each other, resulting in the quenched fluorophore. Once the target nucleic acid hybridize with molecular beacon, the stem structure opens and the fluorescence recovers. AuNPs behave excellently in molecular beacons due to their high quenching efficiency. Dubetret *et al.* used AuNPs to assist molecular beacon to detect DNA targets, they found that sensitivity had improved 100 times compared to conventional molecular beacon.[179] Fan *et al.*[180,181] reported a similar probe to detect three different tumor-suppressor genes (Fig. 7(b)). Additionally, the AuNPs-assisted molecular beacon demonstrate a single-base mismatch selectivity of 25:1, so they hold great potential in multiplexed detection of single-base mismatch.

Generally, small AuNPs (with size less than 20 nm) are more often used in NEST, as absorption is dominant over scattering. While large AuNPs are more often used in PRET as scattering is dominant above 600 nm.[182]

6.2. Surface-Enhanced Raman Scattering

Raman effect originates from the inelastic scattering when light interacts with matter, and it means that an energy exchange occurs between the incident photons and the scattering substance. Like NIR spectrum, the Raman spectrum of different molecules is so unique that it has also been called a "Raman fingerprint".[183] A major limitation of traditional Raman spectroscopy is that its SNR ratio is too low, especially in *in vivo* imaging with short acquisition time. SERs is the most crucial solution to amplify Raman intensity.[184] SERs is a plasmonic effect in which the Raman intensity of a molecule is enhanced enormously (up to 10^{14}-fold) when adsorbed on a rough metal (e.g. Ag or Au) surface with high curvature, such as a nanoparticle.[185,186] Both electromagnetic (EM) enhancement and chemical enhancement are behind the increased inelastic light scattering in SERS. The EM mechanism arise from LSPR of plasmon nanoparticles and is the dominant effect, while chemical enhancement mechanism come from molecular charge-transfer interactions between the molecule and the metallic surface.[187,188] Since the SERs signal intensity is dominant by EM, a common aim among many groups has been to develop substrate with great EM field. These substrates contain single sharp structures like nanocresents, nanostars or couple structure with controllable gap.[189] Graham *et al.*[190] developed a selective SERS detection method to locate Raman dye within the interparticle gap region between plasmonic nanostructures. The appearance of target in this method selectively and reversibly turn the Raman signal on or off. Nam's *et al.*[191] improves these methods by controlling interparticle gap at the nanometer scale. They reported a high-yield synthetic method for preparing SERS-active dimeric gold–silver core–shell with a nanodumbbell structure (Fig. 8(a)). In this structure, nanometer silver shell was formed to engineer the nanogap between AuNPs

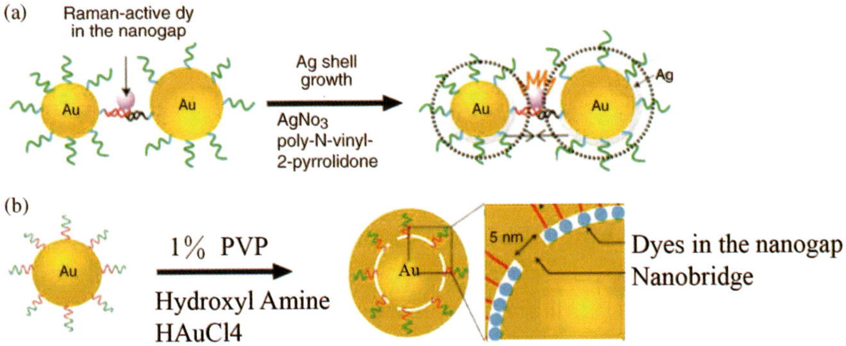

Fig. 8. Schematic SERS-based detection methods: (a) Nanogap-engineerable nanostructures for single molecules detection by SERS. In this method, Ag shell can be controlled at nanometer scale to finely engineer the Raman-signal-amplifying plasmonic nanogap between two core–shell particles. (Reprinted with permission from Ref. 191. Copyright 2009, NPG). (b) A scheme of gold nanostructures containing approximately 1-nm interior nanogap. Raman dyes can be precisely located within the gap and the number of modified dyes can be controlled. Reprinted with permission from Ref. 192. Copyright 2011, NPG.

nanodumbbell. This interparticle nanogap structure offers high enhancement factor (EF) values (>10^{12}) to achieve single-molecule sensitivity. To overcome the limit of silver layer deposition, Nam's *et al.*[192] propose creative DNA-based monomeric gold nanobridged nanogap (Au–NNPs) structures (Fig. 8(b)) with approximately 1-nm interior gap. SERs signals generated by Au–NNPs showed a linear dependence on probe concentration (R_2 > 0.98) and were sensitive down to 10 fM concentrations. Yields of Au–NNPs is higher than 95%, and >90% of Au–NNPs had EFs greater than 10^8 which is sufficient for single-molecular detection.

SERs tag, referred to as SERS-encoded nanoparticles, is composed of the SERs substrate, usually single plasmonic nanoparticles or coupled plasmonic nanoparticles, and reporter, usually a molecular label with high Raman cross-section, and an acceptor which is capable of binding the analyst.[193–195]

SERs tag could be designed to serve as a pH marker of a cellular microenvironment, especially for the tumor microenvironment since acidity is the hallmark for solid tumors due to the high glucose metabolism rate and poor vascular perfusion. Vo-Dinh's *et al.*[196] developed a pH sensing SERS nanoprobe based on gold Nanostructures and para-mercaptobenzoic acid (pMBA) as the reporter, the SERS peak intensity at 1700 cm^{-1} decreases while SERS intensity of peaks at 1014 cm^{-1}, 1136 cm^{-1}, or 1390 cm^{-1} increases when pH changes from 5 to 9. Therefore, pH sensor is also used to probe an image of pH in live cells with subcellular resolution.[197]

A lot of biomolecules (e.g. peptides, proteins, antibodies, antibody fragments, DNA) could be conjugated with SERs substrate for molecular imaging application. For example, anti-EGFR antibody/antibody single-chain variable fragment has been modified on AuNP to detect EGFR in human cancer cells and xenograft tumor models.[198–200] By using fluorescent SERS nanoparticles (F–SERs dots), dynamic physiological process of cells, like apoptosis, could be monitored. F-SERs dots conjugate annexin V, anti-Bad or anti-Bax antibodies were used for multiparametric analysis of apoptosis. It has been proved that F–SERs dots have little toxicity and could analysis apoptosis in both cells and tissues by both fluorescent and Raman signal effectively and simultaneously. Due to the simple spectra, this sensor do not generate enhanced fluorescence background and is commercially available at low cost of small molecular reporter,[201,202] therefore, they are widely used to fabricate multiplex SERs nanoprobes.[203,204] F–SERs dots mentioned above have been employed to detect the expression level of three proteins (CD34, Sca-1, and SP-C) simultaneously in bronchioalveolar stem cells in the murine lung. Follow-up researchers explore the superb multiplexing capability of Raman spectroscopy, and they successfully separated 10 different SERs nanoparticles after subcutaneous injection into a nude mouse.

6.3. Nanocarriers

Gold nanostructures are attractive as nanocarriers due to their non-toxic and high biocompatibility properties, such as tunable control in a variety of shapes and ease of surface modification. Investigation of the acute and subacute oral toxicity of AuNPs in rats[205,206] for 28 days by Pokharkar *et al.* and Zhang *et al.* show no subacute toxicities, such as changes in clinical signs, body weight, food consumption, hematological parameters, organ weights and histopathological changes, were observed at lower doses. However, a decrease in the red blood cells (RBCs) count was observed with increased accumulation in spleen at higher doses. Thus, gold nanostructures could serve as carriers and are carefully considered and prepared to prevent particle aggregation, non-specific protein adsorption, and the uptake of circulating gold nanostructures by the reticuloendothelial system, to prolong circulation in the blood stream and to increase the accumulation of gold nanostructures in tumors through passive targeting. Liang *et al.* demonstrated that small (2 nm) AuNPs functionalized with therapeutic peptides (PMI) and targeting peptides (CRGDK) exhibited strong anticancer activity *in vitro*.[207]

Gold nanostructure are also capable of regulating drug release by outside stimulus, such as intracellular GSH changes, pH changes, light and heat. Vivero-Escoto *et al.* has developed a photo-induced controlled release of pacilitaxel. Upon photoirradiation, the photocleavable linker covalently attached to the gold nanospheres, is protected, was cleaved, resulting in drug release.[208] Prabaharan *et al.* produces a pH-triggered drug release nanocarrier.[209] In this nanocarrier, DOX was covalently conjugated to surface via an acid-cleavable hydrazon linkage, and the release rates of conjugated DOX were greater at pH 5.3 and 6.6 than at pH 7.4.[209] Compared to nanosphere, nanorod significantly shifts their LSPR band by changing the aspect ratios. Therefore, nanorod with different aspect ratios could be applied to selective triggered release of multiple drugs.

In addition to drug loading on the nanoparticle surface, drug encapsulation in the nanoparticle also exhibit high therapeutic efficacy.[210] Gold nanocage have been proposed for controlling drug release. Yavuz *et al.* applied PNiPAAm modified nanocage to control drug release with NIR laser irradiation. PNiPAAm changes this conformation in response to small variations in temperature, when the modified gold nanocages absorb incident light and convert into heat, resulting in an increase in temperature and changing the polymer conformation to expose the pores. This process leads to release of preloaded effectors. Once the light was turned off, the polymer returned to its original state and stopped the drug release.[211] Chan *et al.* developed a modular nanoparticle-delivery system that uses DNA to assemble nanoparticles into "core-satellite" architecture, and DOX inserted into ds-DNA between the core and satellite. They demonstrate that their nanocarrier have 2.1-fold increase in therapeutic effect.[212]

6.4. Photodynamic Therapy/Photothermal Therapy

Due to their ability to accumulate in tumor cells *in vivo* rather than in normal tissues, AuNPs pose great promises in photodynamic therapy (PDT) and PTT.

PDT is a non-invasive, highly selective, effective method for the destruction of unwanted cells and tissues and is widely applied in treating oncological diseases[213–216] such as certain dermal or infectious diseases. Plasmonic particles-assisted PDT involves light-sensitive agents — photosensitizers (PS) adsorbing on the surface of particles, and plasmonic nanoparticles passively targeting tumor cells. After NIR light activation, PS generates highly cytotoxic reactive oxygen species (ROS), particularly singlet oxygen (O_2), which triggers cell apoptosis and necrosis.[217] For instance researchers used AuNRs–AlPCS4 complexes for PDT. AuNRs–AlPCS4 complexes were injected to mouse, and PS were released from the surface of AuNRs and showed highly phototoxic features with irradiation of NIR light in tumor region after passive accumulation by enhancing permeability and retention effect (EPR), while non-phototoxic effects were observed in the blood circulation. The tumor-to-background ratio increased and tumor growth was decreased by 79% with PDT.[218] In addition to the assisted application of plasmon nanoparticles, spherical AuNPs can be directly used as PS. Zhao *et al.* in 2012 reported that 808 nm fs pulsed laser excitation polyvinylpyrrolidone (PVP) coated AuNRs could generate ${}_1O_2$ and induce significant cell death.[219]

PTT employed photothermal agents are applied in selective heating of the local environment to produce cells or for tissue destruction. AuNPs, particularly AuNRs, intensively absorb the NIR light range due to LSPR to heat themselves and warm up local tumor. This kind of PTT involved in plasmon nanoparticle is proposed as plasmonic PTT (PPTT). El-Sayed *et al.*[200] demonstrated with AuNRs for selective PPTT using CW NIR lasers, they found that half the laser energy (10 W/cm^2) was required to damage the cancer cells compared with normal cells (20 W/cm^2). Various work on PPTT-used plasmon nanoparticle demonstrate that the optimal wavelength and intensity of irradiation of NIR light could be tuned by controlling its shape, size and particles components.[220,221] Most notably, the work by Halas *et al.* showed that laser dosages is required to induce tumor damages to drop to 4–10% for traditional PTT, which could minimize the damage to normal tissue.[222,223]

References

1. Anker JN, *et al.* Biosensing with plasmonic nanosensors. *Nat. Mater.* 7(6), 442–453 (2008).
2. Mayer KM, Hafner JH. Localized surface plasmon resonance sensors. *Chem. Rev.* **111**(6), 3828–3857 (2011).

3. Mie G. Beiträge zur Optik trüber Medien, speziell kolloidaler Metallösungen. *Annalen der Physik*. **330**(3), 377–445 (1908).
4. Gans R. Über die Form ultramikroskopischer Goldteilchen. *Annalen der Physik*. **342**(5), 881–900 (1912).
5. Bohren CF, Huffman DR. *Absorption and scattering of light by small particles*. Wiley: New York. **530** (1983).
6. Sardar R, *et al.* Gold nanoparticles: Past, present, and future. *Langmuir* **25**(24), 13840–13851 (2009).
7. Rodriguez-Fernandez J, *et al.* Spectroscopy, imaging, and modeling of individual gold decahedra. *J. Phys. Chem. C.* **113**(43), 18623–18631 (2009).
8. Sheikholeslami S, *et al.* Coupling of optical resonances in a compositionally asymmetric plasmonic nanoparticle dimer. *Nano Lett.* **10**(7), 2655–2660 (2010).
9. Sönnichsen C, *et al.* A molecular ruler based on plasmon coupling of single gold and silver nanoparticles. *Nat. Biotechno.* **23**(6), 741–745 (2005).
10. Brandl DW, Mirin NA, Nordlander P. Plasmon modes of nanosphere trimers and quadrumers. *J. Phys. Chem. B.* **110**(25), 12302–12310 (2006).
11. Brandl DW, Oubre C, Nordlander P. Plasmon hybridization in nanoshell dimers. *J. Chem. Phys.* **123**(2) (2005).
12. Mirin NA, Bao K, Nordlander P. Fano resonances in plasmonic nanoparticle aggregates. *J. Phys. Chem. A.* **113**(16), 4028–4034 (2009).
13. Nordlander P, *et al.* Plasmon hybridizaton in nanoparticle dimers. *Nano Lett.* **4**(5), 899–903 (2004).
14. Reinhard BM, *et al.* Calibration of dynamic molecular rulers based on plasmon coupling between gold nanoparticles. *Nano Lett.* **5**(11), 2246–2252 (2005).
15. Yang LL, *et al.* Calibration of silver plasmon rulers in the 1–25 nm separation range: Experimental indications of distinct plasmon coupling regimes. *J. Phys. Chem. C.* **114**(11), 4901–4908 (2010).
16. Mayer KM, *et al.* A label-free immunoassay based upon localized surface plasmon resonance of gold nanorods. *Acs Nano.* **2**(4), 687–692 (2008).
17. Sun Y, Xia Y,. Increased sensitivity of surface plasmon resonance of gold nanoshells compared to that of gold solid colloids in response to environmental changes. *Anal. Chem.* **74**(20), 5297–5305 (2002).

18. Mock JJ, *et al.* Shape effects in plasmon resonance of individual colloidal silver nanoparticles. *J. Chem. Phys.* **116**(15), 6755–6759 (2002).
19. Chen H, *et al.* Shape- and size-dependent refractive index sensitivity of gold nanoparticles. *Langmuir* **24**(10), 5233–5237 (2008).
20. Sherry LJ, *et al.* Localized surface plasmon resonance spectroscopy of single silver nanocubes. *Nano Lett.* **5**(10), 2034–2038 (2005).
21. Kyeong-Seok Lee and Mostafa A. El-Sayed. Gold and Silver Nanoparticles in Sensing and Imaging: Sensitivity of Plasmon Response to Size, Shape, and Metal Composition. *J. Phys. Chem.* **110**(39), 19220–19225 (2006).
22. Link S, EI-Sayed MA. Size and temperature dependence of the plasmon absorption of colloidal gold nanoparticles. *J. Phys. Chem. B.* **103**(21), 4212–4217 (1999).
23. Bozhevolnyi SI, *et al.* Channel plasmon-polariton guiding by subwavelength metal grooves. *Phys Rev. Lett.* **95**(4), 046802 (2005).
24. Willets KA, Van Duyne RP. Localized surface plasmon resonance spectroscopy and sensing. *Annu. Rev. Phys. Chem.* **58**, 267–297 (2007).
25. Henry AI, *et al.* Correlated structure and optical property studies of plasmonic nanoparticles. *J. Phys. Chem. C.* **115**(19), 9291–9305 (2011).
26. Hong S, *et al.* Au nanodisk-core multishell nanoparticles: Synthetic method for controlling number of shells and intershell distance. *Chem. Mater.* **26**(12), 3618–3623 (2014).
27. Wiley B, *et al.* Shape-controlled synthesis of metal nanostructures: the case of silver. Chemistry (Weinheim an der Bergstrasse, Germany), **11**(2), 454–463 (2005).
28. Wiley BJ, *et al.* Maneuvering the surface plasmon resonance of silver nanostructures through shape-controlled synthesis. *J. Phys. Chem. B.* **110**(32), 15666–15675 (2006).
29. He P, Zhu X. Phospholipid-assisted synthesis of size-controlled gold nanoparticles. *Mater. Res. Bull.* **42**(7), 1310–1315 (2007).
30. Hostetler MJ, *et al.* Alkanethiolate gold cluster molecules with core diameters from 1.5 to 5.2 nm: core and monolayer properties as a function of core size. *Langmuir* **14**(1), 17–30 (1998).
31. Pillai ZS, Kamat PV. What Factors Control the Size and Shape of Silver Nanoparticles in the Citrate Ion Reduction Method? *J. Phys. Chem. B.* **108**(3), 945–951 (2004).

32. Sun Y, *et al.* Seedless and templateless synthesis of rectangular palladium nanoparticles. *Chem. Mater.* **19**(8), 2065–2070 (2007).
33. Haynes CL, *et al.* Angle-resolved nanosphere lithography: Manipulation of nanoparticle size, shape, and interparticle spacing. *J. Phys. Chem. B.* **106**(8), 1898–1902 (2002).
34. Dionne JA, Lezec HJ, Atwater HA. Highly confined photon transport in subwavelength metallic slot waveguides. *Nano Lett.* **6**(9), 1928–1932 (2006).
35. Bozhevolnyi SI, *et al.* Channel plasmon subwavelength waveguide components including interferometers and ring resonators. *Nature* **440**(7083), 508–511 (2006).
36. Yin L, *et al.* Subwavelength focusing and guiding of surface plasmons. *Nano Lett.* **5**(7), 1399–1402 (2005).
37. Ozbay E. Plasmonics: Merging photonics and electronics at nanoscale dimensions. *Science* **311**(5758), 189–193 (2006).
38. Khomutov GB, Koksharov YA. Effects of organic ligands, electrostatic and magnetic interactions in formation of colloidal and interfacial inorganic nanostructures. *Adv. Colloid Int. Sci.* **122**(1–3), 119–147 (2006).
39. Roucoux A, Schulz J, Patin H. Reduced transition metal colloids: A novel family of reusable catalysts? *Chem. Rev.* **102**(10), 3757–3778 (2002).
40. Turkevich J, Stevenson PC, and Hillier J. A Study of the nucleation and growth processes in the synthesis of colloidal gold. *Discussions of the Faraday Society* **1951**(11), 55–58.
41. Frens G. Controlled nucleation for regulation of particle-size in monodisperse gold suspensions. *Nature-Physical Science* **241**(105), 20–22 (1973).
42. Bastus NG, Comenge J, and Puntes V. Kinetically controlled seeded growth synthesis of citrate-stabilized gold nanoparticles of up to 200 nm: size focusing versus Ostwald ripening. *Langmuir* **27**(17), 11098–11105 (2011).
43. Polte J, *et al.* Mechanism of gold nanoparticle formation in the classical citrate synthesis method derived from coupled *in situ* XANES and SAXS evaluation. *J. Am. Chem. Soc.* **132**(4), 1296–1301 (2010).
44. Xia H, *et al.* Synthesis of monodisperse quasi-spherical gold nanoparticles in water via silver(I)-assisted citrate reduction. *Langmuir*: the ACS journal of surfaces and colloids, **26**(5), 3585–3589 (2010).
45. Rodríguez-González B, Mulvaney P, Liz-Marzán Luis M. An electrochemical. model for gold colloid formation via citrate reduction, in *Zeitschrift für Physikalische Chemie*. 415–426 (2007).

46. Ojea-Jiménez I, Bastús NG, and Puntes V. Influence of the sequence of the reagents addition in the citrate-mediated synthesis of gold nanoparticles. *J. Phys. Chem. C.* **115**(32), 15752–15757 (2011).
47. Kimling J, *et al.* Turkevich method for gold nanoparticle synthesis revisited. *J. Phys. Chem. B.* **110**(32), 15700–15707 (2006).
48. Brust M, *et al.* Synthesis of thiol-derivatised gold nanoparticles in a two-phase Liquid-Liquid system. *J. Chem. Soc. Chem. Comm.* **1994**(7), 801–802.
49. Chen S. 4-Hydroxythiophenol-protected gold nanoclusters in aqueous media. *Langmuir* **15**(22), 7551–7557 (1999).
50. Templeton AC, Wuelfing WP, and Murray RW. Monolayer-protected cluster molecules. *Accounts of Chemical Research* **33**(1), 27–36 (2000).
51. Ingram RS, Hostetler MJ, and Murray RW. Poly-hetero-ω-functionalized alkanethiolate-stabilized gold cluster compounds. *J. Am. Chem. Soc.* **119**(39), 9175–9178 (1997).
52. Chen S, and Murray RW. Arenethiolate Monolayer-Protected Gold Clusters. *Langmuir* **15**(3), 682–689 (1999).
53. Hostetler MJ, *et al.* Monolayers in three dimensions: synthesis and electrochemistry. of ω-functionalized alkanethiolate-stabilized gold cluster compounds. *J. Am. Chem. Soc.* **118**(17), 4212–4213 (1996).
54. Templeton AC, *et al.* Water-soluble, isolable gold clusters protected by tiopronin and coenzyme a monolayers. *Langmuir* **15**(1), 66–76 (1999).
55. Lohse SE, Dahl JA, and Hutchison JE. Direct synthesis of large water-soluble functionalized gold nanoparticles using bunte salts as ligand precursors. *Langmuir* **26**(10), 7504–7511 (2010).
56. Itoh H, *et al.* Photochemical assembly of gold nanoparticles utilizing the photodimerization of thymine. *Langmuir* **20**(5), 1972–1976 (2004).
57. Shon Y-S, *et al.* Alkanethiolate-protected gold clusters generated from sodium s-dodecylthiosulfate (*Bunte Salts*). *Langmuir* **16**(16), 6555–6561 (2000).
58. Price RC, and Whetten RL. All-aromatic, nanometer-scale, gold-cluster thiolate complexes. *J. Am. Chem. Soc.* **127**(40), 13750–13751 (2005).
59. Boal AK, *et al.* Self-assembly of nanoparticles into structured spherical and network aggregates. *Nature* **404**(6779), 746–748 (2000).
60. Jana NR, Gearheart L, and Murphy CJ. Seed-mediated growth approach for shape-controlled synthesis of spheroidal and rod-like gold nanoparticles using a surfactant template. *Adv. Mater.* **13**(18), 1389–1393 (2001).

61. Jiang XC, Brioude A, and Pileni MP. Gold nanorods: Limitations on their synthesis and optical properties. *Colloids and Surfaces A: Physicochemical and Engineering Aspects.* **277**(1–3), 201–206 (2006).
62. Nikoobakht B, and El-Sayed MA. Preparation and growth mechanism of gold nanorods (NRs) using seed-mediated growth method. *Chem. Mater* **15**(10), 1957–1962 (2003).
63. Pérez-Juste J, *et al.* Electric-Field-Directed Growth of Gold Nanorods in Aqueous Surfactant Solutions. *Adv. Functional Mater* **14**(6), 571–579 (2004).
64. Sau TK, and Murphy CJ. Seeded high yield synthesis of short Au nanorods in aqueous solution. *Langmuir: The ACS Journal of Surfaces and Colloids* **20**(15), 6414–6420 (2004).
65. Mukherjee S, *et al.* Hot-electron-induced dissociation of H2 on gold nanoparticles supported on SiO2. *J. Am. Chem. Soc.* **136**(1), 64–67 (2014).
66. Liu, and Guyot-Sionnest P. mechanism of silver(i)-assisted growth of gold nanorods and bipyramids. *J. Phys. Chem. B.* **109**(47), 22192–22200 (2005).
67. Sánchez CG, Del Pópolo MG, and Leiva EPM. An embedded atom approach to underpotential deposition phenomena. *Surface Science* **421**(1–2), 59–72 (1999).
68. Rojas MI, *et al.* Erratum to: An embedded atom approach to underpotential deposition phenomena: [Surf. Sci. 421 (1999) 59]. *Surface Science* **453**(1–3), 225–228 (2000).
69. TH H, Koo H-J, and Chung BH. Shape-controlled syntheses of gold nanoprisms and nanorods influenced by specific adsorption of halide ions. *J. Phys. Chem. C.* **111**(3), 1123–1130 (2007).
70. Evanoff DD, and Chumanov G, Size-controlled synthesis of nanoparticles. 2. measurement of extinction, scattering, and absorption cross sections. *J. Phys. Chem. B.* **108**(37), 13957–13962 (2004).
71. Merga G, *et al.* Redox Catalysis on "Naked" Silver Nanoparticles. *J. Phys. Chem. C.* **111**(33), 12220–12226 (2007).
72. Wiley B, *et al.* Shape-controlled synthesis of silver and gold nanostructures. *MRS Bull.* **30**(05), 356–361 (2005).
73. Oliveira MM, *et al.* Influence of synthetic parameters on the size, structure, and stability of dodecanethiol-stabilized silver nanoparticles. *J. Colloid Int. Sci.* **292**(2), 429–435 (2005).
74. Lee PC, and Meisel D. Adsorption and surface-enhanced Raman of dyes on silver and gold sols. *J. Phys. Chem.* **86**(17), 3391–3395 (1982).

75. Koh AL, *et al.* Electron energy-loss spectroscopy (EELS) of surface plasmons in single silver nanoparticles and dimers: influence of beam damage and mapping of dark modes. *Acs Nano.* **3**(10), 3015–3022 (2009).
76. Camden JP, *et al.* Probing the structure of single-molecule surface-enhanced Raman scattering hot spots. *J. Am. Chem. Soc.* **130**(38), 12616–12617 (2008).
77. Wiley BJ, *et al.* Synthesis and electrical characterization of silver nanobeams. *Nano Lett.* **6**(10), 2273–2278 (2006).
78. Xia Y, *et al.* Shape-controlled synthesis of metal nanocrystals: simple chemistry meets complex physics? Angewandte Chemie (International ed. in English), **48**(1), 60–103 (2009).
79. Wiley B, Sun Y, and Xia Y. Synthesis of silver nanostructures with controlled shapes and properties. *Acc. Chem. Res.* **40**(10), 1067–1076 (2007).
80. Tao A, Sinsermsuksakul P, and Yang P. Polyhedral silver nanocrystals with distinct scattering signatures. *Angewandte Chemie* (International ed. in English), **45**(28), 4597–4601 (2006).
81. Tao AR, Habas S, and Yang P. Shape control of colloidal metal nanocrystals. *Small* **4**(3), 310–325 (2008).
82. Sun Y, and Xia Y. ChemInform abstract: Large-scale synthesis of uniform silver nanowires through a soft, self-seeding, polyol process. *Chem. In.form* **33**(36), 833–837 (2002).
83. Silvert P-Y, Herrera-Urbina R, and Tekaia-Elhsissen K, Preparation of colloidal silver dispersions by the polyol process. *J. Mater. Chem.* 7(2), 293–299 (1997).
84. Feldmann C, Polyol-mediated synthesis of nanoscale functional materials. *Solid State Sci.* 7(7), 868–873 (2005).
85. Wiley B, *et al.* Polyol Synthesis of Silver Nanoparticles: Use of Chloride and Oxygen to promote the formation of single-crystal, truncated cubes and tetrahedrons. *Nano Lett.* **4**(9), 1733–1739 (2004).
86. Komarneni S, *et al.* Microwave–Polyol Process for Pt and Ag Nanoparticles. *Langmuir* **18**(15), 5959–5962 (2002).
87. And YW, and Xia Y. Bottom-up and top-down approaches to the synthesis of monodispersed spherical colloids of low melting-point metals. *Nano Lett.* **4**(10), 2047–2050 (2004).
88. Bonet F, *et al.* Synthesis of monodisperse Au, Pt, Pd, Ru and Ir nanoparticles in ethylene glycol. *Nanostructured Mater.* **11**(8), 1277–1284 (1999).

89. Fievet F, *et al.* Controlled nucleation and growth of micrometre-size copper particles prepared by the polyol process. *J. Mater. Chem.* **3**(3), 627–632 (1993).
90. Rycenga M, *et al.* Controlling the synthesis and assembly of silver nanostructures for plasmonic applications. *Chem. Rev.* **111**(6), 3669–3712 (2011).
91. Wang ZL. Transmission electron microscopy of shape-controlled nanocrystals and their assemblies. *J. Phys. Chem. B.* **104**(6), 1153–1175 (2000).
92. SH I, *et al.* Large-scale synthesis of silver nanocubes: the role of HCl in promoting cube perfection and monodispersity. *Angewandte Chemie.* **44**(14), 2154–2157 (2005).
93. Ito T, and Okazaki S. Pushing the limits of lithography. *Nature* **406**(6799), 1027–1031 (2000).
94. Kim S-E, *et al.* One-pot fabrication of various silver nanostructures on substrates using electron beam irradiation. *Nanotechnol.* **21**(7), 75302 (2010).
95. Reyntjens S, and Puers R. A review of focused ion beam applications in microsystem technology. *J. Micromech. Microen.* **11**(4), 287–300 (2001).
96. Stewart ME, *et al.* Nanostructured plasmonic sensors. *Chem. Rev.* **108**(2), 494–521 (2008).
97. Ru ECL, *et al.* Mechanisms of spectral profile modification in surface-enhanced fluorescence. *J. Phys. Chem. C.* **111**(44), 16076–16079 (2007).
98. Wen F, *et al.* Plasmon transmutation: Inducing new modes in nanoclusters by adding dielectric nanoparticles. *Nano Lett.* **12**(9), 5020–5026 (2012).
99. Fan JA, *et al.* Near-normal incidence dark-field microscopy: Applications to nanoplasmonic spectroscopy. *Nano Lett.* **12**(6), 2817–2821 (2012).
100. Luk'Yanchuk B, *et al.* The Fano resonance in plasmonic nanostructures and metamaterials. *Nat. Mater.* **9**(9), 707–715 (2010).
101. Cai Y, *et al.* Fabrication of elliptical nanorings with highly tunable and multiple plasmonic resonances. *Nano Lett.* **12**(9), 4881–4888 (2012).
102. Ament I, *et al.* Single unlabeled protein detection on individual plasmonic nanoparticles. *Nano Lett.* **12**(2), 1092–1095 (2012).
103. Herrmann LO, and Baumberg JJ. Watching single nanoparticles grow in real time through supercontinuum spectroscopy. *Small* **9**, 3743–3747 (2013).

104. Xiao LH, *et al.* Imaging translational and rotational diffusion of single anisotropic nanoparticles with planar illumination microscopy. *J. Am. Chem. Soc.* **133**(27), 10638–10645 (2011).
105. Xiao LH, *et al.* Determining the full three-dimensional orientation of single anisotropic nanoparticles by differential interference contrast microscopy. *Angewandte Chemie (International Edition)* **51**(31), 7734–7738 (2012).
106. Xiao LH, *et al.* Unsynchronized translational and rotational diffusion of nanocargo on a living cell membrane. *Angewandte Chemie-(International Edition)* **51**(17), 4181–4184 (2012).
107. Gu Y, *et al.* Three-dimensional super-localization and tracking of single gold nanoparticles in cells. *Anal. Chem.* **84**(9), 4111–4117 (2012).
108. Marchuk K, and Fang N. Three-dimensional orientation determination of stationary anisotropic nanoparticles with sub-degree precision under total internal reflection scattering microscopy. *Nano Lett.* **13**(11), 5414–5419 (2013).
109. Liu M, *et al.* Dark-field microscopy in imaging of plasmon resonant nanoparticles. *Colloids & Surfaces B. Biointerfaces* **124**, 111–117 (2014).
110. Hall WP, Ngatia SN, and Van Duyne RP. LSPR biosensor signal enhancement using nanoparticle-antibody conjugates. *J. Phys. Chem. C.* **115**(5), 1410–1414 (2011).
111. Nusz GJ, *et al.* Label-free plasmonic detection of biomolecular binding by a single gold nanorod. *Anal. Chem.* **80**(4), 984–989 (2008).
112. Haes AJ, *et al.* Localized surface plasmon resonance spectroscopy near molecular resonances. *J. Am. Chem. Soc.* **128**(33), 10905–10914 (2006).
113. Bellapadrona G, *et al.* Optimization of localized surface plasmon resonance transducers for studying carbohydrate-protein interactions. *Anal. Chem.* **5**(2), 748–760 (2012).
114. Liu GL, *et al.* A nanoplasmonic molecular ruler for measuring nuclease activity and DNA footprinting. *Nat. nanotechnol.* **1**(1), 47–52 (2006).
115. Hall WP, *et al.* A conformation- and ion-sensitive plasmonic biosensor. *Nano Lett.* **11**(3), 1098–1105 (2011).
116. Luo WJ, *et al.* Self-catalyzed, self-limiting growth of glucose oxidase-mimicking gold nanoparticles. *Acs Nano* **4**(12), 7451–7458 (2010).

117. Zheng XX, *et al.* Catalytic Gold Nanoparticles for Nanoplasmonic Detection of DNA Hybridization. *Angewandte Chemie-(International* Edition). **50**(50), 11994–11998 (2011).
118. Liu Q, *et al.* Nanoplasmonic detection of adenosine triphosphate by aptamer regulated self-catalytic growth of single gold nanoparticles. *Chem. Comm.* **48**(77), 9574–9576 (2012).
119. Zayats M, *et al.* Biocatalytic growth of Au nanoparticles: from mechanistic aspects to biosensors design. *Nano Lett.* **5**(1), 21–25 (2005).
120. Xiao Y, *et al.* Catalytic growth of Au nanoparticles by NAD(P)H cofactors: optical sensors for NAD(P)+-dependent biocatalyzed transformations. *Angewandte Chemie (International Edition)*, **43**(34), 4519–4522 (2004).
121. Willner I, Baron R, and Willner B. Growing Metal Nanoparticles by Enzymes. *Adv. Mater.* **18**(9), 1109–1120 (2006).
122. Zhang L, *et al.* Single gold nanoparticles as real-time optical probes for the detection of nadh-dependent intracellular metabolic enzymatic pathways. *Angewandte Chemie (International Edition)*. **50**(30), 6789–6792 (2011).
123. Xu Y, *et al.* Unraveling the role of hydrogen peroxide in α-synuclein aggregation using an ultrasensitive nanoplasmonic probe. *Anal. Chem.* **87**(3), 1968–1973 (2015).
124. Qin WW, *et al.* Visualizing dopamine released from living cells using a nanoplasmonic probe. *Nanoscale* 7(37), 15070–15074 (2015).
125. Peng T, *et al.* Nanoplasmonic imaging of latent fingerprints with explosive RDX residues. *Anal. Chem.* **87**(18), 9403–9407 (2015).
126. Elghanian R, *et al.* Selective colorimetric detection of polynucleotides based on the distance-dependent optical properties of gold nanoparticles. *Science* **277**(5329), 1078–1081 (1997).
127. Storhoff JJ, *et al.* What controls the optical properties of DNA-linked gold nanoparticle assemblies? *J. Am. Chem. Soc.* **122**(19), 4640–4650 (2000).
128. Storhoff JJ, *et al.* One-pot colorimetric differentiation of polynucleotides with single base imperfections using gold nanoparticle probes. *J. Am. Chem. Soc.* **120**(9), 1959–1964 (1998).
129. Wang LH, *et al.* Gold nanoparticle-based optical probes for target-responsive DNA structures. *Gold Bull.* **41**(1), 37–41 (2008).
130. Wang J, *et al.* A gold nanoparticle-based aptamer target binding readout for ATP assay. *Adv. Mater.* **19**(22), 3943–3946 (2007).

131. Li F, *et al.* Adenosine detection by using gold nanoparticles and designed aptamer sequences. *Analyst* **134**(7), 1355–1360 (2009).
132. Li K, *et al.* Nanoplasmonic imaging of latent fingerprints and identification of cocaine. *Angew. Chem.* (*Int Ed Engl*) **52**(44), 11542–11545 (2013).
133. Shi L, *et al.* Plasmon resonance scattering spectroscopy at the single-nanoparticle level: real-time monitoring of a click reaction. *Angew. Chem.* **125**(23), 6127–6130 (2013).
134. Waldeisen JR, *et al.* Disassembly of a core–satellite nanoassembled substrate for colorimetric biomolecular detection. *Acs Nano* **5**(7), 5383–5389 (2011).
135. Sannomiya T, Hafner C, and Voros J. *In situ* Sensing of Single Binding Events by Localized Surface Plasmon Resonance. *Nano Lett.* **8**(10), 3450–3455 (2008).
136. Wang Y, *et al.* Direct detection of microRNA based on plasmon hybridization of nanoparticle dimers. *Analyst.* **140**(4), 1140–1148 (2014).
137. Chen JIL, *et al.* Optical detection of protein in complex media with plasmonic nanoparticle dimers. *Small* 7(14), 1993–1997 (2011).
138. Jain PK, Huang WY, and El-Sayed MA. On the universal scaling behavior of the distance decay of plasmon coupling in metal nanoparticle pairs: A plasmon ruler equation. *Nano Lett.* 7(7), 2080–2088 (2007).
139. Liu GL, *et al.* A nanoplasmonic molecular ruler for measuring nuclease activity and DNA footprinting. *Nat. Nanotechnol.* **1**(1), 47–52 (2006).
140. Chen JIL, Chen Y, and Ginger DS. Plasmonic nanoparticle dimers for optical sensing of DNA in complex media. *J. Am. Chem. Soc.* **132**(28), 9600–9601 (2010).
141. Larsson EM, *et al.* Nanoplasmonic probes of catalytic reactions. *Science* **326**(5956), 1091–1094 (2009).
142. Tang ML, *et al.* Observations of shape-dependent hydrogen uptake trajectories from single nanocrystals. *J. Am. Chem. Soc.* **133**(34), 13220–13223 (2011).
143. Darvill D, Centeno A, and Xie F. Plasmonic fluorescence enhancement by metal nanostructures: shaping the future of bionanotechnology. *Physi. Chem. Chem. Phys.* **15**(38), 15709–15726 (2013).

144. K L, *et al.* DNA-directed assembly of gold nanohalo for quantitative plasmonic imaging of single-particle catalysis. *J. Am. Chem. Soc.* **137**(13), 4292–4295 (2015).
145. Aaron J, *et al.* Dynamic imaging of molecular assemblies in live cells based on nanoparticle plasmon resonance coupling. *Nano Lett.* **9**(10), 3612–3618 (2009).
146. Kumar S, *et al.* Plasmonic nanosensors for imaging intracellular biomarkers in live cells. *Nano Lett.* 7(5), 1338–1343 (2007).
147. Sokolov K, *et al.* Real-time vital optical imaging of precancer using anti-epidermal growth factor receptor antibodies conjugated to gold nanoparticles. *Cancer Res.* 2003, **63**(9), (1999).
148. Rosman C, *et al.* A new approach to assess gold nanoparticle uptake by mammalian cells: combining optical dark-field and transmission electron microscopy. *Small* **8**(23), 3683–3690 (2012).
149. Sayed El-IH, Huang X, and El-Sayed MA. Surface plasmon resonance scattering and absorption of anti-EGFR antibody conjugated gold nanoparticles in cancer diagnostics: applications in oral cancer. *Nano Lett.* **5**(5), 829–834 (2005).
150. Wang J, *et al.* Illuminating epidermal growth factor receptor densities on filopodia through plasmon coupling. *Acs Nano* **5**(8), 6619–6628 (2011).
151. Wang FK, *et al.* PEG-POSS Assisted facile preparation of amphiphilic gold nanoparticles and interface formation of Janus nanoparticles. *Chem. Comm.* **47**(2), 767–769 (2011).
152. Yan B, Boriskina SV, and Reinhard BM. Optimizing gold nanoparticle cluster configurations ($n \leq 7$) for array applications. *J. Phys. Chem. C. Nanomaterials and Interfaces* **115**(11), 4578–4583 (2011).
153. Austin LA, *et al.* Plasmonic imaging of human oral cancer cell communities during programmed cell death by nuclear-targeting silver nanoparticles. *J. Am. Chem. Soc.* **133**(44), 17594–17597 (2011).
154. Dreaden EC, *et al.* Tamoxifen-poly(ethylene glycol)-thiol gold nanoparticle conjugates: enhanced potency and selective delivery for breast cancer treatment. *Bioconjugate Chem.* **20**(12), 2247–2253 (2009).
155. Dreaden EC, *et al.* Beating cancer in multiple ways using nanogold. *Chem. Soc. Rev.* **40**(7), 3391–3404 (2011).
156. Huang B, Babcock H, and Zhuang X. Breaking the diffraction barrier: super-resolution imaging of cells. *Cell* **143**(7), 1047–1058 (2010).

157. Oyelere AK, *et al.* Peptide-conjugated gold nanorods for nuclear targeting. *Bioconjugate Chem.* **18**(5), 1490–1497 (2007).
158. Kang B, Mackey MA, and El-Sayed MA. Nuclear targeting of gold nanoparticles in cancer cells induces DNA damage, causing cytokinesis arrest and apoptosis. *J. Am. Chem. Soc.* **132**(5), 1517–1519 (2010).
159. Yeh HC, *et al.* Single-molecule detection and probe strategies for rapid and ultrasensitive genomic detection. *Current Pharmaceutical Biotechnol.* **6**(6), 453–461 (2005).
160. Radwan SH and Azzazy HM. Gold nanoparticles for molecular diagnostics. Expert Review of Molecular Diagnostics. **9**(5), 511–524 (2009).
161. Qu W-G, *et al.* Plasmonic resonance energy transfer-based nanospectroscopy for sensitive and selective detection of 2,4,6-trinitrotoluene (TNT). *Chem. Comm.* **47**(4), 1237–1239 (2011).
162. Choi Y, *et al.* Selective and sensitive detection of metal ions by plasmonic resonance energy transfer-based nanospectroscopy. *Nat. Nanotechnol.* **4**(11), 742–746 (2009).
163. Choi Y, Kang T, and Lee LP. Plasmon resonance energy transfer (PRET)-based molecular imaging of cytochrome c in living cells. *Nano Lett.* **9**(1), 85–90 (2009).
164. Ray PC, Fortner A, and Darbha GK. Gold nanoparticle based FRET assay for the detection of DNA cleavage. *J. Phys. Chem. B.* **110**(42), 20745–20748 (2006).
165. Gao MX, *et al.* Insight into a reversible energy transfer system. *Nanoscale.* **8**(36) (2016).
166. Lee SE, and Lee LP. Biomolecular plasmonics for quantitative biology and nanomedicine. *Current Opinion in Biotechno.* **21**(4), 489–497 (2010).
167. Liu GL, *et al.* Quantized plasmon quenching dips nanospectroscopy via plasmon resonance energy transfer. *Nat. Methods.* **4**(12), 1015–1017 (2007).
168. Nakayama N, *et al.* Ethanol-induced apoptosis in hepatoma cells proceeds via intracellular Ca(2+) elevation, activation of TLCK-sensitive proteases, and cytochrome c release. *Experimental Cell Res.* **269**(2), 202–213 (2001).
169. Waldeck DH, Alivisatos AP, and Harris CB. Nonradiative damping of molecular electronic excited states by metal surfaces. *Surface Science* **158**(1–3), 103–125 (1985).
170. Prigogine I, and Rice SA. Molecular Fluorescence and Energy *Transfer Near Interfaces* 1589–1595 (1975).

171. Persson BNJ, and Lang ND. Electron-hole-pair quenching of excited states near a metal. *Phys. Rev. B.* **26**(10), 5409–5415 (1982).
172. Kuhnke K, *et al.* C 60 Exciton quenching near metal surfaces. *Phys. Rev. Lett.* **79**(17), 3246–3249 (1997).
173. Seferos DS, *et al.* Nano-flares: probes for transfection and mRNA detection in living cells. *J. Am. Chem. Soc.* **129**(50), 15477–15479 (2011).
174. Prigodich AE, *et al.* Multiplexed nanoflares: mRNA detection in live cells. *Anal. Chem.* **84**(4), 2062–2066 (2012).
175. Prigodich AE, *et al.* Nano-flares for mRNA Regulation and Detection. *Acs Nano* **3**(8), 2147–2152 (2009).
176. Zheng D, *et al.* Aptamer nano-flares for molecular detection in living cells. *Nano Lett.* **9**(9), 3258–3261 (2009).
177. Tyagi S, and Kramer FR. Molecular beacons: probes that fluoresce upon hybridization. *Nat. Biotechno.* **14**(3), 303–308 (1996).
178. Wang K, *et al.* Molecular engineering of DNA: molecular beacons. *Angew. Chemie.* **48**(5), 856–870 (2009).
179. Dubertret B, Calame M, and Libchaber AJ. Single-mismatch detection using gold-quenched fluorescent oligonucleotides. *Nat. Biotechno.* **19**(4), 365–370 (2001).
180. Zhang J, *et al.* Aptamer-based multicolor fluorescent gold nanoprobes for multiplex detection in homogeneous solution. *Small* **6**(2), 201–204 (2010).
181. Song S, *et al.* Gold-nanoparticle-based multicolor nanobeacons for sequence-specific DNA analysis. *Angew. Chem.* **48**(46), 8670–8674 (2009).
182. Lakowicz JR, Radiative decay engineering 5: metal-enhanced fluorescence and plasmon emission. *Anal. Biochem.* **337**(2), 171–194 (2005).
183. Krishnan KS, A New Type of Secondary Radiation. *Nat.* **121**(3048), 501–502 (1927).
184. Zavaleta CL, Kircher MF, and Gambhir SS. Raman's "effect" on molecular imaging. *Journal of Nuclear Medicine Official Publication Society of Nuclear Medicine* **52**(12), 1839–1844 (2011).
185. Lombardi JR, and Birke RL. The theory of surface-enhanced Raman scattering. *J. Chem. Phys.* **136**(136), 2018–2029 (2012).
186. Andreou C, Kishore SA, and Kircher MF. Surface-Enhanced Raman Spectroscopy: A New Modality for Cancer Imaging. *Journal Nuclear Medicine Official Publication Society of Nuclear Medicine* **56**(9), 1295–1299 (2015).

187. Campion A, *et al.* On the mechanism of chemical enhancement in surface-enhanced raman scattering. *J. Am. Chem. Soc.* **117**(47), 11807–11808 (1995).
188. Jensen L, Aikens CM, and Schatz GC. Electronic structure methods for studying surface-enhanced Raman scattering. *Chem. Soc. Rev.* **37**(5), 1061–1073 (2008).
189. Gilroy KD, and Xia Y. Dimerization of Colloidal Particles through Controlled Aggregation for Enhanced Properties and Applications. *Chem. Asian J.* 2341–2351 (2016).
190. Graham D, *et al.* Control of enhanced Raman scattering using a DNA-based assembly process of dye-coded nanoparticles. *Nat. Nanotechnol.* **3**(9), 548–551 (2008).
191. Lim DK, *et al.* Nanogap-engineerable Raman-active nanodumbbells for single-molecule detection. *Nat. Mater.* **9**(1), 60–67 (2010).
192. Lim DK, *et al.* Highly uniform and reproducible surface-enhanced Raman scattering from DNA-tailorable nanoparticles with 1-nm interior gap. *Nat. Nanotechnol.* **6**(7), 452–460 (2011).
193. Jun BH, *et al.* Surface-enhanced Raman scattering-active nanostructures and strategies for bioassays. *Nanomedicine* **6**(8), 1463–1480 (2011).
194. Rodriguezlorenzo L, Fabris L, and Alvarezpuebla RA, Multiplex optical sensing with surface-enhanced Raman scattering: a critical review. *Analytica Chimica Acta* **745**(10), 10–23 (2012).
195. Wang Y, Yan B, and Chen L. SERS Tags: Novel optical nanoprobes for bioanalysis. *Chem. Rev.* **113**(3), 1391–1428 (2012).
196. † YL, *et al.* pH-sensing nanostar probe using surface-enhanced Raman scattering (SERS): theoretical and experimental studies. *J. Raman Spectrosc.* **44**(7), 980–986 (2013).
197. Kneipp J, *et al.* One- and two-photon excited optical ph probing for cells using surface-enhanced Raman and hyper-Raman nanosensors. *Nano Lett.* 7(9), 2819–2823 (2007).
198. Nyagilo J, *et al.* Multi-color Raman nanotags for tumor cell biomarker detection. in Conference: International Conference of the IEEE Engineering in Medicine & Biology Society IEEE Engineering in Medicine & Biology Society Conference. (2009).
199. Qian X, *et al. In vivo* tumor targeting and spectroscopic detection with surface-enhanced Raman nanoparticle tags. *Nat. Biotechnol.* **26**(1), 83–90 (2008).
200. Xiaohua Huang, *et al.* Cancer Cell Imaging and Photothermal Therapy in the Near-Infrared Region by Using Gold Nanorods. *J. Am. Chem. Soc.* **128**(6), 2115–2120 (2006).

201. Qian XM, and Nie SM. ChemInform Abstract: Single-Molecule and Single-Nanoparticle SERS: From Fundamental Mechanisms to Biomedical Applications. *Chem. In. Form.* **37**(31), 912–920 (2008).
202. Zhang Y, Hong H, and Cai W, Imaging with Raman spectroscopy. *Curr. Pharm. Biotechno.* **11**(6), 654–661 (2010).
203. Woo MA, *et al.* Multiplex immunoassay using fluorescent-surface enhanced Raman spectroscopic dots for the detection of bronchioalveolar stem cells in murine lung. *Anal. Chem.* **81**(3), 1008–1015 (2009).
204. Keren S, *et al.* Noninvasive molecular imaging of small living subjects using Raman spectroscopy. *Proceedings of the National Academy of Sciences* **105**(15), 5844–5849 (2008).
205. Pokharkar V, *et al.* Acute and subacute toxicity studies of chitosan reduced gold nanoparticles: a novel carrier for therapeutic agents. *J. Biomed. Nanotechnol.* **5**(3), 233–239 (2009).
206. Zhang XD, *et al.* Toxicologic effects of gold nanoparticles *in vivo* by different administration routes. *Int. J. Nanomedicine* **5**(1), 771–781 (2010).
207. Kumar A, *et al.* Gold nanoparticles functionalized with therapeutic and targeted peptides for cancer treatment. *Biomaterials* **33**(4), 1180–1189 (2012).
208. Vivero-Escoto JL, *et al.* Photoinduced Intracellular Controlled Release Drug Delivery in Human Cells by Gold-Capped Mesoporous Silica Nanosphere. *J. Am. Chem. Soc.* **131**(10), 3462–3463 (2009).
209. Prabaharan M, *et al.* Gold nanoparticles with a monolayer of doxorubicin-conjugated amphiphilic block copolymer for tumor-targeted drug delivery. *Biomater.* **30**(30), 6065–6075 (2009).
210. Wilhelm S, *et al.* Analysis of nanoparticle delivery to tumours. *Nat. Rev. Mater.* **1**(5) (2016).
211. Yavuz MS, *et al.* Gold nanocages covered by smart polymers for controlled release with near-infrared light. *Nat. Mater.* **8**(12), 935–939 (2009).
212. Raeesi V, Chou LYT, and Chan WCW. Tuning the drug loading and release of DNA-assembled gold-nanorod superstructures. *Adv. Mater.* (2016).
213. Liu K, *et al.* Simple Peptide-Tuned Self-Assembly of Photosensitizers towards Anticancer Photodynamic Therapy. *Angew. Chem.* (*Int. Ed.*) **55**(9), 3036–3039 (2016).

214. Theodossiou TA, *et al.* Photochemical internalization of tamoxifens transported by a "Trojan-horse" nanoconjugate into breast-cancer cell lines. *Angew. Chem.* (*Int. Ed.*) **54**(16), 4885–4889 (2015).
215. Wang X, *et al.* Analysis of the *in vivo* and *in vitro* effects of photodynamic therapy on breast cancer by using a sensitizer, sinoporphyrin sodium. *Theranostics* **5**(7), 772–786 (2015).
216. Zhang C, *et al.* Marriage of scintillator and semiconductor for synchronous radiotherapy and deep photodynamic therapy with diminished oxygen dependence. *Angew. Chem.* (*Int. Ed.*) **127**(6), 1790–1794 (2015).
217. Zeng L, *et al.* Resolution of deep angiosperm phylogeny using conserved nuclear genes and estimates of early divergence times. *Nat. Comm.* **5**(4956), 4956–4956 (2014).
218. Jang B, *et al.* Gold nanorod-photosensitizer complex for near-infrared fluorescence imaging and photodynamic/photothermal therapy *in vivo*. *ACS Nano* **5**(2), 1086–1094 (2011).
219. Zhao T, *et al.* Gold nanorods as dual photo-sensitizing and imaging agents for two-photon photodynamic therapy. *Nanoscale* **4**(24), 7712–7719 (2012).
220. Takahashi H, *et al.* Gold Nanorod-sensitized Cell Death: Microscopic Observation of Single Living Cells Irradiated by Pulsed Near-infrared Laser Light in the Presence of Gold Nanorods. *Chem. Lett.* **35**(5), 500–501 (2006).
221. Terry BH, Yan Zhao LT, Matthew NH, Ji-Xin Cheng. Alexander Wei. Hyperthermic effects of gold nanorods on tumor cells. *Nanomedicine* **2**(1), 125–132 (2007).
222. Hirsch LR, *et al.* Nanoshell-Mediated Near-Infrared Thermal Therapy of Tumors under Magnetic Resonance Guidance. *Proceedings of the National Academy of Sciences* **100**(23), 13549–13554 (2003).
223. Loo C, *et al.* Immunotargeted nanoshells for integrated cancer imaging and therapy. *Nano Lett.* **5**(4), 709–711 (2005).

Chapter 7

Delivering Nanoparticles to Cancer Cells

Ronald L. Huston

Department of Mechanical and Materials Engineering
College of Engineering and Applied Sciences
University of Cincinnati
Cincinnati, OH 45221-0072, USA

Abstract: In this chapter, we present a new nanoparticle methodology for treating cancerous tumors. The methodology is based upon a characteristic of cancer cells: that is, cells with an excess of centrioles. Centrioles are small cylindrical organelles lying adjacent to the nucleus. They create the electromagnetic field in a cell. But with an excess of centrioles, the electromagnetic field is enhanced. This enhancement then provides a biomarker for magnetically charged toxic nanoparticles. The chapter provides a basis for these assertions.

Keywords: Cancer, centrioles, microtubules.

1. Introduction

Many people (myself included) believe that the use of nanoparticles to image and destroy cancerous tumors is the greatest envisioned application in all of nanotechnology.[1,2]

In this chapter, we consider a new method for attracting nanoparticles to cancer cells. As difficult as that may seem, the idea is relatively simple; cancer cells are known to have greater electromagnetic (EM) fields than normal cells.[3–12] These enhanced EM fields are then targets for charged nanoparticles.

There are currently three traditional methods for treating cancer: (1) Excision (via surgery); (2) Radiation (focused and repeated); and (3) Chemotherapy (repeated and varied). Clinically, excision has the best outcome if all of the malignant cells are removed. Radiation can also be effective if all the malignant cells are destroyed; but the side effects of radiation can be harmful: Normal cells may also be damaged. Finally, thermotherapy is often a last resort or is used in combination with surgery and radiation. It too can be effective, but often all of the cancer cells are not destroyed. Since cancer tumors grow erratically and usually without sufficient oxygen and nutrients, the tumor recruits blood vessels which expand and also become cancerous, and with erratic growth.[11] With this vascular anomaly, the treating chemicals often do not reach all of the malignant cells. Nevertheless, the tumor is usually shrunk, but over time it recurs, and the recurring cells are usually more resistant to the treating chemicals. But if that were not bad enough, the side effects on normal cells are well known and often devastating.

If we can get nanoparticles to go inside the cancer cells, the particles can be used for accurate imaging of the tumor and more importantly, for killing its cells outright either by toxic deposit, by vibration via external stimulation, or by heating, or also via external stimulation.

The balance of the chapter is divided into six parts with the following part providing some preliminary background information about cell geometry. This is followed by a more detailed description of centriole development and activity within the cell. The next part provides a general description of bionano technology which is then followed by proposed alternatives to traditional nanoparticle therapy. The final part is a discussion with concluding remarks.

2. Preliminary Considerations I: Cell, Centriole, and Microtubule Geometry

To develop the proposed nanotherapy concept it may be helpful to first review some aspects of cell geometry and behavior. Our focus is restricted to human and animal (eukaryotic) cells. Figure 1 shows a cross-section of a typical cell. It consists of a nucleus surrounded by numerous organelles making up the cytoplasm (the cell interior except for the nucleus). Of all these organelles, the centrioles are of greatest interest for studying cell division (mitosis) and overexpressed cell division — as in cancer. For our purposes, if we can prevent cellular overexpression, we have accomplished our goal.

The centrioles are a pair of perpendicular cylinders lying adjacent to the nucleus as seen in Fig. 1. Of all the organelles in the cytoplasm, the centrioles are especially unique: they have precise geometry and

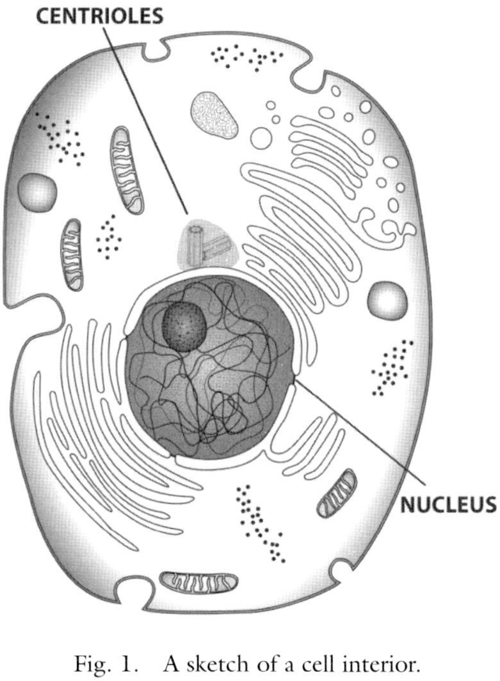

Fig. 1. A sketch of a cell interior.

they have no membrane cover. Also the centrioles duplicate each other in a unique way, and upon duplication the newer of the duplicated pairs moves away and onto the other side of the nucleus to drive the separation and duplication of the cell itself.

From a microscopic perspective, each centriole appears as a hollow cylinder approximately 400–500 nm long, 200 nm in outside diameter and 50 nm in inside diameter. Upon closer examination, the centriole is seen to be composed of parallel hollow tubes known as "microtubules". The microtubules have the length of a centriole (approximately 400–500 nm), but with an outside diameter of only 25 nm. The wall thickness of a microtubule is 5 nm and thus the inside diameter is approximately 10 nm.

Each centriole has at least 27 microtubules, arranged in 9 "blades" of 3 each. Some centrioles also have two microtubules along the centriole axis of the centriole center.

Figure 2 shows a sketch of a centriole pair.

The blades (or triplets) of the microtubules are parallel to each other and in a plane; but the plane is slightly inclined to the radial

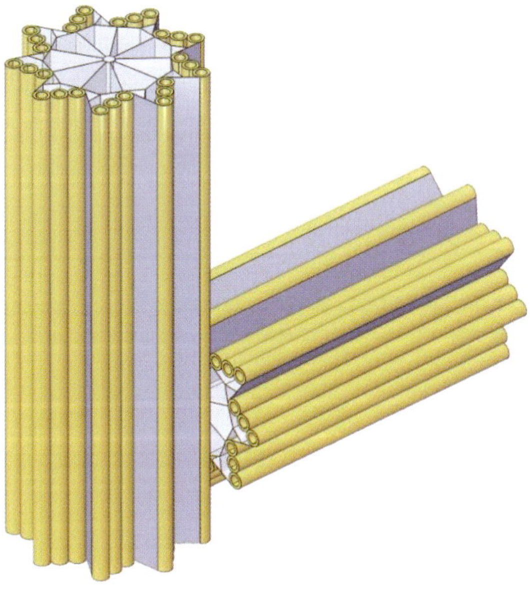

Fig. 2. A centriole pair.

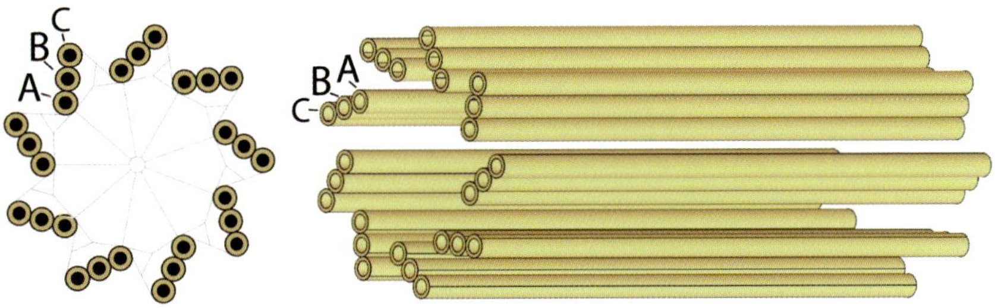

Fig. 3. An end view of a centriole/microtubule blade geometry.

Fig. 4. A side view of a centriole/microtubule blade.

direction. Also the microtubules are not exactly of the same length: the one closest to the center, labeled "A", is the longest; the outer microtubules, labeled "C" are the shortest; and the middle one, with intermediate length, is labeled "B". Figures 3 and 4 provide sketches of this microtubule blade geometry.

A closer look at an individual microtubule shows that it is composed of longitudinal strands (or "filaments") consisting of alternating

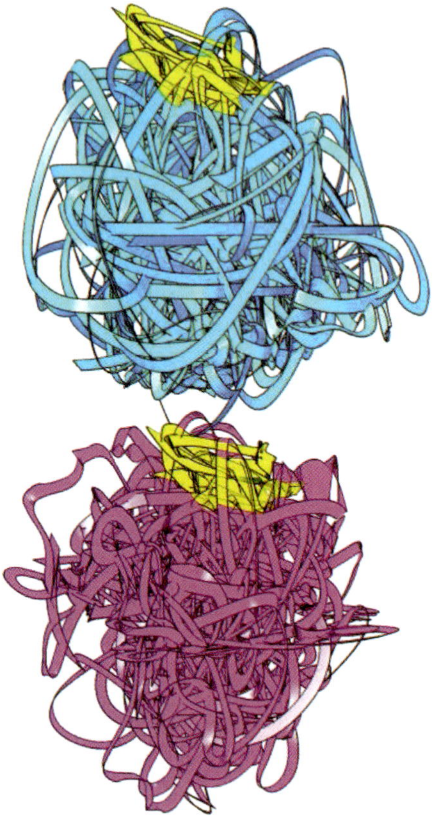

Fig. 5. A sketch of an α/β tubulin dimer.

α and β tubulin proteins. The α and β tubulin have positive and negative charges and taken as a pair (a "dimer"), they may be viewed electrically as a simple battery — as used in a flashlight.

If the α/β dimers are modeled as a short cylinder, the diameter is approximately 5 nm and the length 10 nm. Thus a typical microtubule filament contains approximately 40 α/β dimers.

The microtubule cylinder has 13 parallel filaments. These filaments are believed to be able to move longitudinally, independent of one another. Figures 5–7 provide sketches of an α/β dimer, a filament, and a microtubule cylinder.

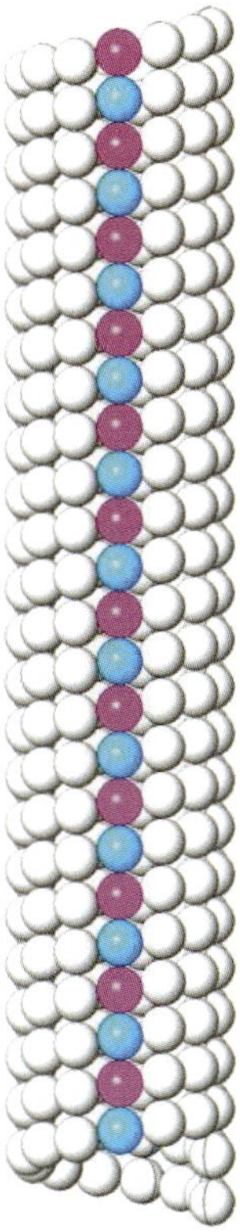

Fig. 6. A typical microtubule filament.

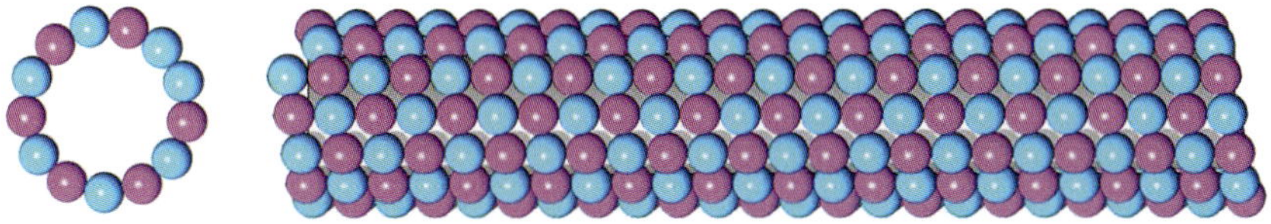

Fig. 7. A microtubule cylinder.

3. Preliminary Consideration II: Cell, Centriole, and Microtubule Development

From a purely mechanical perspective, all division (mitosis) begins with a duplication of the centriole pair. Centriole pair duplication in

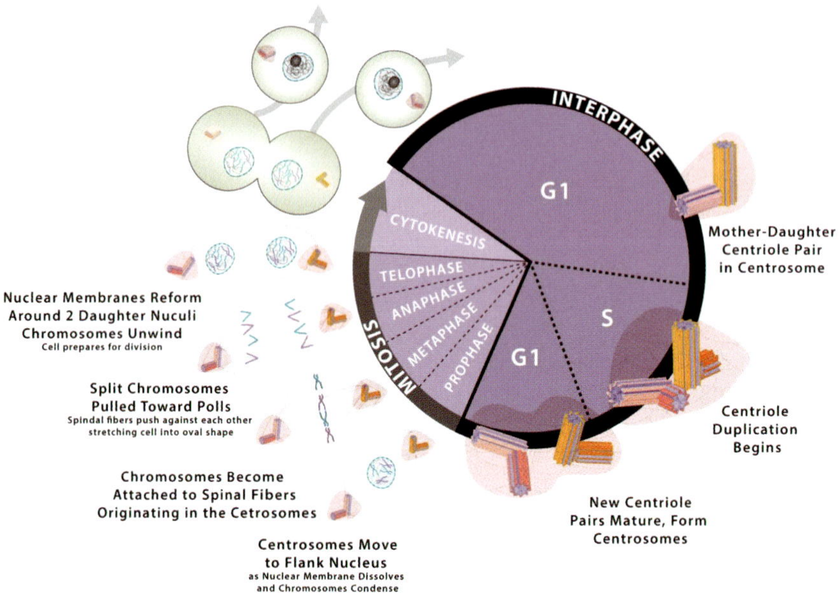

Fig. 8. The mitotic cycle.

turn begins with the development of microtubules on the centriole sides at the base, or proximal, ends of the centriole. This occurs during the "S-phase" of the mitotic cycle. See Fig. 8.

While details of the development process are still being discovered, in general, it occurs as described in the following paragraphs:

The centriole is immersed in a sea of proteins, probably numbering more than 110, known as the centrosome. Tubulins form an abundance of these proteins. The centrosome aside from the centrioles, is known as the microtubule organizing center (MTOC).

During the S-phase of the mitotic cycle, a protein known as "asterless", or "Asl", deposits itself on the outside of a C microtubule, near the base of the proximal end of each centriole. The Asl in turn then recruits another protein: Plk4 (protein-like kinase four) which forms an extended patch over the Asl deposit. Plk4 plays a prominent role in the entire centriole development process both as an enabler and as a regulator.

Once established upon and across the C microtubule the Plk4 recruits SAS6, a protein forming the structural base for the growth of a new centriole: Specifically, the SAS6 protein then emits nine radial spokes angled at 40° to each other where each spoke provides support for the growth of a triplet of microtubules — thus forming a microtubule blade of the newly growing centriole.

The conventional understanding[13–28] for the growth of the microtubules, and thus also of the centrioles, is that the α/β tubulin dimers are recruited to the SAS6 spokes by a γ-tubulin protein. The SAS6 spokes provide platforms for triplets of γ-tubulin rings. These rings have a series of 13 pores through which the α/β dimers are recruited from the MTOC and pushed upward to form microtubule filaments. Figure 9 provides a sketch of the envisioned γ-tubulin ring together with its pores.

The pores used for inserting α/β tubulin are sometimes called the "γ-tubulin small complex" (γ-Tu–SC).

Interestingly, the γ-tubulin ring is not quite planar. Instead, its ends overlap so that it has the form of a lockwasher. This in turn causes the microtubule filaments to be inclined at a small angle.

The centriole host is known as the "mother" and the growing centriole via the growing microtubules is known as the "daughter". In the original mother-daughter centriole pair, each centriole hosts a new centriole. The mother has a new daughter, and the daughter has a daughter of its own — a "granddaughter".

The microtubules of a daughter grow to only about 80% of the length of the mother's microtubule.

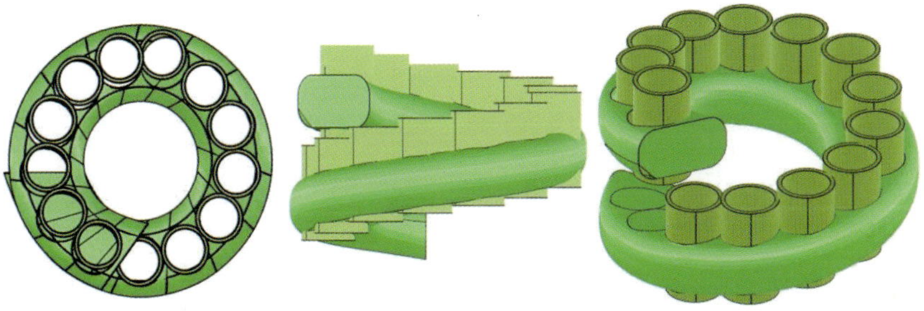

Fig. 9. A γ-Tubulin ring complex (γ-Tu–RC).

Once a second pair of centrioles is formed, the younger pair will break away from the original pair and move to the opposite side of the nucleus — as noted earlier. But during this movement the younger centriole pair becomes more mature, increasing in size so that by the time it is on the opposite side of the nucleus it is of equivalent vitality as the older centriole pair is left behind. Cell separation can then occur by the opposing centriole pairs, extending their microtubules about the nucleus and connecting to the kinetochores at the mid-plane of the nucleus. Once these connections occur the nucleus is pulled apart with each half taking approximately half of the cytoplasm. Thus two cells are formed. See Fig. 10.

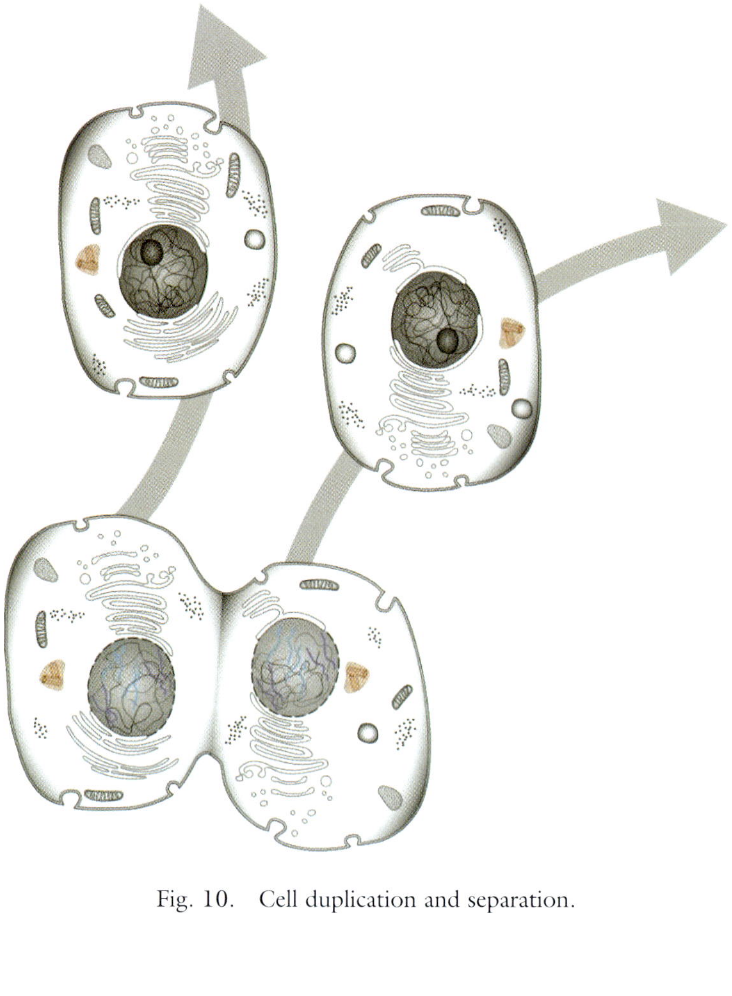

Fig. 10. Cell duplication and separation.

Finally, this process can go awry in several ways — most leading to cell death. But if it happens, that a mother centriole has more than one daughter centriole (i.e. "twins"), the duplication process can be accelerated and tumors can occur.

4. Preliminary Consideration III: Cellular Electromagnetism

Microtubule filaments (see Sec. 1) are composed of approximately 40 α/β dimers connected end-to-end. Each dimer has a positive/negative charge with the net negative charge of the filament series being at the proximal end. For a single dimer, the dipole, a moment (an end charge multiplied by the separation of the charges) is approximately 1000 Debye,[29–34] where a Debye is 10^{-29} Coulomb meters (cm). The dimer length is approximately 10 nm.

We also observed that the microtubule filaments can move relative to one another. Some researchers have suggested oscillation frequencies with values of approximately 465 MHz.[33,35,36] Since a moving charge creates an EM field[29] and an oscillating accelerating charge creates an oscillating EM field.

The evidence for the existence of cellular fields is manifold:

(1) The separation of the centriole pairs just after the younger pair is formed, and then the movement of the younger pair about the nucleus to the opposite side is suggestive of forces exerted at a distance. Aside from gravity, only EM can exert forces at a distance.
(2) In a *Science* article approximately 50 years ago, Paul Schafer[37] reported the maintenance of centriole orientation relative to each other even though separated with no apparent physical connection.
(3) As long ago as 1860, Dubois–Raymond discovered a potential difference across the sides of an open wound.[38]
(4) Potential differences across open wounds have also been investigated and reported by Weiss *et al.* in 1990[39] and by Chang and Zou in 2006.[38]

(5) Cancer cells reportedly have a greater EM field than adjoining normal tissue.[4–7] Cuzick *et al.* (1998) discovered an increased EM field about breast tumors that can be measured at the surface.[8]
(6) It is believed the chromosomes align themselves with the electric field of its cell.[40]
(7) EM fields are also believed to play an important role in blood vessel growth.[41–42]

5. Bio-nano Technology

Despite the many years of studying nanoparticles and various nanotechnology applications, much remains to be discovered and understood before widespread beneficial bioapplications are realized. Nevertheless, great strides have been made and are continuing to be made in the use of nanoparticles to treat illness with the ultimate goal of treating malignant tumors.

A nanoparticle is defined by size: That is a particle is a nanoparticle (NP) if[2]:

$$1\,\text{nm} \leq \text{NP} \leq 100\,\text{nm}, \tag{1}$$

where nm refers to nanometer or 10^{-9} m.

With such minute sizes, nanoparticles can easily penetrate cell membranes.[43] But in spite of their small size, nanoparticles can be composed of powerful toxins which can be released into cancer cells.[11]

If a cancer cell has a penetrating nanoparticle (i.e. magnetic) the particle can be made to vibrate and thus destroy the cancer cell. That is, movement and behavior of a magnetic nanoparticle within a cell can be controlled externally.

To date, therapeutic nanoparticles have primarily been administered by injection into the blood stream — much like chemotherapy. But unlike toxic chemicals nanoparticles can retain their toxicity until they reach a destination — that is, a tumor. The toxin can then be released via external stimulation.[45] Moreover, magnetic nanoparticles can be pulled through the blood vessels to the tumor by external magnets.[45]

In addition to penetrating and destroying tumor cells, nanoparticles within a cell can readily be located externally, either by magnetics or by photons emitted by the particle. In this way, nanoparticles can image tumors and thus assist in surgery and/or radiation.

Imaging with nanoparticles is often conducted using quantum dots and nanoshells. Quantum dots are small particles no more than 10 nm in major dimension and nanoshells are gold covered nanoparticles. Gold nanoparticles can pass through the blood brain barrier.[1]

Small magnetic nanoparticles (<30 nm) can have their magnetism blocked so that they are no longer magnetic, but then become magnetic when in a magnetic field. Such particles are said to be "superparamagnetic".[46] This blocking of the magnetism prevents the particles from clustering together before they reach a tumor target.

Despite the promising observations in the foregoing paragraphs, there is still not an effective and reliable method for getting nanoparticles into the cells of tumors. Although the use of nanoparticles to treat tumors is believed to have some advantages over conventional chemotherapy the difficulties with chemotherapy are also inherited by the injection of therapeutic nanoparticles into the blood stream.

The principal source of the difficulty is the vasculature (blood vessel geometry) leading up to, surrounding, and supplying tumors. Tumor blood vessels have structural abnormalities with evidence of uncontrolled growth. Some have even described the tumor vasculature as "architectural anarchy".[11]

Most tumor cells with their rapid growth are oxygen deprived. Therefore they recruit blood vessels which in turn tend to increase in diameter to accommodate the demand for blood supply. Large diameter blood vessels, however, have high back pressure keeping the supply of therapeutic chemicals and particles from reaching all the tumor cells.[11,47–48]

In addition to these structural deficiencies of the vasculature the blood vessels themselves can become malignant. They can leak,[11] again allowing many malignant cells to escape therapy.

There are other difficulties with blood injected nanoparticles: (1) The particles are quickly removed from the blood vessels[47–48];

(2) Blood can change the characteristics of nanoparticles; and (3) Magnetic nanoparticles tend to cluster together.[49–50] Such clusters can become large so that the nano advantage is lost.

In the following part we consider methods for overcoming these difficulties.

6. Potential Improvements and Technological Advances

The enormous potential of using nanoparticles to treat cancerous tumors will not be realized unless one or more nanoparticles can be imbedded into *all* of the cells within the tumor. Here are a few quotes of prominent researchers:

In the abstract of their paper Azarmi *et al.*[51] state:

> "The targeted delivery of drug molecules to organs of specific sites is one of the most challenging research areas in the pharmaceutical sciences." (2008)

In the Abstract of their paper von Maltzahn, Park, Lin *et al.*[52] state:

> "...our ability to efficiently have these materials (nanoparticles) to regions of disease *in vivo* remains very limited." (2011)

Finally, Sengupta and Sasisekharan[53] state:

> "There are still bottle necks in the existing 'homing' technologies... This emphasizes the need for more targeting strategies." (2007)

Our objective in this chapter is to present such a strategy. The concept is simple: we noted at the end of Sec. 2 that when centriole development goes awry, for whatever reason, multiple (supernumerary) centrioles can be developed. When this happens, the cell division rate can increase without bound and tumors can occur. References.[21,31] provide a listing of papers showing that supernumerary centrioles in a cell are a hallmark of cancer.

We also saw in Sec. 3 that centrioles via their microtubules have an EM field. Supernumerary centrioles in a cancer cell will thus have an overall increased EM field. This increased field can then be a biomarker for magnetized therapeutic nanoparticles.

If the EM charge of the nanoparticle is within a yet to be determined range, the particles will ignore normal cells. Osaka *et al.*[12] reported on the uptake of 40 nm magnetic nanoparticles into breast cancer cells while not invading normal cells (2009).

Lin *et al.*[54] report on the use of carbon nanotubes for targeting tumors in mice (2008) — it is claimed to be the first such application with carbon nanoparticles. In 2011, Wang *et al.*[55] reported the use of composite nanoparticles for targeting breast cancer cells.

As an alternative, along similar lines, is the targeting of tumor blood vessels.[44]

7. Discussion and Concluding Remarks

Success of the proposed methodology depends upon the following presumptions:

(1) Cancer cells have supernumerary centrioles;
(2) These supernumerary centrioles form clusters which create an EM field which is greater than that of normal cells;
(3) This increased EM field of cancer cells creates a biomarker, or target, for magnetically charged nanoparticles;
(4) Magnetically charged nanoparticles can be developed which will be attracted to the biomarker and thereby invade the cancer cells while leaving normal cells unharmed;
(5) External control of magnetically charged nanoparticles can be used to deliver toxins to cancer cells and/or to image a cancerous tumor.

Of all these presumptions, point 4 still requires development, and point 2 requires quantification.

At this time, it appears that the most promising application of the methodology is with aggressive forms of breast cancer, where the tumor is close to the surface. In such instances, the nanoparticles can be directed toward the tumor without prior injection into the blood stream.

This raises an additional area for development: How best to administer the therapeutic nanoparticles?

All of these assertions and issues are likely to provide fertile areas for research and development for the foreseeable future.

Acknowledgment

Support for the research providing the subject matter for this chapter was provided by the Schafer Foundation for Centriolar Research and is sincerely appreciated. Also, the consultation and encouragement of Bardyl Tirana, Roger Adelman, and Donglu Shi is acknowledged. Finally, the assistance of Mary Schartman in preparing the figures is acknowledged.

References

1. Chow EK-H, Ho D. Cancer nanomedicine: From drug delivery to imaging. *Sci. Tran. Med.* **5**(216), 1–12 (2013).
2. Arruebo M, Fernández-Pacheco R, Ibarra MR, Santamaria J. Magnetic nanoparticles for drug delivery. N*anotoday* **2**, 22–32 (2007).
3. Pokorný J, Hasek J, Vanis J, Jelínek F. Biophysical aspects of cancer — eletromagnetic mechanism. *Indian J. Exp. Biol.* **46**(5), 310–321 (2008).
4. McCaig CD, Rajnicek AM, Song B, Zhao M. Controlling cell behavior electrically: Current views and future potential. *Physiol. Rev.* **85**(3), 943–978 (2005).
5. Binggeli R, Weinstein RC. Membrane potentials and sodium channels: hypotheses for growth regulation and cancer formation based on changes in sodium channels and gap junctions. *J. Theor. Biol.* **123**(4), 377–401 (1986).
6. Djamgoz MBA, Mycielska M, Madeja Z, Fraser SP, Korohoda W. Directional movement of rat prostate cancer cells in direct-current electric field: Involvement of voltagegated Na+ channel activity. *J. Cell Sci.* **114**(14), 2697–2705 (2001).
7. Pokorný J. Physical aspects of biological activity and cancer. *AIP Adv.* **2**, 011207–1 to 011207–11 (2012).
8. Cuzick J, Holland R, Barth V, Davies R, Faupel M, Fentiman I, Frischbier HJ, LaMarque JL, Merson M, Sacchini V, Vanel D, Veronesi U. Electropotential measurements as a new diagnostic modality for breast cancer. *Lancet.* **352**(9125), 359–363 (1998).
9. Zhao M, Bai H, Wang E, Forrester JV, McCaig CD. Electrical stimulation directly induced pre-angiogenic responses in vascular

endothelial cells by signaling through YEGF receptors. *J. Cell Sci.* **117**(3), 397–405 (2004).
10. Mycielska ME, Djamgoz MB. Cellular mechanisms of direct-current electric field effects: Galvanotaxis and metastatic disease. *J. Cell Sci.* **177**(9), 1631–1639 (2004).
11. Davis ME, Chen Z, Shin DM. Nanoparticle Therapeutics: an emerging treatment modality for cancer. *Nat. Rev. Drug Discov.* 7, 771–782 (2008).
12. Osaka T, Nakanishi T, Shanmugam S, Takahama S, Zhang H. Effect of surface charge of magnetite nanoparticles on their internalization into breast cancer and umbilical vein endothelial cells. *Colloids and Surfaces B: Biointerfaces* **71**, 325–330 (2009).
13. Alberts B, Bray D, Lewis J, Raff M, Roberts K, Watson JD. *Molecular Biology of the Cell*, 3rd Edition, Garland Publishing, New York, (1994).
14. Hardin J, Bertoni G, Kleinsmith LJ. *Becker's World of the Cell*, 8th Ed., Pearson, (2015).
15. Guest D. *Biology Smart*, The Princeton Review, Random House, New York, 133–135 (1996).
16. Moritz M, Braunfeld MB, Guénebaut V, Heuser J, Agard DA. Structure of the (-tubulin ring complex: A template for microtubule nucleation. *Nat. Cell Biol.* **2**(6), 365–370 (2000).
17. Kollman JM, Polka JK, Zelter A, Davis TN, Agard DA. Microtubule nucleating (-TuSC assembles structures with 13-fold microtubule-like symmetry. *Nature* **466**, 879–883 (2010).
18. Wiese C, Zheng Y. A new function for the (-tubulin ring complex as a microtubule minus-end cap. *Nat. Cell Biol.* **2**, 358–364 (2000).
19. Teixido-Travesa N, Roig J, Luders J. The where, when and how of microtubule nucleation — one ring to rule them all. *J. Cell Sci.* **125**, 4445–4456 (2012).
20. Pokorný J, Hašek J, Jelínek F. Endogenous Electric Field and Organization of Living Matter. *Electromagn. Biol. Med.* **24**(3), 185–197 (2005).
21. Huston RL. A Review of Centriole Activity, and Wrongful Activity, During Cell Division. *Adv. Biosci. Biotechnol.* 7, 169–182 (2016).
22. Strnad P, Gönczy P. Mechanisms of procentriole formation. *Trends in Cell Biol.* **18**(8), 389–396 (2008).
23. Schiebel E. (-tubulin complexes: Binding to the centrosome, regulation and microtubule nucleation. *Curr. Opin. Cell Biol.* **12**, 113–118 (2000).

24. Lin T-C, Neuner A, Schiebel E. Targeting of (-tubulin complexes to microtubule organizing centers: Conservation and divergence. *Trends in Cell Biol.* **25**(5), 296–307 (2015).
25. Oakley BR, Paolillo V, Zheng Y. (-tubulin complexes in microtubule nucleation and beyond. *Mol. Biol. Cell* **26**, 2957–2962 (2015).
26. Bouissou A, *et al.* (-tubulin ring complexes regulate microtubule plus end dynamics. *J. Cell Biol.* **187**(3), 327–334 (2009).
27. Chang P, Giddings TH, Jr Winey M, Stearns T. -Tubulin is required for centriole duplication and microtubule organization. *Nat. Cell Biol.* **5**, 71–76 (2003).
28. Inclán YF, Nogales E. Structural models for the self-assembly and microtubule interactions of (-, *- and g-tubulin. *J. Cell Sci.* **114**, 413–422 (2000).
29. Edminister JA. *Electromagnetics*, Schaum's *Outline Series in Engineering*, McGraw Hill Book Co., New York, Chapter 9, (1979).
30. Griffiths DJ. *Introduction to Electrodynamics*, 2nd Edition, Prentice Hall, Englewood Cliffs, NJ, Chapter 5 (1981).
31. Huston RL. Using the electromagnetics of cancer's centrosome clusters to attract therapeutic nanoparticles. *Adv. Biosci. Biotechnol.* **6**, 172–181 (2015).
32. Pokorný J. Viscous effects on polar vibrations in microtubules. *Electromagn. Biol. Med.* **22**(1), 15–29 (2003).
33. Cifra M, Pokorny J, Havelka D, Kucera O. Electric field generated by axial longitudinal vibration modes of microtubule. *Biosystems.* **100**, 122–131 (2010).
34. Havelka D, Cifra M. Calculation of the electromagnetic field around a microtubule. *Acta Polytech.* **49**, 58–63 (2009).
35. Cifra M, Havelka D, Kucera O. Electric oscillations generated by collective vibration modes of microtubules, in laser applications in life sciences, edited by Matti Kinnamen and Risto Myllyla. *Proc. SPIE*, **7376** (2010).
36. Pokorný J, Vedruccio C, Cifra M, Kučera O. Cancer physics: Diagnosis based on damped cellular elastoelectrical vibrations in microtubules. *Eur. Biophys. J.* **40**(6), 747–759 (2011).
37. Paul W. Schafer. Centrioles of a human cancer: Intercellular order and intracellular disorder. *Science* **164**, 1300–1303 (1969).
38. Cheng K, Zou C. Physics models of centriole replication. *Med. Hypotheses* **67**, 572–577 (2006).
39. Weiss DS, Kirsner R, Eaglstein WH. Electrical stimulation and wound healing. *JAMA Dermatol.* **126**(2), 222–225 (1990).

40. Zhao Y, Zhan Q. Electric oscillation and coupling of chromatin regulate chromosome packaging and transcription in eukaryotic cells. *Theor. Biol. Med. Model.* **9**, 27–38 (2012).
41. Hochwald SN. Molecular-targeted therapy for cancer and nanotechnology. *Methods Mol. Biol.* **624**, 11–23 (2010).
42. Kanno S, Oda N, Abe M, Saito S, Hori K, Handa Y, Tabayashi K, Sato Y. Establishment of a simple and practical procedure applicable to therapeutic angiogenesis. *Circulation* **99**(20), 2682–2687 (1999).
43. Mailänder V, Landfester K. Interaction of nanoparticles with cells. *Biomacromolecules* **10**, 2379–2400 (2009).
44. Biswas S, Torchilin VP. Nanopreparations for organelle-specific delivery in cancer. *Adv. Drug Deliv. Rev.* **66**, 26–41 (2014).
45. Pouponneau P, Leroux JC, Soulez G, Gaboury L, Martel S. Co-encapsulation of magnetic nanoparticles and doxorubicin into biodegradable microcarriers for deep tissue targeting by vascular MRI navigation. *Biomaterials* **32**(13), 3481–3486 (2011).
46. Mannix RJ, Kumar S, Cassiola F, Montoya-Zavala M, Feinstein E, Prentiss M, Ingber DE. Nanomagnetic actuation of receptor-mediated signal transduction. *Nat. Nanotechnol.* **3**, 36–40 (2008).
47. Veiseh O, Gunn JW, Zhang M. Design and fabrication of magnetic nanoparticles for targeted drug delivery and imaging. *Adv. Drug Deliv. Rev.* **62**, 284–304 (2010).
48. Lammers T, Hennink WE, Storm G. Tumour-targeted nanomedicines: Principles and practice. *British J. Cancer* **99**, 392–397 (2008).
49. Owen J, Pankhurst Q, Stride E. Magnetic targeting and ultrasound mediated drug delivery: Benefits, limitations and combination. *Int. J. Hyperthermia* **28**(4), 362–373 (2012).
50. Zhang Y, Sun C, Kohler N, Zhang M. Self-assembled coatings on individual monodisperse magnetite nanoparticles for efficient intracellular uptake. *Biomed. Microdevices* **6**, 33–40 (2004).
51. Azarmi S, Roa WH, Löbenberg R. Targeted delivery of nanoparticles for the treatment of lung diseases. *Adv. Drug Deliv. Rev.* **60**, 863–875 (2008).
52. von Maltzahn G, Park J-H, Lin KY, Singh N, Schwöppe C, Mesters R, Berdel WE, Ruoslahti E, Sailor MJ, Bhatia SN. Nanoparticles that communicate *in vivo* to amplify tumour targeting. *Nat. Mater.* **10**, 545–552 (2011).
53. Sengupta S, Sasisekharan R. Exploiting nanotechnology to target cancer. *Brit. J. Cancer* **96**, 1315–1319 (2007).

54. Liu Z, Kai Chen, Corrine Davis, Sherlock S, Cao Q, Chen X, Dai H. Drug delivery with carbon nanotubes for *in vivo* cancer treatment. *Cancer Res.* **68**, 6652–6660 (2008).
55. Wang F, Pauletti GM, Wang J, Zhang J, Ewing RC, Wang Y, Shi D. Dual surface-functionalized Janus nanocomposites of polystyrene/ $Fe_3O_4@SiO_2$ for simultaneous tumor cell targeting and stimulus-induced drug release. *Adv. Mater. Communications*, www.advmat.de, 1–5 (2013).

Chapter 8

Multifunctional Nanomaterials for Cancer Theranostics

Shun Shen

The Institute for Translational Nanomedicine
Shanghai East Hospital, Tongji University School of Medicine
Shanghai, China

Abstract: Nanomaterials with different morphologies and functions have been widely developed. Among them, the nanotheranostic systems designed and simultaneously used for therapeutics and diagnostics draw particular attention. To achieve the multifunctionality in one nanoformulation, typically four components are included: nanocarriers, targeting ligand, imaging domain and therapeutics. This review is aimed at introducing some examples of possibilities for the different components in a nanotheranostic system.

Keywords: Theranostics, nanocarriers, therapeutic drug, targeting ligand, imaging domain.

1. Introduction

Since the last century, cancer has become a major threat to the survival of mankind.[1] China has nearly 20% of the world's population, but it has 22% of new cancer cases and 27% of the world's cancer deaths.[2–4] Cancer has become a serious threat to the health of our people, the number one killer. The highest incidence of cancer is lung cancer, followed by stomach cancer, liver cancer, esophageal cancer and colorectal cancer.[5,6] Although cancer prevention and control became one of the most important issues Chinese government carried out all over China, there was no significant decrease in cancer mortality in the short term.[7,8] In developed western countries, such as the United States, Britain and other countries, cancer mortality too failed to significantly decline.[9,10] Thus, there is a pressing need to develop novel and effective treatments and approaches that will enhance the survival rates.

Up to now, chemotherapy, surgery and radiotherapy are the first choice of cancer treatment in clinical practices. Although successful in reducing deaths directly related to cancer, unfortunately, their outcome has still been very pessimistic. Furthermore, these modalities can yield many side-effects, such as body weight loss resulting from oral intake disturbance, systemic toxicity and a destructive "bystander" effect to neighboring cells.[11–13] The main reason is that poor pharmacokinetic profiles and non-specific mechanisms of action by chemotherapeutics cause poor biodistribution in the diseased cells of interest, which results in the need for higher and continuous dosages to provide effective treatment.[14] This will inevitably further increase the patient's side reaction. Extensive resection is the best choice of cancer therapy, but such trials must be weighed against the potential survival benefits. Radiation therapy is limited in scope, and accompanied by severe adverse reactions.

In the last 20 years, nanotechnology has been developed rapidly.[15–17] The nanomaterials with different morphologies and functions have been widely reported.[18–20] Among them, the novel nanomaterials designed and used for drug delivery, cancer treatment and diagnosis draw particular attention.[21–27] A large number of animal experiments showed that the systemic toxicity of chemotherapeutic agents was greatly reduced, and the effect of chemotherapy was improved through the drug delivery system of nanomaterials.[28–30] The early diagnosis of cancer is gradually realized using the diagnostic nanomaterials.[31–37] However, there is still a gap between clinical applications. These therapeutic-loaded nanoparticles with poor tumor targeting and single function have revealed limited durable success clinically, which has led to the need to attempt to develop new strategies to improve the efficacy of nanoscale treatments through the addition of functional elements. Active targeting enhancements, as well as contrast agents have been incorporated into the nanomaterials to create truly multifunctional formulations.[38–41]

Nanotheranostics combined with therapeutics and diagnostics into single multifunctional nanoplatform is a promising field which has been proposed recently.[42–48] The theranostic nanomedicine can realize *in vivo* circulation, escape host defenses and deliver the drug and diagnostic agents at the targeted site to diagnose and cure the disease at cellular and molecular level.[47] By taking advantages of powerful theranostic nanomedicines, personalized medicine can be carried out. The concept of personalized medicine is to attempt to tailor medical care to an individual's needs, based on the specific characteristics of each individual patient.[49] Obviously, the nanotheranostics can provide the diagnostic characteristics of each individual, so as to target drug or physical therapy, which may aid in reducing the adverse side effects of treatment and potentially produce better overall results.[47,50,51] In order to achieve the multifunctionality in one nanoformulation, typically four components are included: nanocarriers, targeting ligand, imaging domain and therapeutics.

The diagnostic imaging examination and evaluation is essential before cancer treatment, because imaging can determine the exact nature, phenotype, and stage of the cancer state, which is related to the treatment regimen chosen, drug dosage administration, and prediction of the response to treatment. This capability allows a theranostic nanocompound to act as a guided biomarker that may aid in maximizing drug delivery efficacy and reducing the guesswork associated with development of cancer treatments. As we know, in order to achieve the desired diagnostic imaging of tumor, there is a sufficient dose of contrast agent that needs to accumulate in the tumor tissue. Luckily, the use of nanomaterials platforms has become rather advantageous because the nanomaterials have usually a modified surface, which can be decorated with contrast molecular. Moreover, contrast molecular or agent nanoparticles can be directly loaded or packed into the nanomaterials. Commonly used imaging methods include fluorescence imaging, photoacoustic imaging, Raman imaging, computed tomography (CT), radionuclide imaging (PET and SPECT) and magnetic resonance imaging (MRI).

According to a large number of literature reports, both organic and inorganic nanomaterials can be applied for a nanotheranostic system.[52–54] Common materials used to synthesize or form nanocarriers include polymers, lipids, silicon dioxide, carbon-based materials, gold, and iron oxides *et al.*[53,55–70] In order to ensure *in vivo* circulation escape host defenses and effectively penetrate blood vessels into the tumor tissues, each class of nanomaterials needs to have a size of less than 150 nm in diameter, with highly homogeneous particle distributions, good biocompatibility and sufficient stability *in vivo*. In addition, the nanomaterials must have the functions of modification with imaging contrast, targeting elements and the loading of the anticancer drug, without any changes to its nanomaterials properties. The targeting ligands overexpressed on tumors are attached to the surface of the nanomaterials so that they are bound to the receptors of cancer. Drug loaded nanoparticles should be slow-release before reaching the tumor site. Once entered into the tumor tissue, drug loading nanoparticles have a rapid drug release through physical or chemical action. Compared to the current chemotherapy drugs, these nanocarriers with unique drug-release behavior can enhance the therapeutic effect on cancer, and reduce the drug harm of body and tissues. In the following sections, we briefly introduce some examples of possibilities for the different components in a theranostic system.

2. Nanomaterials for Drug Delivery

According to the needs of drug delivery and tumor therapy, different morphology, size, change and hydrophobicity of nanocarriers can be synthesized. The advantages of the use of nanomaterials as a delivery tool of chemotherapeutic drugs have been widely recognized. The drug delivery vehicle can overcome the following series of limitations associated with traditional drug administration routes. (1) Drug solubility. In order to dissolve the drugs and for convenient clinical use, solubilization agents such as Cremophor EL associated with hypersensitivity reactions are often appended,[71] which further increases the adverse drug reactions. The use of nanocarriers such as polymers, mesoporous silica, carbon nanotubes (CNTs) can provide both a hydrophobic and hydrophilic environment, loading room which enhances solubility and overcomes the side effects of the cosolvent added.[72] (2) Drug burst release and rapid clearance. Without the loading and delivery by nanomaterials, chemotherapy drugs will be immediately absorbed and distributed in the major organs once they enter the body, causing damage to the normal tissues. In addition, the drugs will be soon metabolized and excreted, which subsequently requires repeated high dosages. By the protection of nanomaterials, the encapsulated drugs in delivery vehicles are more stable. Meanwhile, the behavior of slow release is beneficial to reduce the damage caused by the rapid absorption and distribution with traditional drug delivery routes. (3) Poor drug biodistribution. Due to their non-specific nature, only limited chemotherapeutic drugs reach the tumor tissues, and the majority are distributed to other normal tissues. By modifying the targeting ligands on the surface of nanomaterials, the targeting drug delivery systems can significantly be taken by tumor tissues through recognizing the receptors on the surface of the tumor cells, which reduces drugs toxicity to normal tissues and enhances the delivery of drugs to the site of interest. Therefore, the targeting drug delivery nanosystems have unparalleled advantages compared with current chemotherapeutics administration routes as well as great potential for the future.

It should be noted that not any nanocarriers can be achieved *in vivo* drug delivery and cancer therapy. The physicochemical differences of nanomaterials have a significant influence in the *in vivo* distribution and cellular internalization, including their size, shape, charge, etc.[73]

Effect of size. The size of nanomaterials mainly affects the biodistribution, kinetics of release, and cellular uptake.[74–76] The mechanisms of uptake of nanomaterials are known as phagocytosis, diffusion, and fluid phase endocytosis, through passive targeting, adsorptive mechanisms, active targeting and receptor-mediated endocytosis, respectively.[77,78] The nanomaterials with smaller size as a foreign objects are often not recognized by macrophages but microparticles easily taken up by reticuloendothelial systems. It has been found that nanomaterials with the size of ca. 50 nm have a maximum uptake *in vitro*,[79] while a general range of 10–100 nm has been demonstrated for *in vivo* applications.[80] The nanomaterials with less than 10 nm in size or larger size are cleared via glomerular filtration in the kidneys,[81] the Kupffer cells of the liver and through the spleen,[82] respectively. In tumor targeting application, the nanomaterials with the size range between 50 and 150 nm are more appropriate no matter active or passive targeting, while anything larger than 400 nm is considered too large.[83,84]

Effect of shape. The shape of nanomaterials has a significant impact on the internalization interaction between the particles and cells.[85] The selected nanomaterials with a suitable shape are conducive to obtain high effect on the circulation time, *in vivo* distribution, and residency time inside the cancer. Moreover, the effect of targeting the desired cells can be acquired by controlling the shape of nanomaterials. Also, the shape of nanomaterials can also affect their receptor-mediated endocytosis.[86] Generally, the elongated nanocarriers exhibit higher efficiency in adhering to the cells rather than the nanospheres. The rods show higher affinity and efficiency to the endothelial cells in *in vitro* experiments. *In vivo* studies indicated that spheres with the constant forces acting on them often remain in the center of the blood vessel as they circulate.[87] Rods easily accumulate at the vessel walls because of asymmetrical shape susceptible to drag forces and torques as they circulate. Shape is also an important factor affecting circulation time and biodistribution of the nanomaterials. According to the reports, the long circulation time of spheres is shorter than those of non-spherical nanomaterials.[88]

Effect of charge. Negative membrane of cells interacting differently with positively/negatively charged nanomaterials is important in interpretation of mechanism and designing of suitable nanocarriers. Due to electrostatic interaction with the negative cell membrane, the positively charged nanomaterials show higher uptake than negatively charged ones through endocytosis pathways.[89–94] Previous studies showed that internalization of nanoparticles with negative charge surface decreases by increasing of surface charge, but the amount of cellular uptake increases in positively modified particles by rising of charge.[95,96] The permeability of positively charged drug carriers to gastrointestinal mucous barrier is easier than neutral and negatively charged nanoparticles because of the presence of negatively charged proteins in outer surface of gastrointestinal epithelial cells. The anionic nanomaterials can also affect their cellular uptake. For example, by increasing the negative surface charge of carboxymethyl dextran-modified iron oxide nanoparticles, the nanomaterials can be taken increasingly via non-specific pathways.[97]

Besides, the hydrophobic/hydrophilic character of nanomaterials is also an important factor for cell uptake and blood circulation.[98–100] If the hydrophobicity of the nanomaterials is stronger than the cell membrane, the nanomaterials bind more easily to the cell surface through protein adsorption to enhance uptake. However, such nanomaterials cannot be used *in vivo*, especially via intravenous injection. In contrast, hydrophilic nanoparticles will attract fewer of these proteins, which makes nanomaterial *in vivo* circulation adsorb protein hardly. For this reason, nanocarriers are often PEGylated.[101,102] Polyethylene glycol (PEG), a neutral and hydrophilic polymer material, can lead to the steric prevention, low opsonization, high blood circulation time, and escape from reorganization by phagocytic system. PEG could be used to modulate the hydrophilicity of hydrophobic nanomaterials significantly. Obviously, there are many factors that interplay in the design of the nanoparticles for use in theranostic systems. It must be explained that the above effects are only the wealth of information available and reported, for a given system, and each of these factors should be experimentally determined because there is no way to predict the interactions occurring in a given system. The following are some of the mainly introduced and frequently-used theranostics of organic systems and representatives of inorganic systems.[73]

2.1. Polymer

Polymeric nanomaterials are composed of a backbone that is synthesized from a monomer. Anticancer drugs' conjugation with polymeric nanocarriers to form polymer-based prodrugs help circumvent problems associated with solubility of anticancer drugs.[103–105] The polymer-based prodrugs can also enhance bioavailability of drugs, improve chemo-drug pharmacokinetics and pharmacological properties through unique features of passively intratumoral accumulation.[104] Representative of synthetic polymeric nanomaterials are mainly N-(2-hydroxypropyl) Methacrylamide (HPMA), poly(glutamic acid) (PGA), poly(lactic-co-glycolic acid) (PLGA), polyglycolides (PGCA), which are mostly biodegradable and biocompatible with the human body. HPMA, a water soluble, biocompatible and non-immunogenic polymer, provides a large number of pendent functional groups in HPMA copolymers for drugs conjugation.[106–108] Activating or releasing the drugs from the polymer backbone within a targeted region can be carried out by manipulation of their bond, a degradable linker. PGA is naturally synthesized in microbial species. It is a highly anionic polymer which exhibits excellent biocompatibility and non-cytotoxicity. Researchers developed a thiolated PGA nanogel for the delivery of doxorubicin (DOX) to treat MCF-7 cancer cells, showing controlled drug release behavior and higher toxicity as compared to free DOX.[109] PLGA, a Food and Drug Administration (FDA) approved biodegradable and biocompatible copolymer, is synthesized from the two polymers of poly(lactic acid) (PLA) and polyglycolicacid (PGA).[110] Many anticancer drugs can be conjugated or directly loaded with PLGA, and the conjugated or loaded drugs release from the complex, slowly and over time.[111,112] Natural polymers are commonly classified in to three major categories: polysaccharides, polypeptides and polynucleotides. Among them, chitosan is widely applied as the drug delivery systems.[113–115] For example, Nogueira *et al.* produced chitosan nanoparticles loaded with anticancer agent of methotrexate for effective therapy of cancer.[116] Dextran is a water soluble polymeric saccharide. Dextran-based nanocarriers have been explored in *in vitro*, *in vivo* and most recently.[117,118] Pullulan is a natural polysaccharide with biodegradable, non-toxic, non-mutagenic, and non-carcinogenic properties. Due to the low immunogenicity and satisfactory solubility in aqueous and several organic solvents, it has been widely used for cancer drug delivery.[119,120]

2.2. Liposome

The liposome with its capacity to self-assemble into a bilayer structure is composed of both hydrophobic and hydrophilic component, is the most widely used for targeted drug delivery.[121–124] The lipid bilayer can be composed of cationic, anionic, or neutral (phospho)lipids and cholesterol. To improve liposome stability and enhance their circulation times in the blood, some hydrophilic polymers, such as PEG are often chosen for obtaining sterically-stabilized liposome.[125,126] Its unique structure gives the liposome a powerful drug loading function. Hydrophobic drugs can be inserted into the bilayer membrane, with the hydrophilic molecules trapped in the aqueous center. Moreover, owing to the large aqueous center and biocompatible lipid exterior, a variety of macromolecules, such as DNA, proteins and imaging agents can also be delivered by means of the liposome. The use of the liposomes as drug vesicles have been shown to be beneficial for stabilizing therapeutic drugs, overcoming obstacles to cellular and tissue uptake, and improving biodistribution of drug. Moreover, the liposomes are biocompatible, their surface properties can be easily modified according to the need of active targeting to tumor site, and their particle sizes can be precisely controlled. Despite these advantages, the liposome nanocarriers also have shortcomings, including poor stability, low loading efficiency, and poor release profiles.

2.3. Dendrimers

Dendrimers are constituted of successive branched layers with discrete, highly compact and tree-like nanostructures. Their size is well controlled with the increase of the dendrimer generation. Dendrimers as drug carriers also possess a lot of advantages.[127] For instance, because of the excellent monodispersity, the dendrimers loaded anticancer drugs provide a good promise for reproducible pharmacodynamic and pharmacokinetic behaviors.[128] The nanosize and unique structure of dendrimers ensure drug delivery applications *in vivo*.[129,130] The sufficient functional groups on the surface of dendrimers can be modified with various ligands such as targeting moieties, imaging probes and biocompatible ligands for specific therapies. For drug delivery, drug molecules can be encapsulated through the interior pockets of dendrimers via non-covalent interactions such as electrostatic, hydrophobic, and hydrogen-bond interactions, while drug molecules can also be conjugated by the surface functionalities via covalent linkages. Additionally, the cationic dendrimers mainly including polyamidoamine and poly(propyleneimine) with abundant tertiary amine groups on the surface can efficiently condense nucleic acids into nanomaterials for gene delivery.[131–137]

2.4. Polymeric Micelle

In recent years, amphiphilic block copolymers forming self-assembled micellar structure have gained much attention for their application in drug delivery.[138,139] The polymeric micelle has a hydrophobic core consisting of polymer tails and hydrophilic head groups. The micelles are formed through a process of self-assembly that occurs in a concentration-dependent manner, referred to as critical micelle concentration. Like other nanocarriers, the micelle used for drug delivery can improve the drug tissue distribution and reduce off-target cytotoxicity. The following describes some typical micelles. Poloxamer (trade name Pluronics) consists of a central poly(propylene oxide) (PPO) block forming hydrophobic core that is flanked on both sides by two hydrophilic chains of poly(ethylene oxides) (PEO) forming hydrophilic corona, yielding a structure of (PEO)a–(PPO)b–(PEO). Poly(lactic) acid (PLA), approved by the FDA, is used up in citric acid cycle to produce water and carbon dioxide. In order to improve the weak hydrophilicity of PLA, copolymerization of PEG–PLA was developed for the self-assembly of micelles in water with the size range of 10–100 nm.[140,141] The clinically approved micellar preparation of paclitaxel, Genexol-PM has mPEG–PLA as the micelle-forming polymer.[142,143] Another class of amphiphilic polyester polymers, PEGylated poly(caprolactone) (PEG–PCL) is an ideal candidate for drug delivery with its amphiphilic property and ease to synthesis, and exhibits good biocompatibility, biodegradability and low toxicity profile.[144–146] PEG–lipid copolymers are the same as other diblock copolymers. Among them, the most representative nanomaterial is PEG–phosphatidylethanolamine (PEG–PE) micelles, which can be loaded with drugs through two ways, i.e. by the dispersion method and dialysis method.[147,148] Block copolymeric micelles based on PEG–poly(amino acids) (PEG–PAA) utilize polymerized amino acids as hydrophobic core.[149] The advantage of using amino acids as hydrophobic core is their biocompatibility and biodegradability, easy functionalization of terminal amino acid groups with other versatile functional groups.

2.5. Mesoporous Silica Nanoparticles

Mesoporous silica nanoparticles (MSNs) have been promising in the fields of delivery vehicles for drugs with the following features: (1) An ordered pore network, with tunable sizes of 50–300 nm, shape and pore diameters of 2–6 nm, which is very homogeneous in size and allows fine control of the drug load and release kinetics; (2) A high pore volume of 0.6–1 cm^3/g to host the required amount of pharmaceuticals; (3) A high surface area of 700–1000 m^2/g, which implies high potential for drug adsorption; (4) A silanol-containing surface that can be functionalized to allow better control over drug loading and release.[150–159] On the other hand, MSNs can be endocytosed through the control of their surface functionalization, size and morphology by a huge number of mammalian cell lines. *In vitro* cytotoxicity tests indicate that MSNs have no cytotoxicity for different cell lines, even if the nanomaterials concentration was 100 μg/mL, just a transient change in cell metabolism.[160] *In vivo* animal experiments show that the MSNs have good biocompatibility, and almost no toxicity.[161,162] The maximum tolerated dosages can reach 200 mg/kg.[161] The MSNs can be intravenously administrated with good hemocompatibility. Owing to the high specific surface area and tunable mesoporous channel, the MSNs allow the transportation of drug compounds of different nature, from small molecules to proteins among others, either hydrophobic or hydrophilic drugs. Additionally, by the modification of small molecules, polymers, and biomolecules in both the interior (mesoporous channels) and exterior surfaces, the

MSNs for safely carry and deliver anticancer therapeutics in a target-specific, stimuli-responsive control fashion have demonstrated *in vitro* and *in vivo* with a wide variety of cancers. In recent years, an increasing number of researchers are focusing on the use of MSN nanocarriers for the enhancement of chemotherapy efficacy and inhibitions of multidrug resistance (MDR) and cancer cell metastases.[163,164] Besides, in order to develop multifunctional MSNs, various contrast agents for optical, MRI, CT, and PET have been readily incorporated to MSNs.[150,165–170] The combination of the capabilities of MSNs as a therapeutic and diagnostic platform has resulted in theranostic systems.[171,172] The advantages and efficacy of MSN-based theranostic carriers has also been demonstrated in reported studies.[166,173]

2.6. Carbon-based Nanomaterials

Carbon-based nanomaterials can be categorized into fullerenes, CNTs, nanohorns, nanodiamond, nanodots and graphene derivatives according to their morphologies. Here, CNTs and graphene are discussed. According to the numbers of sheets of sp^2 carbon atoms, CNTs can also be generally categorized into single-walled CNTs (SWCNTs) and multiwalled CNTs (MWCNTs). CNTs have great advantages in biomedical applications on antitumor drug delivery, photothermal tumor ablation, or biological imaging due to their unique physical and chemical properties.[174–180] CNTs have ultrahigh area inside and outside that permits tremendous accommodation to efficiently load multiple drug molecules. Thus, they have been intensively investigated as drug carriers, with doxorubicin (DOX) being the most common model drug.[181–185] Supramolecular binding of aromatic molecules such as DOX can be easily achieved by π–π stacking of those molecules onto the polyaromatic surface of nanotubes through non-covalent interactions.[181] The studies indicate the DOX-loaded CNTs have a potent accumulation in different tumor types and the on-demand release of DOX depend on enhanced permeability and retention effect (EPR) and acidic micro-environments of extracellular tissues of tumors and intracellular lysosomes and endosomes.[181] Light-controllable drug release of drug-CNTs complex is also achievable by utilization of CNTs' unique optical properties.[186] When conjugated with a targeting ligand, CNTs can be adopted as tumor-selective delivery vectors. Functionalized MWCNTs modified with a ligand of LRP receptor of angiopep-2 (ANG) were used for the targeted delivery of DOX to glioma. Improved therapeutic efficacy was obtained for DOX–MWCNTs–PEG–ANG when compared with free DOX.[185] Covalent binding of drugs onto CNTs can also be employed for hydrophilic drugs. For example, the anticancer drug of cisplatin and targeting ligand of epidermal growth factor (EGF) were covalently attached to SWCNTs and had enhanced killing efficacy of cancer cells.[187,188] According to the literatures reported, CNTs can also be good gene delivery vector of plasmid DNA and small interfering RNA (siRNA).[189–192] Other categories of carbon nanomaterials are nanosized graphene including graphene oxide (GO), reduced graphene oxide (RGO), and hybrid graphene nanostructures. With unique physical and chemical characteristics of ultra-high surface area, optical absorption, and versatile surface modification, graphene also attracts enormous research interests.[63,193–198] Like CNTs, the presence of reactive functional groups and localized π-electrons, which support the π–π bond with aromatic compounds such as DOX, render graphene a suitable candidate as a drug delivery system.[195,199] Moreover, the multifunctional graphene theranostics has also been developed. For example, DOX was loaded onto magnetic GO to be used for MRI detectable and magnetically targeted drug delivery.[200–202] Following a similar approach, a GO-based T_1 MRI detectable nanoplatform was developed via a GO-gadolinium complex.[203,204] Graphene derivatives have also been used for directing a gene of DNA or RNA, with PEI or chitosan being two primary mediators to improve the DNA/RNA binding and condensation, to the cancerous sites.[205–211]

3. Nanotherapeutic Agents

3.1. Anti-cancer Drugs

Up to now, DOX and paclitaxel (PTX) have been two of the most extensively studied cancer therapeutics that have been encapsulated into nanoparticles for theranostic applications.[73] PEGylated liposomal

DOX (Doxil, Caelyx) is a formulation of DOX in PEG-coated liposomes with a prolonged circulation time and unique toxicity profile used against numerous forms of cancer. Due to its wide range of indications, it is one of the most widely used anti-cancer drugs in the market. It must be said that Doxil also has adverse side effects associated with its administration, which include common symptoms, such as hair loss, vomiting, hand-foot syndrome, and cardiomyopathy.[212,213] The studies show that the mechanism of action of doxorubicin is mainly through interaction with DNA via intercalation.[214] Paclitaxel was discovered and isolated from the bark of the Pacific yew, *Taxus brevifolia*. The mechanism of action of PTX to the cancer cells is mainly through promoting tubulin polymerization and preventing its depolymerization which could block the tumor cells in mitosis phase. Clinical treatment results display that PTX have good therapeutic effect against a series of solid tumors including breast cancer, ovarian cancer, non-small cell lung cancer and head and neck cancer. The water solubility of PTX is very poor, therefore the traditional injection of PTX-taxol uses polyoxyethylene castor oil and ethanol as solubilizing agent to improve the solubility of PTX in water for injection. However, polyoxyethylene castor oil can cause some undesirable side effects such as allergic reaction etc. In 2005, Abraxane (albumin bound PTX) was from the company of Abraxis BioScience approved by Food and Drug Administration (FDA). Abraxane has abandoned traditional solubilizing agent of polyoxyethylene castor oil and taken human serum albumin (HSA) as the carrier and stabilizer. Thanks to good biocompatibility and low toxicity of HAS originating from the human body, the maximum tolerated dose of Abraxane is much higher than that of taxol. In addition, HAS can pass through the vascular endothelial cells through GP60 receptor and into the tumor tissue. Subsequently, HAS could specifically bind to the protein of SPARC, distributed on the surface of tumor cells and deliver PTX into the tumor cells. The clinical curative effect of Abraxane is better than taxol. Nucleic acid-based therapies by siRNA and miRNAs have gained applications in the treatment of various forms of cancers. Because miRNAs play a role in increasing chemoresistance of both breast cancer cells and prostate cancer cells, they can be applied for therapeutic targets through two strategies. The first is to block tumor suppressor miRNA and the other is to target the genes. Thus, the miRNA can be delivered to target cells by using targeted nanocarrier system. siRNA is also utilized in the treatment of cancers. For example, siRNA was successfully delivered to cyclin CD-1 to silence the target in leukocytes by liposomes decorated with monoclonal antibodies.[215] Polymer nanomaterials decorated with the integrin ligand RGD were utilized to deliver siRNA to endothelial cells during angiogenesis.[216]

3.2. Food and Drug Administration-approved Nanodrugs

Because a large number of powerful evidences suggest that the nanocarrier for anticancer drug delivery can improve the tissue distribution of drugs, improve the uptake of drugs in the tumors, enhance the effect of tumor therapy, and reduce the side effects of drugs, the FDA has approved nearly 10 kinds of nanoparticle drug formulations.[73] In 1995, Doxil was approved for the therapy of ovarian cancer and sarcoma. In 1996, DaunoXome was approved for the therapy of sarcoma. In 2002, Eligard (PLGH-polymer and leuprolide formulation) was approved for the therapy of late-stage prostate cancer. In 2005, Abraxane was approved for the therapy of breast cancer. In 2006 and 2007, Oncaspar (PEG-L-asparagine) and Genexol-PM (PLA-PEG micelle with paclitaxel) were approved for the therapy of acute lymphoblastic leukemia and metastatic breast cancer, respectively. Moreover, nanoparticle-based formulations mainly including many liposomal-based drugs are in various stages of clinical development to test their pharmacokinetics and biodistribution profiles. The new drugs of irinotecan SN-38 in phase I/II to treat colorectal cancer and a liposomal-based all-trans-retinoic acid (ATRA) tretinoin in phase II to treat acute promyelocytic leukemia and hormone-refractory prostate cancer are underway. EndoTAG-1, PTX embedded in a cationic liposome, is in phase II trials to treat advanced triple-negative breast cancer and pancreatic cancer. LEP-ETU (liposome- entrapped paclitaxel easy-to-use) as typical another form of PTX, is in phase I/II trials. Additionally, DOPC is a recently developed neutral liposome carrier for siRNA delivery and is currently in clinical testing. Another drug in phase I/II trials is annamycin to treat acute lymphocytic leukemia.

4. Targeting Strategy

4.1. Passive Targeting

Passive targeting is the penetration of nanocarriers through leaky tumor capillary fenestrations into the tumor interstitium and cells by convection or passive diffusion.[217–219] Due to the vascular structure of tumors, selective accumulation of nanocarriers and drugs can be easily carried out by the EPR effect that is now becoming the gold standard in cancer-targeting drug designing. The EPR effect is applicable for almost all solid tumors, and can be observed in almost all human cancers with the exception of hypovascular tumors.[220–222] For the nanocarriers to evade immune surveillance and circulate for a longer period, the EPR effect should be optimal. A large number of studies have shown that higher local concentrations of drug-loaded nanocarriers can be achieved at the tumor site, with ca. 50-fold higher concentration than that in normal tissue within several days.[223] In order to achieve the desired passive targeting effect by the EPR effect, more attention should be paid to at least three properties of nanocarriers. Firstly, the size of the nanocarrier must be strictly controlled. The ideal nanocarrier size is somewhere between 10 and 100 nm.[224–226] The nanocarriers of sizes smaller than 10 nm are easily filtrated by the kidneys, and with size larger than 100 nm, are specifically captured by the liver.[227,228] Secondly, the charge of the nanocarrier is neutral or anionic for efficient avoidance of the renal elimination.[81,229] Last but not least, the nanocarrier should be camouflaged to hide from the reticulo-endothelial system, which destroys all exogenous foreign bodies through opsonization followed by phagocytosis. However, there are shortcomings to solely relying on the EPR effect for passive targeting. The degree of tumor vascularization and angiogenesis affect the effect of the passive targeting. The nanocarrier should be designed according to tumor types and anatomical sites. On the other hand, solid tumors exist high interstitial fluid pressure, which can avoid successful uptake and homogenous distribution of drugs in the tumor. Larger and long-circulating nanocarriers (100 nm) are retained more in the tumor, whereas smaller molecules easily diffuse. Another drawback is that not every type of tumor allows for penetrating accumulation via the EPR effect, including gastric and pancreatic cancers.[230]

4.2. Active Targeting

Although passive targeting has achieved some success, the tumor targeting efficiency needs to be improved. In active targeting, targeting groups that bind to appropriate receptors expressed at the target site are linked at the surface of the nanocarrier. The targeting ligands can actively recognize and bind to a receptor overexpressed by tumor cells or tumor vasculature and not expressed by normal cells.[72,78,231–235] The binding affinity of the ligands affect the tumor penetration because of the "binding-site barrier".[236,237] According to reported studies, there are many types of targeting ligands.

Monoclonal Antibodies (mAbs). mAbs as currently clinically available ligands are widely used for active targeting.[238–241] Clinically available mAbs targeted nanomaterials are mainly rituximab for lymphoma,[242] trastuzumab for breast cancer treatment,[243] bevacizumab for angiogenesis inhibition,[244] and cetuximab for treatment of advanced colorectal cancer.[245] However, these large mAbs have serious limitations. Firstly, most mAbs hardly penetrate into tumors because they are too big to easily leave blood vessels and efficiently diffuse into tissues, which leads mAbs to be trapped by antigens located on the perivascular tumor cells to prevent permeation into the tumor mass. Secondly, slow clearance of mAbs from the body results in high normal tissue exposure. Most mAbs tend to accumulate in the excretory organs and do not reach their targets. In addition, even the humanized mAbs may also induce immune response involving hypersensitivity, neutralizing effects, and changeable pharmacokinetic properties.

Proteins. The use of proteins as targeting ligands that can bind to membrane bound receptors on cancer cells is available for active targeting delivery. Transferrins are iron-binding blood plasma glycoproteins that operate through transferrin receptor-mediated endocytosis, by adjusting the level of free iron in biological fluids. Cancer cells have an overexpression of the transferrin receptor, approximately two to tenfold compared to healthy cells, because they need more iron during growth periods.[246] By utilizing the characteristic of transferrins, Pang *et al.* used biodegradable polymersomes loaded with DOX and conjugated with transferrin to demonstrate significant enhancement of intracellular delivery of DOX in glioma and the chemotherapeutic effect of DOX for glioma rats.[247] EGF is a growth factor that stimulates cell growth, proliferation, and differentiation through binding to its receptor EGFR. Ye *et al.* developed EGF receptor targeted Fe3O4@TiO2 nanoparticles that use the native intracellular trafficking of EGF receptor to improve internalization and nuclear translocation in EGF receptor-expressing HeLa cells.[248] Prostate-specific membrane antigen (PSMA) is a glycoprotein enzyme encoded in humans by the KLK3 gene. Mukherjee *et al.* established a screening method based on humanized anti-PSMA antibody conjugated, silica-coated iron oxide nanoparticles for PSMA-specific cell targeting.[249] Disadvantages of using proteins attached to nanomaterials for active targeting consist of immunogenicity, rapid clearance, and low recognition at target site.

Small Molecules. Compared to mAbs and proteins, advantages of using small molecules linked with nanomaterials for active targeting include rapid accumulation in solid tumors, non-immunogenicity, stability/shelf lives, and cheapness. Folic acid, a classic vitamin, is the one that has been most frequently applied as a biomarker for tumor-targeted drug delivery.[250–254] As we know, all living cells require vitamins for their survival, but cancer cells need them in greater amounts to sustain rapid growth, so vitamin uptake receptors tend to be overexpressed on cancer cells.[255,256] Other than the folic acid receptor, biotin also is a promising vitamin-based targeting ligand for active targeting delivery of drugs or diagnostics.[257–259] Because biotin is required for cell growth, and the biotin demand in tumors is higher than in normal tissues, it is highly overexpressed in several aggressive cancer cell lines including leukemia, ovarian, colon, mastocytoma, lung, renal, and breast cancer cells. The $\alpha v\beta 3$ integrin, highly expressed on neovascular endothelial cells but poorly expressed in resting endothelial cells and most normal organs, is important in the calcium dependent signaling pathway leading to endothelial cell migration.[260,261] Cyclic or linear derivatives of RGD (Arg–Gly–Asp) oligopeptides are the most studied peptides.[262–267] In addition, there are many small molecular compounds for cholecystokin receptor, tropomyosin receptor kinase, glucose transport system, estrogen receptor, progesterone receptor, androgen receptor that will not be discussed here.

5. Imaging Techniques

Imaging technology is an important tool for the diagnosis and prognosis of the disease. The combination of nanomaterials and imaging technology has become a hot research field in nanomedicine, which leads to the development of nanotheranostics. Major imaging techniques are described below:

5.1. Fluorescence Imaging

Fluorescence imaging is one of the most powerful non-invasive imaging techniques that have been widely performed for the detection of disease states.[268–272] The advantages of fluorescence imaging include high sensitivity, real-time observation, lack of non-ionizing radiation, ability to image the spectrum from visible to near-infrared. Moreover, it is inexpensive to detect the photons. The disadvantages are their poor tissue penetration, susceptibility to noise due to scattering, and potential autofluorescence. An ideal fluorescent probe used for nanodiagnosis is expected to show high fluorescence quantum yields, robust photostability, excellent dispersibility in biological conditions, high resistance to degradation in biological systems and favorable biocompatibility. The fluorescent probes include organic dyes, semiconductor quantum dots, organic–inorganic nanohybrids.[273–277] For example, Zhan *et al.* utilized targeting fibronectins PEG–PLA nanoparticles with a near infrared fluorescent dye (DiR) loaded with PTX as a theranostic agent. This system was able to selectively target, treat, and image brain tumors *in vivo*.[278] Kong *et al.* developed ribonuclease-A encapsulated PbS quantum dots which emit in the second near-infrared biological window for ultrasensitive fluorescence *in vivo* imaging. The result showed the prepared quantum dots with high Phi(f) (similar to 17.3%) and peak emission at 4300 nm ensure deep optical penetration to muscle tissues (up to 1.5 cm) and excellent imaging contrast at an extremely low threshold dose similar to 5.2 pmol per mouse.[279] Gianotti *et al.* developed hybrid preparation by loading highly fluorescent dyes of indocyanine inside the pores of MSNs.[280] Uniform monomeric existence of dye inside pores was indicated under diluted dye concentrations. The formation was observed in high concentrations of the dye from the hybrid. *In vitro* fluorescence bioimaging of this hybrid was demonstrated in mammalian cancer cells.[280]

5.2. Photoacoustic Imaging

Photoacoustic imaging, or namely optoacoustic imaging, unlike conventional optical imaging methods, is insensitive to photon scattering within biological tissue and, makes high-resolution optical visualization deep within tissue possible.[281–286] The principle of photoacoustic imaging is that the endogenous molecules and tissues absorb laser energy to generate wave sounds, which are recorded by ultrasound detectors and converted to image. However, due to tissue scattering and negative influence of some endogenous agents, low optical absorption by tissues limits the benefits of photoacoustic imaging. In order to solve the problem for the enhancement of photoacoustic signals, nanoparticle-based contrast agents with high NIR absorption were applied. Among various nanomaterials as contrast agents for photoacoustic imaging, gold-based nanomaterials, CNTs and upconversion nanoparticles have been explored extensively for photoacoustic imaging applications.[64,287–294] For example, Cheng *et al.* developed Prussian blue nanocubes as theranostic agents for photoacoustic imaging guided NIR photothermal therapy (PTT).[295] Cheng *et al.* used PEGylated WS2 nanosheets as a multifunctional theranostic agent for *in vivo* dual-modal CT/photoacoustic imaging guided PTT.[296]

5.3. Computed Tomography

CT is a diagnostic imaging technique based on the fundamental principle that the density of the tissue passed by the X-ray beam measured from the calculation of the attenuation coefficient. Thus, anatomical information of a 3D image with high spatial resolution can be clearly provided by CT owing to the differences in X-ray attenuation of biological tissues such as bone, muscle, fat, water, and air. Because the atomic number of noble metal nanomaterials is higher than that of non-ionic iodine preparation, the CT diagnostic effect by gold nanoparticles and so on is better. Wei *et al.* reported a simple non-injection method to fabricate uniformly sized pure Bi nanoparticles. The Bi nanoparticles obtained had highly uniform size, excellent monodispersity, and impressive antioxidant capacity. After being modified with oligosaccharide, the "sweet" Bi nanoprobe with comfortable patient experience and favorable biocompatibility was successfully used in CT visualization of gastrointestinal tract in detail.[297] Motiei *et al.* developed a nanoparticle-based approach, utilizing glucose-functionalized gold nanoparticles as a metabolically targeted CT contrast agent.[298] The approach showed specific tumor targeting and had successfully distinguished between cancer and inflammatory processes in a combined tumor-inflammation mouse model, due to dissimilarities in angiogenesis occurring under different pathologic conditions.[298]

5.4. Radionuclide Imaging

Radionuclide imaging technique mainly includes PET and SPECT. PET utilizes more expensive radionuclides and their emitted γ rays with a shorter half-life to produce images. Commonly used radionuclides have ^{111}In, ^{64}Cu, and ^{18}F. Compared with SPECT, the advantages of using PET are that a more sensitive 3D image with low concentration of radionuclides it can be provided by PET. The disadvantages of using PET contain spatial resolution limitations, cost of the equipment, it can only image one radionuclide at a time, and radiation concerns are present. Pang *et al.* reported a facile synthesis of radiolabeled PdCu@Au core–shell tripods for use in PET and image-guided photothermal cancer treatment by directly incorporating radioactive ^{64}Cu atoms into the crystal lattice. The result demonstrated the use of radioactive PdCu@Au tripods for effective photothermal treatment in the 4T1 tumor model as guided by PET imaging. The efficacy of treatment was confirmed by the significant reduction in tumor metabolic activity revealed through the use of ^{18}F-fluorodeoxyglucose PET/CT imaging.[299] SPECT utilizes a camera to detect a dosage of γ radiation that is emitted from the tissue of interest upon administration. Commonly used radionuclides have ^{99m}Tc, ^{131}I, and ^{67}Ga and so on. Compared with PET, the advantages of using SPECT are that it can allow for precise measurements, have little background interference, require no signal amplification, less expensive, long half-lives of radionuclides, high sensitivity. The disadvantages of using SPECT include a low spatial resolution, high costs, and use of radiation and relatively cumbersome size of detection equipment. Zhao *et al.* synthesized Au nanoparticles doped with ^{199}Au atoms for targeted SPECT tumor imaging in a mouse triple negative breast cancer model.[199] By directly incorporating ^{199}Au atoms into the crystal lattice of each Au nanoparticle, the stability of the radiolabel could be ensured. When conjugated with D-Ala1-peptide T-amide, the Au nanoparticles doped with ^{199}Au atoms can serve as a C–C chemokine receptor 5 (CCR5)-targeted nanoprobe for the sensitive and specific detection of both triple negative breast cancer and its metastasis in a mouse tumor model.[300] Zhao *et al.* reported chlorotoxin-conjugated multifunctional dendrimers labeled with radionuclide ^{131}I for SPECT and radiotherapy of gliomas. The result indicated the ^{131}I labeling render the dendrimer platform with an ability to be used for targeted SPECT imaging and radiotherapy of an MMP2-overexpressing glioma model *in vivo*.[301]

5.5. Magnetic Resonance Imaging

MRI is a non-invasive diagnostic method, which is the most commonly used in current medical examination for getting high-resolution and 3D anatomical images in the body through differing relaxation times between physiological tissues. The parameters of longitudinal ($T1$) or transverse ($T2$) represent the differences in these relaxation times with the use of image contrasts. DTPA–Gd (Magnevist®) is a commercial MRI Gd-based contrast agent. In order to develop more efficient contrast agents, more and more attention is focused on nanotechnology. Currently the most studied nanosized contrast agents mainly include, Fe_3O_4-based T_2 contrast agent, manganese oxide nanoparticles-based T_1 contrast agent.[302–304] For example, Xu *et al.* developed a premix membrane emulsification method to prepare uniform PEGylated PLGA microcapsules with superparamagnetic Fe_3O_4 nanoparticles embedded in the shell for ultrasound/magnetic resonance bimodal imaging.[305] However, spherical superparamagnetic Fe_3O_4 nanoparticles exhibit relatively low transverse relaxivity. In order to solve this deficiency, Zhao *et al.* reported a new strategy to achieve high transverse relaxivity by controlling the morphology of iron oxide nanoparticles.[306] Size-controllable octapod Fe_3O_4 nanoparticles with an ultrahigh transverse relaxivity value were successfully fabricated. These octapod Fe_3O_4 nanoparticles were much more effective T_2 contrast agents for *in vivo* imaging and small tumor detection in comparison with conventional Fe_3O_4 nanoparticles. Detection of brain gliomas at the earliest stage is of great importance to improve outcomes, but it remains a most challenging task. Chen *et al.* prepared folic acid modified manganese oxide nanoparticles for the early diagnosis of brain gliomas.[307] T_1–T_2 dual modal MRI can offer complementary diagnostic information, leading to more precise diagnosis. Li *et al.* reported novel hybrids of nanocubes with both iron and Gd elements as T_1–T_2 dual modal contrast agents.[308] *In vivo* MRI of rats showed both T_1-positive and T_2-negative contrast enhancements in the livers, which were about two folds of those of Gd_2O_3 nanoparticles and Fe_3O_4 nanocubes, respectively.

6. Therapy

In this section, several therapeutic methods including PTT, photodynamic therapy as well as combined therapy are mainly discussed. Controlled drug delivery in the previous section has been mentioned.

6.1. Photothermal Therapy

PTT is a therapeutic strategy in which photon energy is converted into heat for the treatment of many medical conditions, including cancer. It is known that NIR light is harmless to the human body. So, ideal photothermal agents should exhibit strong absorbance in the NIR region, and should efficiently transfer the absorbed NIR optical energy into heat. Various types of NIR-absorbing nanomaterials, such as carbon-based nanomaterials, gold-based nanomaterials, copper sulfide nanoparticles as well as organic polymers have been widely explored.[309–318] For example, CNTs have been developed as effective photothermal conversion agents for the PTT of cancer.[319–321] The results indicated that PEGylated CNTs exhibited excellent biocompatibility, low cost, non-toxic and stabilization.[185,322,323] Due to their strong NIR absorption, gold nanotubes have also been extensively explored in PTT.[324,325] Even so, the potential toxicity of carbon-based, gold-based nanomaterials, is still an unresolved debate, which will inevitably limit future clinic applications of PTT. These non-degradable or slowly degradable nanoparticles easily accumulate in bodily organs, resulting in increased oxidative stress, inflammatory cytokine production and cell death. Fe_3O_4 nanoparticles have received tremendous attention for their excellent magnetic, biocompatible and potentially non-toxic properties. Iron is a nutrient and is readily metabolized by cellular regulation using the transferrin pathway. Thus, Fe_3O_4 nanoparticles are easily degradable and pass in and out of cells across the plasma membrane. Shen *et al.* firstly reported photothermal effect of nano-Fe_3O_4 clusters for the photothermal ablation of tumors. Compared with individual magnetic Fe_3O_4 nanoparticles, clustered Fe_3O_4 nanoparticles can result in a significant increase in the NIR absorption.[326]

6.2. Photodynamic Therapy

PDT is a procedure that uses a photosensitizing drug to apply light therapy selectively to lesion locations. The wavelength of the light source needs to be appropriate for exciting the photosensitizer to produce radicals and/or reactive oxygen species. Its advantages lessen the need for delicate surgery and lengthy recuperation and minimal formation of scar tissue and disfigurement. A side effect is the associated photosensitisation of skin tissue. However, conventional PDT strategies result in ineffective therapeutic response for deep-seated tumors because the wavelength of illumination light of most clinically approved photosensitizers is located in the UV/VIS range that possesses limited tissue penetration ability. The combination of PDT and nanotechnology is becoming a promising approach to fight against deep tumors.[327] The maximum absorption of clinically approved photosensitizers at the UV–VIS region can be obtained with the help of nanomaterials for deep-seated tumor treatment. Firstly, these nanomaterials act as carriers for photosensitizers to enhance their long circulation capability and active/passive targeting. Secondly, they are used as antennae to capture photons of illumination light with good penetrability and then emit light to excite the photosensitizers nearby via efficient energy or electron transfer process. To strive for the best therapeutic response, the energy transfer efficiency between antenna nanomaterials and photosensitizers must be optimized, which is highly dependent on the photosensitizers loading approach. It is generally known that there are three major approaches for the combination of photosensitizers and antenna nanoparticles: colocation, physical loading and covalent conjugation.[328]

6.3. Combined Therapy and Theranostics

Current clinical therapy investigation has shown that combined therapy techniques such as thermo-chemotherapy or chemo-photodynamic therapy could result in additionally enhanced anticancer efficacy. However, for traditional treatment techniques, the synergistic effects of combined therapy are difficult to realize *in vivo* because codelivery of different agents or tools together to the target tissues is still challenging. The rapid development of nanotechnology provides a new impetus for cancer therapy due to its unique physical properties. For example, mesoporous silica-encapsulated gold nanorods (GNRs@ $mSiO_2$) as a NIR-light-absorbing agent and a drug delivery system for *in vivo* chemo-photothermal destruction of tumor were developed.[329] In order to target the tumor and increase the therapeutic efficacy, RGD peptides were conjugated on the terminal groups of PEG on GNRs@$mSiO_2$. DOX was also successfully loaded into pGNRs@ $mSiO_2$-RGD nanomaterials. The results proved that the synergistic effect of DOX–pGNRs@$mSiO_2$–RGD for the efficacious treatment of tumor were better than that of the chemotherapy or photothermal therapy alone. The side effects of antitumor drugs and low treatment efficacy are two major challenges of current chemotherapy. To address these issues, a new kind of smart nanocarriers based on upconversion nanoparticle (UCN)-loaded folate-conjugated polymeric lipid vesicles (UFPLVs) that combine pH-responsive chemotherapy and NIR-light-triggered photodynamic therapy were constructed. Merocyanine 540 (MC540), as a photosensitizer, was loaded in the UFPLVs; DOX was conjugated to the surfaces of UFPLVs by pH-sensitive hydrazone bonds. *In vitro* and *in vivo* combination treatments evidenced the high antitumor effects of MC540@DOX-UFPLVs under NIR-light-irradiation.[330] Checkpoint inhibitors are among the most promising immunotherapies for patients with advanced colon cancer, but their durable response rate remains low. He *et al.* reported the use of immunogenic nanoparticles to augment the antitumor efficacy of PD-L1 antibody-mediated cancer immunotherapy. Nanoscale coordination polymer core–shell nanoparticles carry oxaliplatin in the core and the photosensitizer pyropheophorbide-lipid conjugate in the shell for effective chemotherapy and PDT.[331]

Theranostics, the integration of therapeutics and diagnostics into a system, is a recently proposed concept.[42,47,332–334] It can provide real-time monitoring of the treatment process, and improve therapeutic efficiency with better planning and prognosis. The treatment methods include chemo/gene/PTT/PDT, and diagnostic technologies with fluorescence/bioluminescence, MRI, PET, SPECT, photoacoustic,

Raman imaging, etc. Therefore, there are many different combinations of nanoscale theranostic agents. For example, for better therapeutic planning, many research teams have developed imaging-guided PTT/PDT with the help of nanotheranostic agents.[335–338] Recently, graphene oxides modified with iron oxide nanoparticles (GO–IONP) as a nanotheranostic agent have been developed to diagnose and treat regional lymph nodes (RLN) metastasis of pancreatic cancer.[339] The GO–IONP showed powerful ability of dual-modality mapping of regional lymphatic system by MRI, as well as dark color of agent providing valuable information that was instrumental for the surgeon in making the preoperative plan before operation and to intraoperatively distinguish RLN from surrounding tissue. Under the guidance of dual-modality mapping, the metastatic lymph nodes including abdominal nodes could be effectively ablated by NIR irradiation with an incision operation. Similar nanotheranostic platforms constructed by manganese oxide-coated CNTs as dual-modality lymph mapping agents was applied for PTT of tumor metastasis.[340] Nanoscale ultrasound contrast agents with drugs, genes or functional nanomaterials as payloads could serve as efficient theranostics systems. For instance, DOX loaded perfluoropentane (PFP) nanoemulsion using a biodegradable block copolymerpoly PEG–PLLA was constructed.[341] Upon intravenous injection into mice, DOX–PFP nanoemulsions could selectively extravasate into the tumor sites, producing a strong durable ultrasound contrast via PFP evaporation to form nanomicrobubbles. In the meantime, DOX was released from those nanomicrobubble under ultrasound irradiation to offer significant chemotherapeutic efficacy.

7. Conclusion

At present, the rapid development of nanotechnology provides powerful means for the cancer theranostic. The pre-clinical animal experiments have demonstrated that nanotheranostic systems can lead to an enhancement of treatment options, leading to truly personalized medicine. Despite all of the advantages of nanomaterial-based systems, there is always room for further improvements in this field. The toxicity of the nanomaterial carriers is always a major concern when they are used in real patients. Reducing the dosage of nanocarriers and enhancing their biodegradability *in vivo* are the most promising solutions.

References

1. Ferrari M. Cancer nanotechnology: Opportunities and challenges, Nature reviews. *Cancer* **5**, 161–171 (2005).
2. Ma X, Lin C, Zhen W. Cancer care in China: A general review. *Biomed. Imaging Int. J.* **4**, e39 (2008).
3. Xu GW. Gastric cancer in China: A review. *J. R. Soc. Med.* **74**, 210–211 (1981).
4. Yu KD, Di GH, Wu J, Lu JS, Shen KW, Shen ZZ, Shao ZM. Development and trends of surgical modalities for breast cancer in China: A review of 16-year data. *Ann. Surgi. Oncol.* **14**, 2502–2509 (2007).
5. Chen W, Zheng R, Zeng H, Zhang S. Epidemiology of lung cancer in China. *Thoracic Cancer* **6**, 209–215 (2015).
6. Zhou C. Lung cancer molecular epidemiology in China: Recent trends. *Transl. Lung Cancer. Res.* **3**, 270–279 (2014).
7. Gao H. Chinese government admits existence of cancer villages. *Lancet. Oncol.* **14**, 284 (2013).
8. Wen C. Chinese government urged to tame cancer threat. *Lancet Oncol.* **6**, 453–455 (2005).
9. Moore MA, Kunimoto T, Park CB, Takasuka N, Tsuda H. Cross-country comparisons of colon and rectal cancer mortality suggest the existence of differences in risk factors in eastern and western Europe. *Eur. J. Cancer Prevention* (ECP). **8**, 67–71 (1999).
10. Cole P, Rodu B. Declining cancer mortality in the United States. *Cancer* **78**, 2045–2048 (1996).
11. Tao JJ, Visvanathan K, Wolff AC. Long term side effects of adjuvant chemotherapy in patients with early breast cancer. *Breast* **24**(2), S149–S153 (2015).
12. Wujcik D. Current research in side effects of high-dose chemotherapy. *Sem. Oncol. Nur.* **8**, 102–112 (1992).
13. Herrera FG, Cruz OS, Achtari C, Bourhis J, Ozsahin M. Long-term outcome and late side effects in endometrial cancer patients treated with surgery and postoperative radiation therapy. *Ann. Surg. Oncol.* **21**, 2390–2397 (2014).

14. Krukiewicz K, Zak JK. Biomaterial-based regional chemotherapy: Local anticancer drug delivery to enhance chemotherapy and minimize its side-effects. *Mater. Sci. Eng. C Mater. Biol. Appl.* **62**, 927–942 (2016).
15. Banerjee HN, Verma M. Use of nanotechnology for the development of novel cancer biomarkers. *Expert. Rev. Mol. Diagn.* **6**, 679–683 (2006).
16. Kozako T, Arima N, Yoshimitsu M, Honda SI, Soeda S. Liposomes and nanotechnology in drug development: focus on oncotargets. *Int. J. Nanomedicine* 7, 4943–4951 (2012).
17. Viswanath B, Kim S, Lee K. Recent insights into nanotechnology development for detection and treatment of colorectal cancer. *Int. J. Nanomedicine* **11**, 2491–2504 (2016).
18. Walt DR. Nanomaterials: Top-to-bottom functional design. *Nat. Mater.* **1**, 17–18 (2002).
19. Cheng L, Wang C, Feng L, Yang K, Liu Z. Functional nanomaterials for phototherapies of cancer. *Chem. Rev.* **114**, 10869–10939 (2014).
20. Yin Y, Talapin D. The chemistry of functional nanomaterials. *Chem. Soc. Rev.* **42**, 2484–2487 (2013).
21. Fernandez-Fernandez A, Manchanda R, McGoron AJ. Theranostic applications of nanomaterials in cancer: Drug delivery, image-guided therapy, and multifunctional platforms. *Appl. Biochem. Biotechnol.* **165**, 1628–1651 (2011).
22. Hammond PT. Virtual issue on nanomaterials for drug delivery. *ACS Nano.* **5**, 681–684 (2011).
23. Hubbell JA, Chilkoti A. Chemistry. Nanomaterials for drug delivery. *Science* **337**, 303–305 (2012).
24. Habibi N, Kamaly N, Memic A, Shafiee H. Self-assembled peptide-based nanostructures: Smart nanomaterials toward targeted drug delivery. *Nano Today* **11**, 41–60 (2016).
25. Oliveira ON, Jr. Iost RM, Siqueira JR, Jr. Crespilho FN, Caseli L. Nanomaterials for diagnosis: Challenges and applications in smart devices based on molecular recognition. *ACS. Appl. Mater. Int.* **6**, 14745–14766 (2014).
26. Pietronave S, Iafisco M, Locarno D, Rimondini L, Maria Prat M. Functionalized nanomaterials for diagnosis and therapy of cancer. *J. Appl. Biomater. Biomech.* 7, 77–89 (2009).
27. Rosenblum LT, Kosaka N, Mitsunaga M, Choyke PL, Kobayashi H. *In vivo* molecular imaging using nanomaterials: General *in vivo* characteristics of nano-sized reagents and applications for cancer diagnosis. *Mol. Membr. Biol.* **27**, 274–285 (2010).

28. Khaliq NU, Sandra FC, Park DY, Lee JY, Oh KS, Kim D, Byun Y, Kim IS, Kwon IC, Kim SY, Yuk SH. Doxorubicin/heparin composite nanoparticles for caspase-activated prodrug chemotherapy. *Biomatrials* **101**, 131–142 (2016).
29. Li M, Luo Z, Zhao Y. Hybrid nanoparticles as drug carriers for controlled chemotherapy of cancer. *Chemical. Record* **16**, 1833–1851 (2016).
30. Li SY, Sun R, Wang HX, Shen S, Liu Y, Du XJ, Zhu YH, Jun W. Combination therapy with epigenetic-targeted and chemotherapeutic drugs delivered by nanoparticles to enhance the chemotherapy response and overcome resistance by breast cancer stem cells. *J. Control. Release* **205**, 7–14 (2015).
31. Alexiou A, Vairaktarakis C, Tsiamis V, Ashraf GM. Application of efficient nanoparticles for early diagnosis and treatment of cancer. *Curr. Drug Metab.* **16**, 662–675 (2015).
32. Zhang C, Yan Y, Zou Q, Chen J, Li C. Superparamagnetic iron oxide nanoparticles for MR imaging of pancreatic cancer: Potential for early diagnosis through targeted strategies. *Asia Pac. J. Clini. Oncol.* **12**, 13–21 (2016).
33. Laurentius LB, Owens NA, Park J, Crawford AC, Porter MD. Advantages and limitations of nanoparticle labeling for early diagnosis of infection. *Expert Rev. Mol. Diagn.* **16**, 883–895 (2016).
34. Cho H, Yeh EC, Sinha R, Laurence TA, Bearinger JP, Lee LP. Single-step nanoplasmonic VEGF165 aptasensor for early cancer diagnosis. *ACS Nano.* **6**, 7607–7614 (2012).
35. Cheng K, Cheng Z. Near infrared receptor-targeted nanoprobes for early diagnosis of cancers. *Curr. Med. Chem.* **19**, 4767–4785 (2012).
36. Xifre-Perez E, Guaita-Esteruelas S, Baranowska M, Pallares J, Masana L, Marsal LF. *In vitro* biocompatibility of surface-modified porous alumina particles for hepg2 tumor cells: Toward early diagnosis and targeted treatment. *ACS. Appl. Mater. Int.* 7, 18600–18608 (2015).
37. Wallnofer EA, Thurner GC, Abdelmoez AA, Rohr I, Klammsteiner N, Talasz H, Kremser C, Jaschke W, Debbage P. MRI molecular imaging with nanoparticles: A Technical platform for early diagnosis of cancer. *Int. J. Clin. Pharmacol. Ther.* **49**, 73–74 (2011).
38. Jia Z, Song L, Zang F, Song J, Zhang W, Yan C, Xie J, Ma Z, Ma M, Teng G, Gu N, Zhang Y. Active-target T1-weighted MR imaging of tiny

hepatic tumor via RGD modified ultra-small Fe3O4 nanoprobes. *Theranostics* **6**, 1780–1791 (2016).
39. Zhang X, Zheng Y, Wang Z, Huang S, Chen Y, Jiang W, Zhang H, Ding M, Li Q, Xiao X, Luo X, Wang Z, Qi H. Methotrexate-loaded PLGA nanobubbles for ultrasound imaging and synergistic targeted therapy of residual tumor during HIFU ablation. *Biomaterials* **35**, 5148–5161 (2014).
40. Shi J, Wang L, Zhang J, Ma R, Gao J, Liu Y, Zhang C, Zhang Z. A tumor-targeting near-infrared laser-triggered drug delivery system based on GO@Ag nanoparticles for chemo-photothermal therapy and X-ray imaging. *Biomaterials* **35**, 5847–5861 (2014).
41. Owens EA, Henary M, Ei Fakhri G, Choi HS. Tissue-specific near-infrared fluorescence imaging. *Acc. Chem. Res.* **49**, 1731–1740 (2016).
42. Zhang P, Hu C, Ran W, Meng J, Yin Q, Li Y. Recent progress in light-triggered nanotheranostics for cancer treatment. *Theranostics* **6**, 948–968 (2016).
43. Jo SD, Ku SH, Won YY, Kim SH, Kwon IC. Targeted nanotheranostics for future personalized medicine: Recent progress in cancer therapy. *Theranostics* **6**, 1362–1377 (2016).
44. Kunjachan S, Ehling J, Storm G, Kiessling F, Lammers T. Noninvasive imaging of nanomedicines and nanotheranostics: Principles, progress, and prospects. *Chem. Rev.* **115**, 10907–10937 (2015).
45. Hu B, Dai F, Fan Z, Ma G, Tang Q, Zhang X. Nanotheranostics: Congo red/rutin-MNPs with enhanced magnetic resonance imaging and H2O2-responsive therapy of alzheimer's disease in APPswe/PS1dE9 transgenic mice. *Adv. Mater.* **27**, 5499–5505 (2015).
46. Sharma S. Nanotheranostics in evidence based personalized medicine. *Curr. Drug Targets* **15**, 915–930 (2014).
47. Muthu MS, Leong DT, Mei L, Feng SS. Nanotheranostics — application and further development of nanomedicine strategies for advanced theranostics. *Theranostics* **4**, 660–677 (2014).
48. Mura S, Couvreur P. Nanotheranostics for personalized medicine. *Adv. Drug Deliv. Rev.* **64**, 1394–1416 (2012).
49. Snyderman R, Meade C, Drake C. Value of Personalized medicine. *JAMA* **315**, 613 (2016).
50. Muthu MS, Mei L, Feng SS. Nanotheranostics: Advanced nanomedicine for the integration of diagnosis and therapy. *Nanomedicine* **9**, 1277–1280 (2014).

51. Roy Chowdhury M, Schumann C, Bhakta-Guha D, Guha G. Cancer nanotheranostics: Strategies, promises and impediments. *Biomed. Pharmacothera* **84**, 291–304 (2016).
52. Gautier J, Allard-Vannier E, Munnier E, Souce M, Chourpa I. Recent advances in theranostic nanocarriers of doxorubicin based on iron oxide and gold nanoparticles. *J. Control. Release* **169**, 48–61 (2013).
53. Li Y, Duong HT, Laurent S, MacMillan A, Whan RM, Elst LV, Muller RN, Hu J, Lowe A, Boyer C, Davis TP. Nanoparticles based on star polymers as theranostic vectors: endosomal-triggered drug release combined with MRI sensitivity. *Adv. Health. Mater.* **4**, 148–156 (2015).
54. Li Y, Ma J, Zhu H, Gao X, Dong H, Shi D. Green synthetic, multifunctional hybrid micelles with shell embedded magnetic nanoparticles for theranostic applications. *ACS Appl. Mater. Int.* **5**, 7227–7235 (2013).
55. Dong Y, Chen H, Chen C, Zhang X, Tian X, Zhang Y, Shi Z, Liu Q. Polymer-lipid hybrid theranostic nanoparticles co-delivering ultrasmall superparamagnetic iron oxide and paclitaxel for targeted magnetic resonance imaging and therapy in atherosclerotic plaque. *J. Biomed. Nanotechnol.* **12**, 1245–1257 (2016).
56. Liu Y, Li J, Liu F, Feng L, Yu D, Zhang N. Theranostic polymeric micelles for the diagnosis and treatment of hepatocellular carcinoma. *J. Biomed. Nanotechnol.* **11**, 613–622 (2015).
57. Kim YD, Park TE, Singh B, Maharjan S, Choi YJ, Choung PH, Arote RB, Cho CS. Nanoparticle-mediated delivery of siRNA for effective lung cancer therapy. *Nanomedicine* **10**, 1165–1188 (2015).
58. Zeng Y, Zhang D, Wu M, Liu Y, Zhang X, Li L, Li Z, Han X, Wei X, Liu X. Lipid-AuNPs@PDA nanohybrid for MRI/CT imaging and photothermal therapy of hepatocellular carcinoma. *ACS Appl. Mater. Int.* **6**, 14266–14277 (2014).
59. Paliwal SR, Paliwal R, Vyas SP. A review of mechanistic insight and application of pH-sensitive liposomes in drug delivery. *Drug Delivery* **22**, 231–242 (2015).
60. Li Z, Barnes JC, Bosoy A, Stoddart JF, Zink JI. Mesoporous silica nanoparticles in biomedical applications. *Chem. Soc. Rev.* **41**, 2590–2605 (2012).
61. Vivero-Escoto JL, Huxford-Phillips RC, Lin W. Silica-based nanoprobes for biomedical imaging and theranostic applications. *Chem. Soc. Rev.* **41**, 2673–2685 (2012).
62. Chan MH, Lin HM. Preparation and identification of multifunctional mesoporous silica nanoparticles for *in vitro* and *in vivo* dual-mode

imaging, theranostics, and targeted tracking. *Biomaterials* **46**, 149–158 (2015).

63. Yang K, Feng L, Shi X, Liu Z. Nano-graphene in biomedicine: Theranostic applications. *Chem. Soc. Rev.* **42**, 530–547 (2013).
64. Chen D, Dougherty CA, Zhu K, Hong H. Theranostic applications of carbon nanomaterials in cancer: Focus on imaging and cargo delivery. *J. Control. Release* **210**, 230–245 (2015).
65. Bartelmess J, Quinn SJ, Giordani S. Carbon nanomaterials: Multifunctional agents for biomedical fluorescence and Raman imaging. *Chem. Soc. Rev.* **44**, 4672–4698 (2015).
66. Song J, Wang F, Yang X, Ning B, Harp MG, Culp SH, Hu S, Huang P, Nie L, Chen J, Chen X. Gold Nanoparticle coated carbon nanotube ring with enhanced raman scattering and photothermal conversion property for theranostic applications. *J. Am. Chem. Soc.* **138**, 7005–7015 (2016).
67. Kim JY, Ryu JH, Schellingerhout D, Sun IC, Lee SK, Jeon S, Kim J, Kwon IC, Nahrendorf M, Ahn CH, Kim K, Kim DE. Direct imaging of cerebral thromboemboli using computed tomography and fibrin-targeted gold nanoparticles. *Theranostics* **5**, 1098–1114 (2015).
68. Chen M, Tang S, Guo Z, Wang X, Mo S, Huang X, Liu G, Zheng N. Core-shell Pd@Au nanoplates as theranostic agents for *in vivo* photoacoustic imaging, CT imaging, and photothermal therapy. *Adv. Mater.* **26**, 8210–8216 (2014).
69. Bollhorst T, Shahabi S, Worz K, Petters C, Dringen R, Maas M, Rezwan K. Bifunctional submicron colloidosomes coassembled from fluorescent and superparamagnetic nanoparticles. *Angewandte Chemie* **54**, 118–123 (2015).
70. Yang K, Hu L, Ma X, Ye S, Cheng L, Shi X, Li C, Li Y, Liu Z. Multimodal imaging guided photothermal therapy using functionalized graphene nanosheets anchored with magnetic nanoparticles. *Adv. Mater.* **24**, 1868–1872 (2012).
71. Gelderblom H, Verweij J, Nooter K, Sparreboom A. Cremophor EL: the drawbacks and advantages of vehicle selection for drug formulation. *Eur. J. Cancer* **37**, 1590–1598 (2001).
72. Cho K, Wang X, Nie S, Chen ZG, Shin DM. Therapeutic nanoparticles for drug delivery in cancer. *Clin. Cancer Res.* **14**, 1310–1316 (2008).
73. Cole JT, Holland NB. Multifunctional nanoparticles for use in theranostic applications. *Drug Deliv. Transl. Res.* **5**, 295–309 (2015).

74. Yu H, Tang Z, Zhang D, Song W, Zhang Y, Yang Y, Ahmad Z, Chen X. Pharmacokinetics, biodistribution and *in vivo* efficacy of cisplatin loaded poly(L-glutamic acid)-g-methoxy poly(ethylene glycol) complex nanoparticles for tumor therapy. *J. Control. Release* **205**, 89–97 (2015).
75. Arami H, Khandhar A, Liggitt D, Krishnan KM. In vivo delivery, pharmacokinetics, biodistribution and toxicity of iron oxide nanoparticles. *Chem. Soc. Rev.* **44**, 8576–8607 (2015).
76. Perez-Campana C, Gomez-Vallejo V, Puigivila M, Martin A. Calvo-Fernandez T, Moya SE, Ziolo RF, Reese T, Llop J. Biodistribution of different sized nanoparticles assessed by positron emission tomography: A general strategy for direct activation of metal oxide particles. *ACS Nano.* 7, 3498–3505 (2013).
77. Sahay G, Alakhova DY, Kabanov AV. Endocytosis of nanomedicines. *J. Control. Release* **145**, 182–195 (2010).
78. Elkhodiry MA, Momah CC, Suwaidi SR, Gadalla D, Martins AM, Vitor RF, Husseini GA. Synergistic nanomedicine: Passive, active, and ultrasound-triggered drug delivery in cancer treatment. *J. Nanosci. Nanotechnol.* **16**, 1–18 (2016).
79. Gratton SE, Ropp PA, Pohlhaus PD, Luft JC, Madden VJ, Napier ME, DeSimone JM. The effect of particle design on cellular internalization pathways. *Proc. Nat. Acad. Sci. USA.* **105**, 11613–11618 (2008).
80. Wang J, Byrne JD, Napier ME, DeSimone JM. More effective nanomedicines through particle design. *Small* 7, 1919–1931 (2011).
81. Choi HS, Liu W, Misra P, Tanaka E, Zimmer JP, Itty Ipe B, Bawendi MG, Frangioni JV. Renal clearance of quantum dots. *Nat. Biotechnol.* **25**, 1165–1170 (2007).
82. Moghimi SM, Szebeni J. Stealth liposomes and long circulating nanoparticles: critical issues in pharmacokinetics, opsonization and protein-binding properties. *Prog. Lipid. Res.* **42**, 463–478 (2003).
83. Kamaly N, Xiao Z, Valencia PM, Radovic-Moreno AF, Farokhzad OC. Targeted polymeric therapeutic nanoparticles: Design, development and clinical translation. *Chem. Soc. Rev.* **41**, 2971–3010 (2012).
84. Wang Q, Sun Y, Zhang Z, Duan Y. Targeted polymeric therapeutic nanoparticles: Design and interactions with hepatocellular carcinoma. *Biomaterials* **56**, 229–240 (2015).
85. Toy R, Peiris PM, Ghaghada KB, Karathanasis E. Shaping cancer nanomedicine: The effect of particle shape on the *in vivo* journey of nanoparticles. *Nanomedicine* **9**, 121–134 (2014).

86. Shi W, Wang J, Fan X, Gao H. Size and shape effects on diffusion and absorption of colloidal particles near a partially absorbing sphere: Implications for uptake of nanoparticles in animal cells. *Phy. Rev. E. Stat. Nonlin. Soft. Matter. Phys.* **78**, 061914 (2008).
87. Toy R, Bauer L, Hoimes C, Ghaghada KB, Karathanasis E. Targeted nanotechnology for cancer imaging. *Adv. Drug. Delivery. Rev.* **76**, 79–97 (2014).
88. Decuzzi P, Godin B, Tanaka T, Lee SY, Chiappini C, Liu X, Ferrari M. Size and shape effects in the biodistribution of intravascularly injected particles. *J. Control. Rel.* **141**, 320–327 (2010).
89. Calatayud MP, Sanz B, Raffa V, Riggio C, Ibarra MR, Goya GF. The effect of surface charge of functionalized Fe3O4 nanoparticles on protein adsorption and cell uptake. *Biomaterials* **35**, 6389–6399 (2014).
90. Zhang Y, Yang M, Park JH, Singelyn J, Ma H, Sailor MJ, Ruoslahti E, Ozkan M, Ozkan C. A surface-charge study on cellular-uptake behavior of F3-peptide-conjugated iron oxide nanoparticles. *Small* **5**, 1990–1996 (2009).
91. Nam HY, Kwon SM, Chung H, Lee SY, Kwon SH, Jeon H, Kim Y, Park JH, Kim J, Her S, Oh YK, Kwon IC, Kim K, Jeong SY. Cellular uptake mechanism and intracellular fate of hydrophobically modified glycol chitosan nanoparticles. *J. Control. Rel.* **135**, 259–267 (2009).
92. Ge Y, Zhang Y, Xia J, Ma M, He S, Nie F, Gu N. Effect of surface charge and agglomerate degree of magnetic iron oxide nanoparticles on KB cellular uptake *in vitro*, colloids and surfaces. *B. Biointerfaces* **73**, 294–301 (2009).
93. Win KY, Feng SS. Effects of particle size and surface coating on cellular uptake of polymeric nanoparticles for oral delivery of anticancer drugs. *Biomaterials* **26**, 2713–2722 (2005).
94. Jambhrunkar S, Qu Z, Popat A, Yang J, Noonan O, Acauan L, Ahmad Nor Y, Yu C, Karmakar S. Effect of surface functionality of silica nanoparticles on cellular uptake and cytotoxicity. *Mol. Pharm.* **11**, 3642–3655 (2014).
95. Yue ZG, Wei W, L PP, Yue H, Wang LY, Su ZG, Ma GH. Surface charge affects cellular uptake and intracellular trafficking of chitosan-based nanoparticles. *Biomacromolecules* **12**, 2440–2446 (2011).
96. Frohlich E. The role of surface charge in cellular uptake and cytotoxicity of medical nanoparticles. *Int. J. Nanomedicine* 7, 5577–5591 (2012).

97. Hill A, Payne CK. Impact of serum proteins on MRI contrast agents: Cellular binding and T2 relaxation. *RSC Adv.* **4**, 31735–31744 (2014).
98. Liu Z, Niu D, Zhang J, Zhang W, Yao Y, Li P, Gong J. Amphiphilic core-shell nanoparticles containing dense polyethyleneimine shells for efficient delivery of microRNA to Kupffer cells. *Int. J. Nanomedicine* **11**, 2785–2797 (2016).
99. Raja MA, Zeenat S, Arif M, Liu C. Self-assembled nanoparticles based on amphiphilic chitosan derivative and arginine for oral curcumin delivery. *Int. J. Nanomedicine* **11**, 4397–4412 (2016).
100. Gu L, Faig A, Abdelhamid D, Uhrich K. Sugar-based amphiphilic polymers for biomedical applications: From nanocarriers to therapeutics. *Acc. Chem. Res.* **47**, 2867–2877 (2014).
101. Guo X, Huang L. Recent advances in nonviral vectors for gene delivery. *Acc. Chem. Res.* **45**, 971–979 (2012).
102. Cui L, Lin Q, Jin CS, Jiang W, Huang H, Ding L, Muhanna N, Irish JC, Wang F, Chen J, Zheng G. A PEGylation-free biomimetic porphyrin nanoplatform for personalized cancer theranostics. *ACS Nano.* **9**, 4484–4495 (2015).
103. Dehaini D, Fang RH, Luk BT, Pang Z, Hu CM, Kroll AV, Yu CL, Gao W, Zhang L. Ultra-small lipid-polymer hybrid nanoparticles for tumor-penetrating drug delivery. *Nanoscale* **8**, 14411–14419 (2016).
104. Karimi M, Ghasemi A, Sahandi Zangabad P, Rahighi R, Moosavi Basri SM, Mirshekari H, Amiri M, Shafaei Pishabad Z, Aslani A, Bozorgomid M, Ghosh D, Beyzavi A, Vaseghi A, Aref AR, Haghani L, Bahrami S, Hamblin MR. Smart micro/nanoparticles in stimulus-responsive drug/gene delivery systems. *Chem. Soc. Rev.* **45**, 1457–1501 (2016).
105. Khandare J, Calderon M, Dagia NM, Haag R. Multifunctional dendritic polymers in nanomedicine: Opportunities and challenges. *Chem. Soc. Rev.* **41**, 2824–2848 (2012).
106. Zhang R, Yang J, Sima M, Zhou Y, Kopecek J. Sequential combination therapy of ovarian cancer with degradable N-(2-hydroxypropyl)methacrylamide copolymer paclitaxel and gemcitabine conjugates. *Proc. Nat. Acad. Sci. USA.* **111**, 12181–12186 (2014).
107. Yang Y, Zhou Z, He S, Fan T, Jin Y, Zhu X, Chen C, Zhang ZR, Huang Y. Treatment of prostate carcinoma with (galectin-3)-targeted HPMA copolymer-(G3-C12)-5-Fluorouracil conjugates. *Biomaterials* **33**, 2260–2271 (2012).
108. Oberoi HS, Nukolova NV, Kabanov AV, Bronich TK. Nanocarriers for delivery of platinum anticancer drugs. *Adv. Drug Delivery Rev.* **65**, 1667–1685 (2013).

109. Bae HH, Cho MY, Hong JH, Poo H, Sung MH, Lim YT. Bio-derived poly(gamma-glutamic acid) nanogels as controlled anticancer drug delivery carriers. *J. Microbiol. Biotechnol.* **22**, 1782–1789 (2012).
110. Makadia HK, Siegel SJ. Poly Lactic-co-glycolic acid (PLGA) as biodegradable controlled drug delivery carrier. *Polymers* **3**, 1377–1397 (2011).
111. Pakulska MM, Miersch S, Shoichet MS. Designer protein delivery: From natural to engineered affinity-controlled release systems. *Science* **351**, aac4750 (2016).
112. Zhu H, Chen H, Zeng X, Wang Z, Zhang X, Wu Y, Gao Y, Zhang J, Liu K, Liu R, Cai L, Mei L, Feng SS. Co-delivery of chemotherapeutic drugs with vitamin E TPGS by porous PLGA nanoparticles for enhanced chemotherapy against multi-drug resistance. *Biomaterials* **35**, 2391–2400 (2014).
113. Huang WT, Larsson M, Wang YJ, Chiou SH, Lin HY, Liu DM. Demethoxycurcumin-carrying chitosan-antibody core-shell nanoparticles with multitherapeutic efficacy toward malignant A549 lung tumor: from *in vitro* characterization to *in vivo* evaluation. *Mole. Phar.* **12**, 1242–1249 (2015).
114. Obara K, Ishihara M, Ozeki Y, Ishizuka T, Hayashi T, Nakamura S, Saito Y, Yura H, Matsui T, Hattori H, Takase B, Ishihara M, Kikuchi M, Maehara T. Controlled release of paclitaxel from photocrosslinked chitosan hydrogels and its subsequent effect on subcutaneous tumor growth in mice. *J. Control. Rel.* **110**, 79–89 (2005).
115. Verma NK, Purohit MP, Equbal D, Dhiman N, Singh A, Kar AK, Shankar J, Tehlan S, Patnaik S. Targeted smart pH and thermoresponsive N,O-carboxymethyl chitosan conjugated nanogels for enhanced therapeutic efficacy of DOX in MCF-7 breast cancer cells. *Bioconjug. Chem.* **27**, 2605–2619 (2016).
116. Nogueira DR, Scheeren LE, Macedo LB, Marcolino AI, Pilar Vinardell M, Mitjans M, Rosa Infante M, Farooqi AA, Rolim CM. Inclusion of a pH-responsive amino acid-based amphiphile in methotrexate-loaded chitosan nanoparticles as a delivery strategy in cancer therapy. *Amino Acids* **48**, 157–168 (2016).
117. Cheung RY, Ying Y, Rauth AM, Marcon N, Yu Wu X. Biodegradable dextran-based microspheres for delivery of anticancer drug mitomycin C. *Biomaterials* **26**, 5375–5385 (2005).
118. Dai T, Zhou S, Yin C, Li S, Cao W, Liu W, Sun K, Dou H, Cao Y, Zhou G. Dextran-based fluorescent nanoprobes for sentinel lymph node mapping. *Biomaterials* **35**, 8227–8235 (2014).

119. Zhang HZ, Li XM, Gao FP, Liu LR, Zhou ZM, Zhang QQ. Preparation of folate-modified pullulan acetate nanoparticles for tumor-targeted drug delivery. *Drug Delivery* **17**, 48–57 (2010).
120. Liu Y, Wang Y, Zhang C, Zhou P, Liu Y, An T, Sun D, Zhang N, Wang Y. Core-shell nanoparticles based on pullulan and poly(beta-amino) ester for hepatoma-targeted codelivery of gene and chemotherapy agent. *ACS. Appl. Mater. Int.* **6**, 18712–18720 (2014).
121. Assanhou AG, Li W, Zhang L, Xue L, Kong L, Sun H, Mo R, Zhang C. Reversal of multidrug resistance by co-delivery of paclitaxel and lonidamine using a TPGS and hyaluronic acid dual-functionalized liposome for cancer treatment. *Biomaterials* **73**, 284–295 (2015).
122. Choi KM, Kwon IC, Ahn HJ. Self-assembled amphiphilic DNA-cholesterol/DNA-peptide hybrid duplexes with liposome-like structure for doxorubicin delivery. *Biomaterials* **34**, 4183–4190 (2013).
123. Ekdawi SN, Stewart JM, Dunne M, Stapleton S, Mitsakakis N, Dou YN, Jaffray DA, Allen C. Spatial and temporal mapping of heterogeneity in liposome uptake and microvascular distribution in an orthotopic tumor xenograft model. *J. Control. Rel.* **207**, 101–111 (2015).
124. Liu Z, Xiong M, Gong J, Zhang Y, Bai N, Luo Y, Li L, Wei Y, Liu Y, Tan X, Xiang R. Legumain protease-activated TAT-liposome cargo for targeting tumours and their microenvironment. *Nat. Commun.* **5**, 4280 (2014).
125. Fan Y, Sahdev P, Ochyl LJ, J JA, Moon JJ. Cationic liposome-hyaluronic acid hybrid nanoparticles for intranasal vaccination with subunit antigens. *J. Control. Rel.* **208**, 121–129 (2015).
126. Wu WD, Yi XL, Jiang LX, Li YZ, Gao J, Zeng Y, Yi RD, Dai LP, Li W, Ci XY, Si DY, Liu CX. The Targeted-liposome Delivery System of Antitumor Drugs. *Curr. Drug Metab.* **16**, 894–910 (2015).
127. Namazi H, Adeli M. Dendrimers of citric acid and poly (ethylene glycol) as the new drug-delivery agents. *Biomaterials* **26**, 1175–1183 (2005).
128. Spataro G, Malecaze F, Turrin CO, Soler V, Duhayon C, Elena PP, Majoral JP, Caminade AM. Designing dendrimers for ocular drug delivery. *Eur. J. Med. Chem.* **45**, 326–334 (2010).
129. Dhanikula RS, Hildgen P. Synthesis and evaluation of novel dendrimers with a hydrophilic interior as nanocarriers for drug delivery. *Bioconjug. Chem.* **17**, 29–41 (2006).
130. Svenson S. Dendrimers as versatile platform in drug delivery applications. *Eur. J. Pharm. Biopharm.* **71**, 445–462 (2009).

131. Pisani MJ, Wheate NJ, Keene FR, Aldrich-Wright JR, Collins JG. Anionic PAMAM dendrimers as drug delivery vehicles for transition metal-based anticancer drugs. *J. Inorg. Biochem.* **103**, 373–380 (2009).
132. Jiang YY, Tang GT, Zhang LH, Kong SY, Zhu SJ, Pei YY. PEGylated PAMAM dendrimers as a potential drug delivery carrier: *In vitro* and in vivo comparative evaluation of covalently conjugated drug and non-covalent drug inclusion complex. *J. Drug Targeting* **18**, 389–403 (2010).
133. Yellepeddi VK, Kumar A, Palakurthi S. Biotinylated poly(amido)amine (PAMAM) dendrimers as carriers for drug delivery to ovarian cancer cells in vitro. *Anticancer Res.* **29**, 2933–2943 (2009).
134. Dutta T, Agashe HB, Garg M, Balakrishnan P, Kabra M, Jain NK. Poly (propyleneimine) dendrimer based nanocontainers for targeting of efavirenz to human monocytes/macrophages *in vitro*. *J. Drug Targeting* **15**, 89–98 (2007).
135. Patel HK, Gajbhiye V, Kesharwani P, Jain NK. Ligand anchored poly(propyleneimine) dendrimers for brain targeting: Comparative *in vitro* and *in vivo* assessment. *J. Colloid. Int. Sci.* **482**, 142–150 (2016).
136. Kuo JH, Jan MS, Lin YL. Interactions between U-937 human macrophages and poly(propyleneimine) dendrimers. *J Control. Rel.* **120**, 51–59 (2007).
137. Kesharwani P, Tekade RK, Jain NK. Generation dependent cancer targeting potential of poly(propyleneimine) dendrimer. *Biomaterials* **35**, 5539–5548 (2014).
138. Li W, Huang L, Ying X, Jian Y, Hong Y, Hu F, Du Y. Antitumor drug delivery modulated by a polymeric micelle with an upper critical solution temperature. *Angewandte Chemie* **54**, 3126–3131 (2015).
139. Lu HL, Syu WJ, Nishiyama N, Kataoka K, Lai PS. Dendrimer phthalocyanine-encapsulated polymeric micelle-mediated photochemical internalization extends the efficacy of photodynamic therapy and overcomes drug-resistance *in vivo*. *J. Control. Rel.* **155**, 458–464 (2011).
140. Mei H, Shi W, Pang Z, Wang H, Lu W, Jiang X, Deng J, Guo T, Hu Y. EGFP-EGF1 protein-conjugated PEG-PLA nanoparticles for tissue factor targeted drug delivery. *Biomaterials* **31**, 5619–5626 (2010).
141. Dou S, Yang XZ, Xiong MH, Sun CY, Yao YD, Zhu YH, Wang J. ScFv-decorated PEG-PLA-based nanoparticles for enhanced siRNA delivery to Her2(+) breast cancer. *Adv. Health. Mater.* **3**, 1792–1803 (2014).

142. Kim TY, Kim DW, Chung JY, Shin SG, Kim SC, Heo DS, Kim NK, Bang YJ. Phase I and pharmacokinetic study of Genexol-PM, a cremophor-free, polymeric micelle-formulated paclitaxel, in patients with advanced malignancies. *Clin. Cancer Res.* **10**, 3708–3716 (2004).
143. Lee KS, Chung HC, Im SA, Park YH, Kim CS, Kim SB, Rha SY, Lee MY, Ro J. Multicenter phase II trial of Genexol-PM, a Cremophor-free, polymeric micelle formulation of paclitaxel, in patients with metastatic breast cancer. *Breast Cancer Res. Treat.* **108**, 241–250 (2008).
144. Endres T, Zheng M, Beck-Broichsitter M, Kissel T. Lyophilised ready-to-use formulations of PEG-PCL-PEI nano-carriers for siRNA delivery. *Int. J. Pharm.* **428**, 121–124 (2012).
145. Xin H, Sha X, Jiang X, Zhang W, Chen L, Fang X. Anti-glioblastoma efficacy and safety of paclitaxel-loading Angiopep-conjugated dual targeting PEG-PCL nanoparticles. *Biomaterials* **33**, 8167–8176 (2012).
146. Chi Y, Zhu S, Wang C, Zhou L, Zhang L, Li Z, Dai Y. Glioma homing peptide-modified PEG-PCL nanoparticles for enhanced anti-glioma therapy. *J. Drug Targeting* **24**, 224–232 (2016).
147. Wang Y, Chen L, Tan L, Zhao Q, Luo F, Wei Y, Qian Z. PEG-PCL based micelle hydrogels as oral docetaxel delivery systems for breast cancer therapy. *Biomaterials* **35**, 6972–6985 (2014).
148. Diao YY, Li HY, Fu YH, Han M, Hu YL, Jiang HL, Tsutsumi Y, Wei QC, Chen DW, Gao JQ. Doxorubicin-loaded PEG-PCL copolymer micelles enhance cytotoxicity and intracellular accumulation of doxorubicin in adriamycin-resistant tumor cells. *Int. J. Nanomedicine* **6**, 1955–1962 (2011).
149. Kakizawa Y, Furukawa S, Kataoka K. Block copolymer-coated calcium phosphate nanoparticles sensing intracellular environment for oligodeoxynucleotide and siRNA delivery. *J. Control. Rel.* **97**, 345–356 (2004).
150. Kim J, Kim HS, Lee N, Kim T, Kim H, Yu T, Song IC, Moon WK, Hyeon T. Multifunctional uniform nanoparticles composed of a magnetite nanocrystal core and a mesoporous silica shell for magnetic resonance and fluorescence imaging and for drug delivery. *Angewandte Chemie* **47**, 8438–8441 (2008).
151. Zhao YL, Li Z, Kabehie S, Botros YY, Stoddart JF, Zink JI. pH-operated nanopistons on the surfaces of mesoporous silica nanoparticles. *J. Am. Chem. Soc.* **132**, 13016–13025 (2010).
152. Luo Z, Cai K, Hu Y, Zhao L, Liu P, Duan L, Yang W. Mesoporous silica nanoparticles end-capped with collagen: redox-responsive

nanoreservoirs for targeted drug delivery. *Angewandte Chemie* **50**, 640–643 (2011).
153. Muhammad F, Guo M, Qi W, Sun F, Wang A, Guo Y, Zhu G. pH-Triggered controlled drug release from mesoporous silica nanoparticles via intracelluar dissolution of ZnO nanolids. *J. Am. Chem. Soc.* **133**, 8778–8781 (2011).
154. Agostini A, Mondragon L, Bernardos A, Martinez-Manez R, Marcos MD, Sancenon F, Soto J, Costero A, Manguan-Garcia C, Perona R, Moreno-Torres M, Aparicio-Sanchis R, Murguia JR. Targeted cargo delivery in senescent cells using capped mesoporous silica nanoparticles. *Angewandte Chemie* **51**, 10556–10560 (2012).
155. Li ZY, Liu Y, Wang XQ, Liu LH, Hu JJ, Luo GF, Chen WH, Rong L, Zhang XZ. One-pot construction of functional mesoporous silica nanoparticles for the tumor-acidity-activated synergistic chemotherapy of glioblastoma. *ACS Appl. Mater. Int.* **5**, 7995–8001 (2013).
156. Liu J, Bu W, Pan L, Shi J. NIR-triggered anticancer drug delivery by upconverting nanoparticles with integrated azobenzene-modified mesoporous silica. *Angewandte Chemie* **52**, 4375–4379 (2013).
157. Ahmadi E, Dehghannejad N, Hashemikia S, Ghasemnejad M, Tabebordbar H. Synthesis and surface modification of mesoporous silica nanoparticles and its application as carriers for sustained drug delivery. *Drug Delivery* **21**, 164–172 (2014).
158. Zhang B, Luo Z, Liu J, Ding X, Li J, Cai K. Cytochrome c end-capped mesoporous silica nanoparticles as redox-responsive drug delivery vehicles for liver tumor-targeted triplex therapy *in vitro* and *in vivo*. *J. Control. Rel.* **192**, 192–201 (2014).
159. Ma M, Huang Y, Chen H, Jia X, Wang S, Wang Z, Shi J. Bi2S3-embedded mesoporous silica nanoparticles for efficient drug delivery and interstitial radiotherapy sensitization. *Biomaterials* **37**, 447–455 (2015).
160. Slowing II, Vivero-Escoto JL, Wu CW, Lin VS. Mesoporous silica nanoparticles as controlled release drug delivery and gene transfection carriers. *Adv. Drug Delivery Rev.* **60**, 1278–1288 (2008).
161. Liu T, Li L, Teng X, Huang X, Liu H, Chen D, Ren J, He J, Tang F. Single and repeated dose toxicity of mesoporous hollow silica nanoparticles in intravenously exposed mice. *Biomaterials* **32**, 1657–1668 (2011).
162. Wu SH, Hung Y, Mou CY. Mesoporous silica nanoparticles as nanocarriers. *Chem. Commun.* **47**, 9972–9985 (2011).

163. Shen J, He Q, Gao Y, Shi J, Li Y. Mesoporous silica nanoparticles loading doxorubicin reverse multidrug resistance: Performance and mechanism. *Nanoscale* **3**, 4314–4322 (2011).
164. Pan L, Liu J, He Q, Wang L, Shi J. Overcoming multidrug resistance of cancer cells by direct intranuclear drug delivery using TAT-conjugated mesoporous silica nanoparticles. *Biomaterials* **34**, 2719–2730 (2013).
165. Lee JE, Lee N, Kim H, Kim J, Choi SH, Kim JH, Kim T, Song IC, Park SP, Moon WK, Hyeon T. Uniform mesoporous dye-doped silica nanoparticles decorated with multiple magnetite nanocrystals for simultaneous enhanced magnetic resonance imaging, fluorescence imaging, and drug delivery. *J. Am. Chem. Soc.* **132**, 552–557 (2010).
166. Lee JE, Lee N, Kim T, Kim J, Hyeon T. Multifunctional mesoporous silica nanocomposite nanoparticles for theranostic applications. *Acc. Chem. Res.* **44**, 893–902 (2011).
167. Chen Y, Yin Q, Ji X, Zhang S, Chen H, Zheng Y, Sun Y, Qu H, Wang Z, Li Y, Wang X, Zhang K, Zhang L, Shi J. Manganese oxide-based multifunctionalized mesoporous silica nanoparticles for pH-responsive MRI, ultrasonography and circumvention of MDR in cancer cells. *Biomaterials* **33**, 7126–7137 (2012).
168. Xue S, Wang Y, Wang M, Zhang L, Du X, Gu H, Zhang C. Iodinated oil-loaded, fluorescent mesoporous silica-coated iron oxide nanoparticles for magnetic resonance imaging/computed tomography/fluorescence trimodal imaging. *Int. J. Nanomed.* **9**, 2527–2538 (2014).
169. Chen F, Hong H, Zhang Y, Valdovinos HF, Shi S, Kwon GS, Theuer CP, Barnhart TE, Cai W. *In vivo* tumor targeting and image-guided drug delivery with antibody-conjugated, radiolabeled mesoporous silica nanoparticles. *ACS Nano.* 7, 9027–9039 (2013).
170. Chen F, Nayak TR, Goel S, Valdovinos HF, Hong H, Theuer CP, Barnhart TE, Cai W. *In vivo* tumor vasculature targeted PET/NIRF imaging with TRC105(Fab)-conjugated, dual-labeled mesoporous silica nanoparticles. *Mol. Pharm.* **11**, 4007–4014 (2014).
171. Kempen PJ, Greasley S, Parker KA, Campbell JL, Chang HY, Jones JR, Sinclair R, Gambhir SS, Jokerst JV. Theranostic mesoporous silica nanoparticles biodegrade after pro-survival drug delivery and ultrasound/magnetic resonance imaging of stem cells. *Theranostics* **5**, 631–642 (2015).
172. Chen Y, Chen H, Shi J. *In vivo* bio-safety evaluations and diagnostic/therapeutic applications of chemically designed mesoporous silica nanoparticles. *Adv. Mater.* **25**, 3144–3176 (2013).

173. Mekaru H, Lu J, Tamanoi F. Development of mesoporous silica-based nanoparticles with controlled release capability for cancer therapy. *Adv. Drug Delivery Rev.* **95**, 40–49 (2015).
174. Tian B, Wang C, Zhang S, Feng L, Liu Z. Photothermally enhanced photodynamic therapy delivered by nano-graphene oxide. *ACS Nano.* **5**, 7000–7009 (2011).
175. Wang X, Wang C, Cheng L, Lee ST, Liu Z. Noble metal coated single-walled carbon nanotubes for applications in surface enhanced Raman scattering imaging and photothermal therapy. *J. Am. Chem. Soc.* **134**, 7414–7422 (2012).
176. Chu M, Peng J, Zhao J, Liang S, Shao Y, Wu Q. Laser light triggered-activated carbon nanosystem for cancer therapy. *Biomaterials* **34**, 1820–1832 (2013).
177. Liang C, Diao S, Wang C, Gong H, Liu T, Hong G, Shi X, Dai H, Liu Z. Tumor metastasis inhibition by imaging-guided photothermal therapy with single-walled carbon nanotubes. *Adv. Mater.* **26**, 5646–5652 (2014).
178. Lim DJ, Sim M, Oh L, Lim K, Park H. Carbon-based drug delivery carriers for cancer therapy. *Arch. Phar. Res.* **37**, 43–52 (2014).
179. Liang C, Xu L, Song G, Liu Z. Emerging nanomedicine approaches fighting tumor metastasis: Animal models, metastasis-targeted drug delivery, phototherapy, and immunotherapy. *Chem. Soc. Rev.* (2016).
180. Yang K, Feng L, Liu Z. Stimuli responsive drug delivery systems based on nano-graphene for cancer therapy. *Adv. Drug Delivery Rev.* **105**, 228–241 (2016).
181. Liu Z, Sun X, Nakayama-Ratchford N, Dai H. Supramolecular chemistry on water-soluble carbon nanotubes for drug loading and delivery. *ACS Nano.* **1**, 50–56 (2007).
182. Ali-Boucetta H, Al-Jamal KT, McCarthy D, Prato M, Bianco A, Kostarelos K. Multiwalled carbon nanotube-doxorubicin supramolecular complexes for cancer therapeutics. *Chem. Commun.* 459–461 (2008).
183. Liu Z, Fan AC, Rakhra K, Sherlock S, Goodwin A, Chen X, Yang Q, Felsher DW, Dai H. Supramolecular stacking of doxorubicin on carbon nanotubes for *in vivo* cancer therapy. *Angewandte Chemie* **48**, 7668–7672 (2009).
184. Zhang X, Meng L, Lu Q, Fei Z, Dyson PJ. Targeted delivery and controlled release of doxorubicin to cancer cells using modified single wall carbon nanotubes. *Biomaterials* **30**, 6041–6047 (2009).

185. Ren J, Shen S, Wang D, Xi Z, Guo L, Pang Z, Qian Y, Sun X, Jiang X. The targeted delivery of anticancer drugs to brain glioma by PEGylated oxidized multi-walled carbon nanotubes modified with angiopep-2. *Biomaterials* **33**, 3324–3333 (2012).
186. Del Canto E, Natali M, Movia D, Giordani S. Photo-controlled release of zinc metal ions by spiropyran receptors anchored to single-walled carbon nanotubes. *Phys. Chem. Chem. Phys.* **14**, 6034–6043 (2012).
187. Bhirde AA, Patel V, Gavard J, Zhang G, Sousa AA, Masedunskas A, Leapman RD, Weigert R, Gutkind JS, Rusling JF. Targeted killing of cancer cells *in vivo* and *in vitro* with EGF-directed carbon nanotube-based drug delivery. *ACS Nano.* **3**, 307–316 (2009).
188. Wong BS, Yoong SL, Jagusiak A, Panczyk T, Ho HK, Ang WH, Pastorin G. Carbon nanotubes for delivery of small molecule drugs. *Adv. Drug Delivery Rev.* **65**, 1964–2015 (2013).
189. Pantarotto D, Singh R, McCarthy D, Erhardt M, Briand JP, Prato M, Kostarelos K, Bianco A. Functionalized carbon nanotubes for plasmid DNA gene delivery. *Angewandte Chemie* **43**, 5242–5246 (2004).
190. Singh R, Pantarotto D, McCarthy D, Chaloin O, Hoebeke J, Partidos CD, Briand JP, Prato M, Bianco A, Kostarelos K. Binding and condensation of plasmid DNA onto functionalized carbon nanotubes: toward the construction of nanotube-based gene delivery vectors. *J. Am. Chem. Soc.* **127**, 4388–4396 (2005).
191. Liu Z, Winters M, Holodniy M, Dai H. siRNA delivery into human T cells and primary cells with carbon-nanotube transporters. *Angewandte Chemie* **46**, 2023–2027 (2007).
192. Herrero MA, Toma FM, Al-Jamal KT, Kostarelos K, Bianco A, Da Ros T, Bano F, Casalis L, Scoles G, Prato M. Synthesis and characterization of a carbon nanotube-dendron series for efficient siRNA delivery. *J. Am. Chem. Soc.* **131**, 9843–9848 (2009).
193. Shen H, Zhang L, Liu M, Zhang Z. Biomedical applications of graphene. *Theranostics* **2**, 283–294 (2012).
194. Kim H, Namgung R, Singha K, Oh IK, Kim WJ. Graphene oxide-polyethylenimine nanoconstruct as a gene delivery vector and bioimaging tool. *Bioconjug Chem.* **22**, 2558–2567 (2011).
195. Sun X, Liu Z, Welsher K, Robinson JT, Goodwin A, Zaric S, Dai H. Nano-graphene oxide for cellular imaging and drug delivery. *Nano Res.* **1**, 203–212 (2008).
196. Liu Z, Liang XJ. Nano-carbons as theranostics. *Theranostics* **2**, 235–237 (2012).

197. Muthoosamy K, Bai RG, Manickam S. Graphene and graphene oxide as a docking station for modern drug delivery system. *Curr. Drug Delivery* **11**, 701–718 (2014).
198. Chen YW, Su YL, Hu SH, Chen SY. Functionalized graphene nanocomposites for enhancing photothermal therapy in tumor treatment. *Adv. Drug Delivery Rev.* **105**, 190–204 (2016).
199. Zhang L, Xia J, Zhao Q, Liu L, Zhang Z. Functional graphene oxide as a nanocarrier for controlled loading and targeted delivery of mixed anticancer drugs. *Small* **6**, 537–544 (2010).
200. Sherlock SP, Tabakman SM, Xie L, Dai H. Photothermally enhanced drug delivery by ultrasmall multifunctional FeCo/graphitic shell nanocrystals. *ACS Nano.* **5**, 1505–1512 (2011).
201. Wang Y, Huang R, Liang G, Zhang Z, Zhang P, Yu S, Kong J. MRI-visualized, dual-targeting, combined tumor therapy using magnetic graphene-based mesoporous silica. *Small* **10**, 109–116 (2014).
202. Zhang M, Cao Y, Chong Y, Ma Y, Zhang H, Deng Z, Hu C, Zhang Z. Graphene oxide based theranostic platform for T1-weighted magnetic resonance imaging and drug delivery. *ACS Appl. Mater. Int.* **5**, 13325–13332 (2013).
203. Yang HW, Huang CY, Lin CW, Liu HL, Huang CW, Liao SS, Chen PY, Lu YJ, Wei KC, Ma CC. Gadolinium-functionalized nanographene oxide for combined drug and microRNA delivery and magnetic resonance imaging. *Biomaterials* **35**, 6534–6542 (2014).
204. Gizzatov A, Keshishian V, Guven A, Dimiev AM, Qu F, Muthupillai R, Decuzzi P, Bryant RG, Tour JM, Wilson LJ. Enhanced MRI relaxivity of aquated Gd3+ ions by carboxyphenylated water-dispersed graphene nanoribbons. *Nanoscale* **6**, 3059–3063 (2014).
205. Feng L, Zhang S, Liu Z. Graphene based gene transfection. *Nanoscale* **3**, 1252–1257 (2011).
206. Feng L, Yang X, Shi X, Tan X, Peng R, Wang J, Liu Z. Polyethylene glycol and polyethylenimine dual-functionalized nano-graphene oxide for photothermally enhanced gene delivery. *Small* **9**, 1989–1997 (2013).
207. Teimouri M, Nia AH, Abnous K, Eshghi H, Ramezani M. Graphene oxide-cationic polymer conjugates: Synthesis and application as gene delivery vectors. *Plasmid* 84–**85**, 51–60 (2016).
208. Bao H, Pan Y, Ping Y, Sahoo NG, Wu T, Li L, Li J, Gan LH. Chitosan-functionalized graphene oxide as a nanocarrier for drug and gene delivery. *Small* 7, 1569–1578 (2011).

209. Zhang L, Lu Z, Zhao Q, Huang J, Shen H, Zhang Z. Enhanced chemotherapy efficacy by sequential delivery of siRNA and anticancer drugs using PEI-grafted graphene oxide. *Small* 7, 460–464 (2011).
210. Choi HY, Lee TJ, Yang GM, Oh J, Won J, Han J, Jeong GJ, Kim J, Kim JH, Kim BS, Cho SG. Efficient mRNA delivery with graphene oxide-polyethylenimine for generation of footprint-free human induced pluripotent stem cells. *J. Control. Rel.* **235**, 222–235 (2016).
211. Dong H, Ding L, Yan F, Ji H, Ju H. The use of polyethylenimine-grafted graphene nanoribbon for cellular delivery of locked nucleic acid modified molecular beacon for recognition of microRNA. *Biomaterials* **32**, 3875–3882 (2011).
212. Hubert A, Lyass O, Pode D, Gabizon A. Doxil (Caelyx): An exploratory study with pharmacokinetics in patients with hormone-refractory prostate cancer. *Anti-cancer Drugs* **11**, 123–127 (2000).
213. Orditura M, Quaglia F, Morgillo F, Martinelli E, Lieto E, De Rosa G, Comunale D, Diadema MR, Ciardiello F, Catalano G, De Vita F. Pegylated liposomal doxorubicin: pharmacologic and clinical evidence of potent antitumor activity with reduced anthracycline-induced cardiotoxicity (review). *Oncol. Rep.* **12**, 549–556 (2004).
214. Goto S, Ihara Y, Urata Y, Izumi S, Abe K, Koji T, Kondo T. Doxorubicin-induced DNA intercalation and scavenging by nuclear glutathione S-transferase pi. *FASEB Journal* **15**, 2702–2714 (2001).
215. Peer D, Park EJ, Morishita Y, Carman CV, Shimaoka M. Systemic leukocyte-directed siRNA delivery revealing cyclin D1 as an anti-inflammatory target. *Science* **319**, 627–630 (2008).
216. Schiffelers RM, Storm G. ICS-283: A system for targeted intravenous delivery of siRNA. *Expert Opin. Drug Delivery* **3**, 445–454 (2006).
217. Torchilin VP. Drug targeting. *Eur. J. Phar. Sci.* **11**(2), S81–S91 (2000).
218. Lammers T, Kiessling F, Hennink WE, Storm G. Drug targeting to tumors: principles, pitfalls and (pre-) clinical progress. *J. Control. Rel.* **161**, 175–187 (2012).
219. Lopez-Davila V, Seifalian AM, Loizidou M. Organic nanocarriers for cancer drug delivery. *Curr. Opin. Pharm.* **12**, 414–419 (2012).
220. Maeda H, Nakamura H, Fang J. The EPR effect for macromolecular drug delivery to solid tumors: Improvement of tumor uptake, lowering of systemic toxicity, and distinct tumor imaging *in vivo*. *Adv. Drug Delivery Rev.* **65**, 71–79 (2013).

221. Schleich N, Po C, Jacobs D, Ucakar B, Gallez B, Danhier F, Preat V. Comparison of active, passive and magnetic targeting to tumors of multifunctional paclitaxel/SPIO-loaded nanoparticles for tumor imaging and therapy. *J. Control. Rel.* **194**, 82–91 (2014).
222. Theek B, Gremse F, Kunjachan S, Fokong S, Pola R, Pechar M, Deckers R, Storm G, Ehling J, Kiessling F, Lammers T. Characterizing EPR-mediated passive drug targeting using contrast-enhanced functional ultrasound imaging. *J. Control. Rel.* **182**, 83–89 (2014).
223. Hirsjarvi S, Passirani C, Benoit JP. Passive and active tumour targeting with nanocarriers. *Curr. Drug Discovery Technol.* **8**, 188–196 (2011).
224. Meng H, Xue M, Xia T, Ji Z, Tarn DY, Zink JI, Nel AE. Use of size and a copolymer design feature to improve the biodistribution and the enhanced permeability and retention effect of doxorubicin-loaded mesoporous silica nanoparticles in a murine xenograft tumor model. *ACS Nano.* **5**, 4131–4144 (2011).
225. Aslan B, Ozpolat B, Sood AK, Lopez-Berestein G. Nanotechnology in cancer therapy. *J. Drug Target.* **21**, 904–913 (2013).
226. Hirsjarvi S, Dufort S, Gravier J, Texier I, Yan Q, Bibette J, Sancey L, Josserand V, Passirani C, Benoit JP, Coll JL. Influence of size, surface coating and fine chemical composition on the *in vitro* reactivity and *in vivo* biodistribution of lipid nanocapsules versus lipid nanoemulsions in cancer models. *Nanomedicine* **9**, 375–387 (2013).
227. Agarwal R, Jurney P, Raythatha M, Singh V, Sreenivasan SV, Shi L, Roy K. Effect of shape, size, and aspect ratio on nanoparticle penetration and distribution inside solid tissues using 3D spheroid models. *Adv. Health. Mater.* **4**, 2269–2280 (2015).
228. Wojnicki M, Luty-Blocho M, Bednarski M, Dudek M, Knutelska J, Sapa J, Zygmunt M, Nowak G, Fitzner K. Tissue distribution of gold nanoparticles after single intravenous administration in mice. *Pharmacol. Rep.* **65**, 1033–1038 (2013).
229. Kaminskas LM, Boyd BJ, Karellas P, Henderson SA, Giannis MP, Krippner GY, Porter CJ. Impact of surface derivatization of poly-L-lysine dendrimers with anionic arylsulfonate or succinate groups on intravenous pharmacokinetics and disposition. *Mole. Pharm.* **4**, 949–961 (2007).
230. Maeda H. Vascular permeability in cancer and infection as related to macromolecular drug delivery, with emphasis on the EPR effect for tumor-selective drug targeting, *Proceedings of the Japan Academy. Series B. Phys. Biol. Sci.* **88**, 53–71 (2012).

231. Greish K. Enhanced permeability and retention of macromolecular drugs in solid tumors: A royal gate for targeted anticancer nanomedicines. *J. Drug Target.* **15**, 457–464 (2007).
232. Jabr-Milane LS, van Vlerken LE, Yadav S, Amiji MM. Multi-functional nanocarriers to overcome tumor drug resistance. *Cancer Treat. Rev.* **34**, 592–602 (2008).
233. Godin B, Tasciotti E, Liu X, Serda RE, Ferrari M. Multistage nanovectors: from concept to novel imaging contrast agents and therapeutics. *Acc. Chem. Res.* **44**, 979–989 (2011).
234. Ruoslahti E. Peptides as targeting elements and tissue penetration devices for nanoparticles. *Adv. Mater.* **24**, 3747–3756 (2012).
235. Wang Z, Yu Y, Dai W, Lu J, Cui J, Wu H, Yuan L, Zhang H, Wang X, Wang J, Zhang X, Zhang Q. The use of a tumor metastasis targeting peptide to deliver doxorubicin-containing liposomes to highly metastatic cancer. *Biomaterials* **33**, 8451–8460 (2012).
236. Mukherjee B. Nanosize drug delivery system. *Curr. Pharm. Biotechnol.* **14**, 1221 (2013).
237. Beduneau A, Saulnier P, Benoit JP. Active targeting of brain tumors using nanocarriers. *Biomaterials* **28**, 4947–4967 (2007).
238. Gupta B, Torchilin VP. Monoclonal antibody 2C5-modified doxorubicin-loaded liposomes with significantly enhanced therapeutic activity against intracranial human brain U-87 MG tumor xenografts in nude mice. *Cancer Immunol. Immunother* **56**, 1215–1223 (2007).
239. Nie S, Xing Y, Kim GJ, Simons JW. Nanotechnology applications in cancer. *Ann. Rev. Biomed. Eng.* **9**, 257–288 (2007).
240. Ding B, Wu X, Fan W, Wu Z, Gao J, Zhang W, Ma L, Xiang W, Zhu Q, Liu J, Ding X, Gao S. Anti-DR5 monoclonal antibody-mediated DTIC-loaded nanoparticles combining chemotherapy and immunotherapy for malignant melanoma: Target formulation development and *in vitro* anticancer activity. *Int. J. Nanomed.* **6**, 1991–2005 (2011).
241. Al-Ahmady ZS, Chaloin O, Kostarelos K. Monoclonal antibody-targeted, temperature-sensitive liposomes: *In vivo* tumor chemotherapeutics in combination with mild hyperthermia. *J. Control. Rel.* **196**, 332–343 (2014).
242. Maloney DG, Grillo-Lopez AJ, White CA, Bodkin D, Schilder RJ, Neidhart JA, Janakiraman N, Foon KA, Liles TM, Dallaire BK, Wey K, Royston I, Davis T, Levy R. IDEC-C2B8 (Rituximab) anti-CD20 monoclonal antibody therapy in patients with relapsed low-grade non-Hodgkin's lymphoma. *Blood* **90**, 2188–2195 (1997).

243. Baselga J, Norton L, Albanell J, Kim YM, Mendelsohn J. Recombinant humanized anti-HER2 antibody (Herceptin) enhances the antitumor activity of paclitaxel and doxorubicin against HER2/neu overexpressing human breast cancer xenografts. *Cancer Res.* **58**, 2825–2831 (1998).
244. Hurwitz H, Fehrenbacher L, Novotny W, Cartwright T, Hainsworth J, Heim W, Berlin J, Baron A, Griffing S, Holmgren E, Ferrara N, Fyfe G, Rogers B, Ross R, Kabbinavar F. Bevacizumab plus irinotecan, fluorouracil, and leucovorin for metastatic colorectal cancer. *N. Engl. J. Med.* **350**, 2335–2342 (2004).
245. Cunningham D, Humblet Y, Siena S, Khayat D, Bleiberg H, Santoro A, Bets D, Mueser M, Harstrick A, Verslype C, Chau I, Van Cutsem E. Cetuximab monotherapy and cetuximab plus irinotecan in irinotecan-refractory metastatic colorectal cancer. *N. Engl. J. Med.* **351**, 337–345 (2004).
246. Cavanaugh PG, Jia L, Zou Y, Nicolson GL. Transferrin receptor overexpression enhances transferrin responsiveness and the metastatic growth of a rat mammary adenocarcinoma cell line. *Breast Cancer Res. Treat.* **56**, 203–217 (1999).
247. Pang Z, Gao H, Yu Y, Guo L, Chen J, Pan S, Ren J, Wen Z, Jiang X. Enhanced intracellular delivery and chemotherapy for glioma rats by transferrin-conjugated biodegradable polymersomes loaded with doxorubicin. *Bioconjugate Chem.* **22**, 1171–1180 (2011).
248. Yuan Y, Chen S, Paunesku T, Gleber SC, Liu WC, Doty CB, Mak R, Deng J, Jin Q, Lai B, Brister K, Flachenecker C, Jacobsen C, Vogt S, Woloschak GE. Epidermal growth factor receptor targeted nuclear delivery and high-resolution whole cell X-ray imaging of Fe3O4@TiO2 nanoparticles in cancer cells. *ACS Nano.* 7, 10502–10517 (2013).
249. Mukherjee A, Darlington T, Baldwin R, Holz C, Olson S, Kulkarni P, DeWeese TL, Getzenberg RH, Ivkov R, Lupold SE. Development and screening of a series of antibody-conjugated and silica-coated iron oxide nanoparticles for targeting the prostate-specific membrane antigen. *Chem. Med. Chem.* **9**, 1356–1360 (2014).
250. Hong G, Yuan R, Liang B, Shen J, Yang X, Shuai X. Folate-functionalized polymeric micelle as hepatic carcinoma-targeted, MRI-ultrasensitive delivery system of antitumor drugs. *Biomed. Microdevices* **10**, 693–700 (2008).
251. Kularatne SA, Low PS. Targeting of nanoparticles: Folate receptor. *Methods Mol. Biol.* **624**, 249–265 (2010).

252. Pawar H, Surapaneni SK, Tikoo K, Singh C, Burman R, Gill MS, Suresh S. Folic acid functionalized long-circulating co-encapsulated docetaxel and curcumin solid lipid nanoparticles: *In vitro* evaluation, pharmacokinetic and biodistribution in rats. *Drug Deliv.* **23**, 1453–1468 (2016).
253. Liu Y, Li K, Pan J, Liu B, Feng SS. Folic acid conjugated nanoparticles of mixed lipid monolayer shell and biodegradable polymer core for targeted delivery of Docetaxel. *Biomaterials* **31**, 330–338 (2010).
254. Lin JJ, Chen JS, Huang SJ, Ko JH, Wang YM, Chen TL, Wang LF. Folic acid-Pluronic F127 magnetic nanoparticle clusters for combined targeting, diagnosis, and therapy applications. *Biomaterials* **30**, 5114–5124 (2009).
255. Liu J, Kolar C, Lawson TA, Gmeiner WH. Targeted drug delivery to chemoresistant cells: Folic acid derivatization of FdUMP[10]. enhances cytotoxicity toward 5-FU-resistant human colorectal tumor cells. *J. Org. Chem.* **66**, 5655–5663 (2001).
256. Low PS, Henne WA, Doorneweerd DD. Discovery and development of folic-acid-based receptor targeting for imaging and therapy of cancer and inflammatory diseases. *Acc. Chem. Res.* **41**, 120–129 (2008).
257. Cheng C, Wei H, Zhu JL, Chang C, Cheng H, Li C, Cheng SX, Zhang XZ, Zhuo RX. Functionalized thermoresponsive micelles self-assembled from biotin-PEG-b-P(NIPAAm-co-HMAAm)-b-PMMA for tumor cell target. *Bioconjugate Chem.* **19**, 1194–1201 (2008).
258. Yang W, Cheng Y, Xu T, Wang X, Wen LP. Targeting cancer cells with biotin-dendrimer conjugates. *Eur. J. Med. Chem.* **44**, 862–868 (2009).
259. Maiti S, Park N, Han JH, Jeon HM, Lee JH, Bhuniya S, Kang C, Kim JS. Gemcitabine-coumarin-biotin conjugates: A target specific theranostic anticancer prodrug. *J. Am. Chem. Soc.* **135**, 4567–4572 (2013).
260. Wang Y, Wang X, Zhang Y, Yang S, Wang J, Zhang X, Zhang Q. RGD-modified polymeric micelles as potential carriers for targeted delivery to integrin-overexpressing tumor vasculature and tumor cells. *J. Drug Target.* **17**, 459–467 (2009).
261. Fani M, Mueller A, Tamma ML, Nicolas G, Rink HR, Cescato R, Reubi JC, Maecke HR. Radiolabeled bicyclic somatostatin-based analogs: a novel class of potential radiotracers for SPECT/PET of neuroendocrine tumors. *J. Nuc. Med.* **51**, 1771–1779 (2010).
262. Iyer AK, Singh A, Ganta S, Amiji MM. Role of integrated cancer nanomedicine in overcoming drug resistance. *Adv. Drug Deliv. Rev.* **65**, 1784–1802 (2013).

263. Kulhari H, Pooja D, Shrivastava S, Telukutala SR, Barui AK, Patra CR, Naidu Vegi GM, Adams DJ, Sistla R. Cyclic-RGDfK peptide conjugated succinoyl-TPGS nanomicelles for targeted delivery of docetaxel to integrin receptor over-expressing angiogenic tumours. *Nanomedicine* **11**, 1511–1520 (2015).
264. Melemenidis S, Jefferson A, Ruparelia N, Akhtar AM, Xie J, Allen D, Hamilton A, Larkin JR, Perez-Balderas F, Smart SC, Muschel RJ, Chen X, Sibson NR, Choudhury RP. Molecular magnetic resonance imaging of angiogenesis in vivo using polyvalent cyclic RGD-iron oxide microparticle conjugates. *Theranostics* **5**, 515–529 (2015).
265. Danhier F, Vroman B, Lecouturier N, Crokart N, Pourcelle V, Freichels H, Jerome C, Marchand-Brynaert J, Feron O, Preat V. Targeting of tumor endothelium by RGD-grafted PLGA-nanoparticles loaded with paclitaxel. *J. Control. Rel.* **140**, 166–173 (2009).
266. Wang X, Li S, Shi Y, Chuan X, Li J, Zhong T, Zhang H, Dai W, He B, Zhang Q. The development of site-specific drug delivery nanocarriers based on receptor mediation. *J. Control. Rel.* **193**, 139–153 (2014).
267. Arosio D, Casagrande C. Advancement in integrin facilitated drug delivery. *Adv. Drug Deliv. Rev.* **97**, 111–143 (2016).
268. Luo G, Long J, Zhang B, Liu C, Ji S, Xu J, Yu X, Ni Q. Quantum dots in cancer therapy. *Expert Opin. Drug Delivery* **9**, 47–58 (2012).
269. Chi C, Du Y, Ye J, Kou D, Qiu J, Wang J, Tian J, Chen X. Intraoperative imaging-guided cancer surgery: from current fluorescence molecular imaging methods to future multi-modality imaging technology. *Theranostics* **4**, 1072–1084 (2014).
270. Li Y, Lin TY, Luo Y, Liu Q, Xiao W, Guo W, Lac D, Zhang H, Feng C, Wachsmann-Hogiu S, Walton JH, Cherry SR, Rowland DJ, Kukis D, Pan C, Lam KS. A smart and versatile theranostic nanomedicine platform based on nanoporphyrin. *Nat. Commun.* **5**, 4712 (2014).
271. Martelli C, Lo Dico A, Diceglie C, Lucignani G, Ottobrini L. Optical imaging probes in oncology. *Oncotarget* 7, 48753–48787 (2016).
272. Kolodziej A, Krajewski W, Matuszewski M, Tupikowski K. Review of current optical diagnostic techniques for non-muscle-invasive bladder cancer. *Cent. European J. Urol.* **69**, 150–156 (2016).
273. Neefjes J, Dantuma NP. Fluorescent probes for proteolysis: Tools for drug discovery. *Nat. Rev. Drug Discov.* **3**, 58–69 (2004).
274. Kim JS, Quang DT. Calixarene-derived fluorescent probes. *Chem. Rev.* **107**, 3780–3799 (2007).

275. Fernandez-Suarez M, Ting AY. Fluorescent probes for super-resolution imaging in living cells. *Nat. Rev. Mole. Cell Biol.* **9**, 929–943 (2008).
276. Goncalves MS. Fluorescent labeling of biomolecules with organic probes. *Chem. Rev.* **109**, 190–212 (2009).
277. Pak YL, Swamy KM, Yoon J. Recent Progress in Fluorescent Imaging Probes. *Sensors* **15**, 24374–24396 (2015).
278. Zhan C, Gu B, Xie C, Li J, Liu Y, Lu W. Cyclic RGD conjugated poly(ethylene glycol)-co-poly(lactic acid) micelle enhances paclitaxel anti-glioblastoma effect. *J. Control. Rel.* **143**, 136–142 (2010).
279. Kong Y, Chen J, Fang H, Heath G, Wo Y, Wang W, Li Y, Guo Y, Evans SD, Chen S, Zhou D. Highly Fluorescent Ribonuclease-A-Encapsulated Lead Sulfide Quantum Dots for Ultrasensitive Fluorescence in Vivo Imaging in the Second Near-Infrared Window. *Chem. Mater.* **28**, 3041–3050 (2016).
280. Gianotti E, Bertolino CA, Benzi C, Nicotra G, Caputo G, Castino R, Isidoro C, Coluccia S. Photoactive hybrid nanomaterials: Indocyanine immobilized in mesoporous MCM-41 for "in-cell" bioimaging. *ACS Appl. Mater. Int.* **1**, 678–687 (2009).
281. Jansen K, van Soest G, van der Steen AF. Intravascular photoacoustic imaging: a new tool for vulnerable plaque identification. *Ultrasound Med. Biol.* **40**, 1037–1048 (2014).
282. Li K, Liu B. Polymer-encapsulated organic nanoparticles for fluorescence and photoacoustic imaging. *Chem. Soc. Rev.* **43**, 6570–6597 (2014).
283. Pekkanen AM, DeWitt MR, Rylander MN. Nanoparticle enhanced optical imaging and phototherapy of cancer. *J. Biomed. Nanotechnol* **10**, 1677–1712 (2014).
284. Xia J, Yao J, Wang LV. Photoacoustic tomography: Principles and advances. *Electromagnetic Waves* **147**, 1–22 (2014).
285. Zackrisson S, van de Ven SM, Gambhir SS. Light in and sound out: emerging translational strategies for photoacoustic imaging. *Cancer Res.* **74**, 979–1004 (2014).
286. Li J, Cheng F, Huang H, Li L, Zhu JJ. Nanomaterial-based activatable imaging probes: From design to biological applications. *Chem. Soc. Rev.* **44**, 7855–7880 (2015).
287. Li W, Chen X. Gold nanoparticles for photoacoustic imaging. *Nanomedicine* **10**, 299–320 (2015).
288. Chen WH, Yang CX, Qiu WX, Luo GF, Jia HZ, Lei Q, Wang XY, Liu G, Zhuo RX, Zhang XZ. Multifunctional theranostic nanoplatform for

cancer combined therapy based on gold nanorods. *Adv. Health. Mater.* **4**, 2247–2259 (2015).

289. Song J, Yang X, Jacobson O, Lin L, Huang P, Niu G, Ma Q, Chen X. Sequential drug release and enhanced photothermal and photoacoustic effect of hybrid reduced graphene oxide-loaded ultrasmall gold nanorod vesicles for cancer therapy. *ACS Nano.* **9**, 9199–9209 (2015).
290. Galanzha EI, Shashkov EV, Kelly T, Kim JW, Yang L, Zharov VP. In vivo magnetic enrichment and multiplex photoacoustic detection of circulating tumour cells. *Nat. Nanotechnol.* **4**, 855–860 (2009).
291. Moon H, Kumar D, Kim H, Sim C, Chang JH, Kim JM, Kim H, Lim DK. Amplified photoacoustic performance and enhanced photothermal stability of reduced graphene oxide coated gold nanorods for sensitive photoacoustic imaging. *ACS Nano.* **9**, 2711–2719 (2015).
292. Xie L, Wang G, Zhou H, Zhang F, Guo Z, Liu C, Zhang X, Zhu L. Functional long circulating single walled carbon nanotubes for fluorescent/photoacoustic imaging-guided enhanced phototherapy. *Biomaterials* **103**, 219–228 (2016).
293. Maji SK, Sreejith S, Joseph J, Lin M, He T, Tong Y, Sun H, Yu SW, Zhao Y. Upconversion nanoparticles as a contrast agent for photoacoustic imaging in live mice. *Adv. Mater.* **26**, 5633–5638 (2014).
294. Rieffel J, Chen F, Kim J, Chen G, Shao W, Shao S, Chitgupi U, Hernandez R, Graves SA, Nickles RJ, Prasad PN, Kim C, Cai W, Lovell JF. Hexamodal imaging with porphyrin-phospholipid-coated upconversion nanoparticles. *Adv. Mater.* **27**, 1785–1790 (2015).
295. Cheng L, Gong H, Zhu W, Liu J, Wang X, Liu G, Liu Z. PEGylated Prussian blue nanocubes as a theranostic agent for simultaneous cancer imaging and photothermal therapy. *Biomaterials* **35**, 9844–9852 (2014).
296. Cheng L, Liu J, Gu X, Gong H, Shi X, Liu T, Wang C, Wang X, Liu G, Xing H, Bu W, Sun B, Liu Z. PEGylated WS(2) nanosheets as a multifunctional theranostic agent for in vivo dual-modal CT/photoacoustic imaging guided photothermal therapy. *Adv. Mater.* **26**, 1886–1893 (2014).
297. Wei B, Zhang X, Zhang C, Jiang Y, Fu YY, Yu C, Sun SK, Yan XP. Facile synthesis of uniform-sized bismuth nanoparticles for CT visualization of gastrointestinal tract *in vivo*. *ACS Appl. Mater. Int.* **8**, 12720–12726 (2016).

298. Motiei M, Dreifuss T, Betzer O, Panet H, Popovtzer A, Santana J, Abourbeh G, Mishani E, Popovtzer R. Differentiating between cancer and inflammation: A metabolic-based method for functional computed tomography imaging. *ACS Nano.* **10**, 3469–3477 (2016).
299. Pang B, Zhao Y, Luehmann H, Yang X, Detering L, You M, Zhang C, Zhang L, Li ZY, Ren Q, Liu Y, Xia Y. (6)(4)Cu-Doped PdCu@Au Tripods: A multifunctional nanomaterial for positron emission tomography and image-guided photothermal cancer treatment. *ACS Nano.* **10**, 3121–3131 (2016).
300. Zhao Y, Pang B, Luehmann H, Detering L, Yang X, Sultan D, Harpstrite S, Sharma V, Cutler CS, Xia Y, Liu Y. Gold nanoparticles doped with (199) Au atoms and their use for targeted cancer imaging by SPECT. *Adv. Health Mater.* **5**, 928–935 (2016).
301. Zhao L, Zhu J, Cheng Y, Xiong Z, Tang Y, Guo L, Shi X, Zhao J. Chlorotoxin-conjugated multifunctional dendrimers labeled with radionuclide 131I for single photon emission computed tomography imaging and radiotherapy of gliomas. *ACS Appl. Mater. Int.* 7, 19798–19808 (2015).
302. Xu C, Sun S. New forms of superparamagnetic nanoparticles for biomedical applications. *Adv. Drug Deliv. Rev.* **65**, 732–743 (2013).
303. Couto D, Freitas M, Carvalho F, Fernandes E. Iron oxide nanoparticles: An insight into their biomedical applications. *Curr. Medi. Chem.* (2015).
304. Long NV, Yang Y, Teranishi T, Thi CM, Cao Y, Nogami M. Biomedical applications of advanced multifunctional magnetic nanoparticles. *J. Nanosci. Nanotechnol.* **15**, 10091–10107 (2015).
305. Xu S, Yang F, Zhou X, Zhuang Y, Liu B, Mu Y, Wang X, Shen H, Zhi G, Wu D. Uniform PEGylated PLGA microcapsules with embedded Fe3O4 nanoparticles for US/MR dual-modality imaging. *ACS Appl. Mater. Int.* 7, 20460–20468 (2015).
306. Zhao Z, Zhou Z, Bao J, Wang Z, Hu J, Chi X, Ni K, Wang R, Chen X, Chen Z, Gao J. Octapod iron oxide nanoparticles as high-performance T(2) contrast agents for magnetic resonance imaging. *Nat. Commun.* **4**, 2266 (2013).
307. Chen N, Shao C, Qu Y, Li S, Gu W, Zheng T, Ye L, Yu C. Folic acid-conjugated MnO nanoparticles as a T1 contrast agent for magnetic resonance imaging of tiny brain gliomas. *ACS Appl. Mater. Int.* **6**, 19850–19857 (2014).

308. Li F, Zhi D, Luo Y, Zhang J, Nan X, Zhang Y, Zhou W, Qiu B, Wen L, Liang G. Core/shell Fe3O4/Gd2O3 nanocubes as T1-T2 dual modal MRI contrast agents. *Nanoscale* **8**, 12826–12833 (2016).
309. Huang HC, Barua S, Sharma G, Dey SK, Rege K. Inorganic nanoparticles for cancer imaging and therapy. *J. Control. Rel.* **155**, 344–357 (2011).
310. Fang J, Chen YC. Nanomaterials for photohyperthermia: A review. *Curr. Pharm. Design.* **19**, 6622–6634 (2013).
311. Singh R, Torti SV. Carbon nanotubes in hyperthermia therapy. *Adv. Drug Deliv. Rev.* **65**, 2045–2060 (2013).
312. Shanmugam V, Selvakumar S, Yeh CS. Near-infrared light-responsive nanomaterials in cancer therapeutics. *Chem. Soc. Rev.* **43**, 6254–6287 (2014).
313. Dykman L, Khlebtsov N. Gold nanoparticles in biomedical applications: recent advances and perspectives. *Chem. Soc. Rev.* **41**, 2256–2282 (2012).
314. Chen W, Zhang S, Yu Y, Zhang H, He Q. Structural-engineering rationales of gold nanoparticles for cancer theranostics. *Adv. Mater.* **28**, 8567–8585 (2016).
315. Zha Z, Wang S, Zhang S, Qu E, Ke H, Wang J, Dai Z. Targeted delivery of CuS nanoparticles through ultrasound image-guided microbubble destruction for efficient photothermal therapy. *Nanoscale* **5**, 3216–3219 (2013).
316. Wang S, Riedinger A, Li H, Fu C, Liu H, Li L, Liu T, Tan L, Barthel MJ, Pugliese G, De Donato F, Scotto D'Abbusco M, Meng X, Manna L, Meng H, Pellegrino T. Plasmonic copper sulfide nanocrystals exhibiting near-infrared photothermal and photodynamic therapeutic effects. *Acs Nano.* **9**, 1788–1800 (2015).
317. Wang C, Xu H, Liang C, Liu Y, Li Z, Yang G, Cheng L, Li Y, Liu Z. Iron oxide @ polypyrrole nanoparticles as a multifunctional drug carrier for remotely controlled cancer therapy with synergistic antitumor effect. *Acs Nano.* 7, 6782–6795 (2013).
318. Liu Z, Ye B, Jin M, Chen H, Zhong H, Wang X, Guo Z. Dye-free near-infrared surface-enhanced Raman scattering nanoprobes for bioimaging and high-performance photothermal cancer therapy. *Nanoscale* 7, 6754–6761 (2015).
319. Robinson JT, Welsher K, Tabakman SM, Sherlock SP, Wang H, Luong R, Dai H. High performance *in vivo* Near-IR (>1 mum) imaging and photothermal cancer therapy with carbon nanotubes. *Nano Res.* **3**, 779–793 (2010).

320. Elhissi AM, Ahmed W, Hassan IU, Dhanak VR, D'Emanuele A. Carbon nanotubes in cancer therapy and drug delivery. *J. Drug Deliv.* **2012**, 837327 (2012).
321. Wang L, Shi J, Hao Y, Zhang P, Zhao Y, Meng D, Li D, Chang J, Zhang Z. Magnetic multi-walled carbon nanotubes for tumor theranostics. *J. Biomed. Nanotechnol.* **11**, 1653–1661 (2015).
322. Hadidi N, Kobarfard F, Nafissi-Varcheh N, Aboofazeli R. PEGylated single-walled carbon nanotubes as nanocarriers for cyclosporin A delivery. *AAPS Pharm. Sci. Tech.* **14**, 593–600 (2013).
323. Wen S, Zhao Q, An X, Zhu J, Hou W, Li K, Huang Y, Shen M, Zhu W, Shi X. Multifunctional PEGylated multiwalled carbon nanotubes for enhanced blood pool and tumor MR imaging. *Adv. Healthc. Mater.* **3**, 1568–1577, 1525 (2014).
324. Gormley AJ, Greish K, Ray A, Robinson R, Gustafson JA, Ghandehari H. Gold nanorod mediated plasmonic photothermal therapy: A tool to enhance macromolecular delivery. *Int. J. Pharm.* **415**, 315–318 (2011).
325. Okuno T, Kato S, Hatakeyama Y, Okajima J, Maruyama S, Sakamoto M, Mori S, Kodama T. Photothermal therapy of tumors in lymph nodes using gold nanorods and near-infrared laser light. *J. Control. Rel.* **172**, 879–884 (2013).
326. Shen S, Wang S, Zheng R, Zhu X, Jiang X, Fu D, Yang W. Magnetic nanoparticle clusters for photothermal therapy with near-infrared irradiation. *Biomaterials* **39**, 67–74 (2015).
327. Master A, Livingston M, Sen Gupta A. Photodynamic nanomedicine in the treatment of solid tumors: Perspectives and challenges. *J. Control. Rel.* **168**, 88–102 (2013).
328. Hu J, Tang Y, Elmenoufy AH, Xu H, Cheng Z, Yang X. Nanocomposite-based photodynamic therapy strategies for deep tumor treatment. *Small* **11**, 5860–5887 (2015).
329. Shen S, Tang H, Zhang X, Ren J, Pang Z, Wang D, Gao H, Qian Y, Jiang X, Yang W. Targeting mesoporous silica-encapsulated gold nanorods for chemo-photothermal therapy with near-infrared radiation. *Biomaterials* **34**, 3150–3158 (2013).
330. Wang S, Yang W, Cui J, Li X, Dou Y, Su L, Chang J, Wang H, Li X, Zhang B. pH- and NIR light responsive nanocarriers for combination treatment of chemotherapy and photodynamic therapy. *Biomater. Sci.* **4**, 338–345 (2016).

331. He C, Duan X, Guo N, Chan C, Poon C, Weichselbaum RR, Lin W. Core-shell nanoscale coordination polymers combine chemotherapy and photodynamic therapy to potentiate checkpoint blockade cancer immunotherapy. *Nat. Commun.* 7, 12499 (2016).
332. Diou O, Tsapis N, Fattal E. Targeted nanotheranostics for personalized cancer therapy. *Expert Opin. Drug Deliv.* **9**, 1475–1487 (2012).
333. Wang LS, Chuang MC, Ho JA. Nanotheranostics — a review of recent publications. *Int. J. Nanomedicine* 7, 4679–4695 (2012).
334. Baetke SC, Lammers T, Kiessling F. Applications of nanoparticles for diagnosis and therapy of cancer. *B. J. Radiol.* **88**, 20150207 (2015).
335. Gao F, Li L, Fu C, Nie L, Chen D, Tang F. LHRH-PE40 fusion protein tethered silica nanorattles for imaging-guided tumor-specific drug delivery and bimodal therapy. *Adv. Mater.* **25**, 5508–5513 (2013).
336. Jia X, Cai X, Chen Y, Wang S, Xu H, Zhang K, Ma M, Wu H, Shi J, Chen H. Perfluoropentane-encapsulated hollow mesoporous prussian blue nanocubes for activated ultrasound imaging and photothermal therapy of cancer. *ACS Appl. Mater. Int.* 7, 4579–4588 (2015).
337. Rong P, Huang P, Liu Z, Lin J, Jin A, Ma Y, Niu G, Yu L, Zeng W, Wang W, Chen X. Protein-based photothermal theranostics for imaging-guided cancer therapy. *Nanoscale* 7, 16330–16336 (2015).
338. Wang K, Fan X, Zhao L, Zhang X, Zhang X, Li Z, Yuan Q, Zhang Q, Huang Z, Xie W, Zhang Y, Wei Y. Aggregation induced emission fluorogens based nanotheranostics for targeted and imaging-guided chemo-photothermal combination Therapy. *Small* **12**, 6568–6575 (2016).
339. Wang S, Zhang Q, Luo XF, Li J, He H, Yang F, Di Y, Jin C, Jiang XG, Shen S, Fu de L. Magnetic graphene-based nanotheranostic agent for dual-modality mapping guided photothermal therapy in regional lymph nodal metastasis of pancreatic cancer. *Biomaterials* **35**, 9473–9483 (2014).
340. Wang S, Zhang Q, Yang P, Yu X, Huang LY, Shen S, Cai S. Manganese oxide-coated carbon nanotubes as dual-modality lymph mapping agents for photothermal therapy of tumor metastasis. *ACS Appl. Mater. Int.* **8**, 3736–3743 (2016).
341. Du L, Jin Y, Zhou W, Zhao J. Ultrasound-triggered drug release and enhanced anticancer effect of doxorubicin-loaded poly(D,L-lactide-co-glycolide)-methoxy-poly(ethylene glycol) nanodroplets. *Ultrasound Med. Biol.* **37**, 1252–1258 (2011).

Index

D

E

J

L

M

N

O

P

Frontiers in Nanobiomedical Research

(Continuation of series card page)

Vol. 6 *Handbook of Immunological Properties of Engineered Nanomaterials*
Second Edition
(In 3 Volumes)
Volume 1: Key Considerations for Nanoparticle Characterization Prior to Immunotoxicity Studies
Volume 2: Haematocompatibility of Engineered Nanomaterials
Volume 3: Engineered Nanomaterials and the Immune Cell Function
Second Edition
edited by Marina A Dobrovolskaia (Leidos Biomedical Research Inc., USA) and Scott E McNeil (Leidos Biomedical Research Inc., USA)

Vol. 5 *The World Scientific Encyclopedia of Nanomedicine and Bioengineering I: Nanotechnology for Translational Medicine: Tissue Engineering, Biological Sensing, Medical Imaging, and Therapeutics*
(A 4-Volume Set)
editor-in-chief: Donglu Shi (East Hospital, Tongji University School of Medicine, China & University of Cincinnati, USA)
Volume 1: Noble Metal Nanoparticles for Biomedical Applications
edited by Yu Cheng (East Hospital, Tongji University School of Medicine, China)
Volume 2: Application of Nano-structured Materials and Devices in Chemical and Biological Sensing
edited by Jia Huang (School of Materials Science and Engineering Tongji University, China)
Volume 3: Nanotechnology in Gene Delivery
Yarong Liu (University of Southern California, USA)
edited by Pin Wang (University of Southern California, USA)
Volume 4: Nano Imaging: From Fundamental Principles to Translational Medical Applications
edited by Bingbo Zhang (Tongji University School of Medicine, China)

Vol. 4 *Bio-Inspired Nanomaterials and Applications: Nano Detection, Drug/Gene Delivery, Medical Diagnosis and Therapy*
edited by Donglu Shi (University of Cincinnati, USA)

Vol. 3 *Handbook of Nanobiomedical Research: Fundamentals, Applications and Recent Developments*
(In 4 Volumes)
Volume 1: Materials for Nanomedicine
Volume 2: Applications in Therapy
Volume 3: Applications in Diagnostics
Volume 4: Biology, Safety and Novel Concepts in Nanomedicine
edited by Vladimir Torchilin (Northeastern University, USA)

Vol. 2 *Tissue Regeneration: Where Nano-Structure Meets Biology*
edited by Qing Liu (3D Biotek, USA & Tongji University, China) and Hongjun Wang (Stevens Institute of Technology, USA)

Vol. 1 *Handbook of Immunological Properties of Engineered Nanomaterials*
edited by Marina A. Dobrovolskaia (SAIC-Frederick, Inc., USA) and Scott E. McNeil (SAIC-Frederick, Inc., USA)